T0270957

ALGEBRAIC IDENTIFICATION AND ESTIMATION METHODS IN FEEDBACK CONTROL SYSTEMS

ALGEBRAIC IDENTIFICATION AND ESTIMATION METHODS IN FEEDBACK CONTROL SYSTEMS

Hebertt Sira-Ramírez
CINVESTAV-IPN, Mexico

Carlos García-Rodríguez
Technological University of the Mixteca, Mexico

John Cortés-Romero
National University of Colombia, Colombia

Alberto Luviano-Juárez
UPIITA – Instituto Politécnico Nacional, Mexico

This edition first published 2014
© 2014 John Wiley & Sons Ltd

Registered office
John Wiley & Sons Ltd, The Atrium, Southern Gate, Chichester, West Sussex, PO19 8SQ, United Kingdom

For details of our global editorial offices, for customer services and for information about how to apply for permission to reuse the copyright material in this book please see our website at www.wiley.com.

The right of the author to be identified as the author of this work has been asserted in accordance with the Copyright, Designs and Patents Act 1988.

All rights reserved. No part of this publication may be reproduced, stored in a retrieval system, or transmitted, in any form or by any means, electronic, mechanical, photocopying, recording or otherwise, except as permitted by the UK Copyright, Designs and Patents Act 1988, without the prior permission of the publisher.

Wiley also publishes its books in a variety of electronic formats. Some content that appears in print may not be available in electronic books.

Designations used by companies to distinguish their products are often claimed as trademarks. All brand names and product names used in this book are trade names, service marks, trademarks or registered trademarks of their respective owners. The publisher is not associated with any product or vendor mentioned in this book.

Limit of Liability/Disclaimer of Warranty: While the publisher and author have used their best efforts in preparing this book, they make no representations or warranties with respect to the accuracy or completeness of the contents of this book and specifically disclaim any implied warranties of merchantability or fitness for a particular purpose. It is sold on the understanding that the publisher is not engaged in rendering professional services and neither the publisher nor the author shall be liable for damages arising herefrom. If professional advice or other expert assistance is required, the services of a competent professional should be sought.

Library of Congress Cataloging-in-Publication Data

Sira Ramírez, Hebertt J.
  Algebraic identification and estimation methods in feedback control systems / Hebertt Sira-Ramirez, Carlos Garcia-Rodriguez, John Alexander Cortes-Romero, Alberto Luviano-Juarez.
    pages cm – (Wiley series in dynamics and control of electromechanical systems)
  Includes bibliographical references and index.
  ISBN 978-1-118-73060-7 (hardback)
  1. Feedback control systems – Mathematical models. 2. Control theory – Mathematics. 3. Differential algebra.
I. Title.
  TJ216.S467 2014
  629.8'301512 – dc23

2013049876

Typeset in 10/12pt TimesLTStd by Laserwords Private Limited, Chennai, India

1 2014

This book is dedicated to our families, friends, colleagues, and students. Also, to our beloved countries: Venezuela, Mexico, and Colombia

Contents

Series Preface

Electromechanical Systems permeate the engineering and technology fields in aerospace, automotive, mechanical, biomedical, civil/structural, electrical, environmental, and industrial systems. The Wiley Book Series on dynamics and control of electromechanical systems will cover a broad range of engineering and technology these fields. As demand increases for innovation in these areas, feedback control of these systems is becoming essential for increased productivity, precision operation, load mitigation, and safe operation. Furthermore, new applications in these areas require a reevaluation of existing control methodologies to meet evolving technological requirements. An example involves distributed control of energy systems. The basics of distributed control systems are well documented in several textbooks, but the nuances of its use for future applications in the evolving area of energy system applications, such as wind turbines and wind farm operations, solar energy systems, smart grids, and energy generation, storage and distribution, require an amelioration of existing distributed control theory to specific energy system needs. The book series serves two main purposes: 1) a delineation and explication of theoretical advancements in electromechanical system dynamics and control, and 2) a presentation of application driven technologies in evolving electromechanical systems.

This book series will embrace the full spectrum of dynamics and control of electrome-chanical systems from theoretical foundations to real world applications. The level of the presentation should be accessible to senior undergraduate and first-year graduate students, and should prove especially well-suited as a self-study guide for practicing professionals in the fields of mechanical, aerospace, automotive, biomedical, and civil/structural engineering. The aim is an interdisciplinary series ranging from high-level undergraduate/graduate texts, explanation and dissemination of science and technology and good practice, through to impor-tant research that is immediately relevant to industrial development and practical applications. It is hoped that this new and unique perspective will be of perennial interest to students, schol-ars, and employees in aforementioned engineering disciplines. Suggestions for new topics and authors for the series are always welcome.

Mark J. Balas
John L. Crassidis
Florian Holzapfel
Series Editors

Preface

This work has been made possible thanks to Professor Michel Fliess's professional mathematical vision of real engineering problems. Without his convincing and precise mathematical formulation of fundamental problems in control theory of uncertain systems and signal processing, this book would never have existed.

The quest for an algebraic approach to parameter identification started one lovely summer afternoon in 2002, while having lunch and lively discussions on automatic control matters at Ma Bourgogne restaurant, La Place des Vosges, Paris. I was in the company of Michel and Richard Marquez, an outstanding doctoral student of Michel's in Paris, who had only recently defended his thesis and who had also been a superb Master's student of mine a few years back in Mérida (Venezuela). The discussion ended later that night at Michel's apartment with the distinctive feeling that a "can of crazy worms" had just been opened. Michel and myself worked feverishly over the next few months and years. Sometimes across the Atlantic Ocean, via internet; sometimes in Paris, and on other occasions in the gardens of Cinvestav, in Mexico City. Our colleagues Mamadou Mboup, Hugues Mounier, Cedric Join, Joachim Rudolph, and Johann Reger joined Michel's efforts and quickly found applications and outstanding results of the innovative theory in new and challenging areas such as communications systems, failure detection, and chaotic systems synchronization. As set out by Michel from the beginning, the theory, of course, does not need probability theory and for that reason neither do we.

The approach to parameter estimation, state estimation, and perturbation rejection adopted in this book is radically different from existing approaches in three main respects: (1) it is not based on asymptotic approaches, (2) it does not require a probabilistic setting, and (3) it does not elude the need to compute iterated time derivatives of actual noise-corrupted signals. The fact that the computations do not lead to asymptotic schemes is buried deep in the algebraic nature of the approach. We exploit the system model in performing valid algebraic manipulations, leading to sensible schemes yielding parameters, states, or external perturbations. Naturally, our scheme rests on the category of *model-based* methods. We should point out, however, that the power of the algebraic approach is of such a nature that it also allows complete reformulation of non-model-based control schemes. One of the crucial assumptions that allows us to free ourselves from probabilistic considerations is that of "high-frequency" noises, or, more precisely, "rapidly varying perturbations." A complete theory exists nowadays, based on non-standard analysis, for the rigorous characterization and derivation of the fundamental properties of such noises. White noises, characteristic of the existing literature, are known to constitute a "worst-case" idealization, which allows for a comfortable mathematical treatment of the expressions but one that is devoid of any physical reality.

The research work contained in this book has primarily been supported by the Centro de Investigación y Estudios Avanzados del Instituto Politécnico Nacional (Cinvestav-IPN), Mexico City and the generous financial assistance of the Consejo Nacional de Ciencia y Tecnología (CONACYT), Mexico, under Research Projects No. 42231-Y and 60877-Y. Generous financial assistance of the CNRS (France) and of the Stix and Gage Laboratories of the École Polytechnique is gratefully acknowledged. The first author would like to thank Marc Giusti, Emmanuel Delaleau, and Joachim Rudolph for their kind invitations to, respectively, Palaisseau, Brest, and Dresden on several enjoyable occasions. Back at Cinvestav, the generous friendship of Dr. Gerardo Silva-Navarro is sincerely acknowledged and thanked, in many laboratory undertakings by students and challenging academic discussions. We also acknowledge his special administrative skills to produce "ways and means," as materialized in equipment and infrastructure.

The work gathered by the first author over the years has benefited enormously from the wisdom, patience, and determination of the three co-authors of this book, who set out to clean examples from many mistakes, perform the required computer simulations, and carry out successful laboratory experiments. Carlos García Rodríguez is credited for having obtained, for the first time in the world, an actual experimental application of parameter and derivative estimation, from an algebraic standpoint, in the control of an oscillatory mechanical system. His initial contribution in ordering of the material, finding useful variants, and his remarkable ability to recreate lost simulation files and carry out experiments put us on the trail of pursuing the writing of a book on the subject of algebraic parameter and state estimation. Alberto Luviano-Juárez and John Cortés-Romero, two extraordinary PhD students, joined the venture, giving the available material a definite positive push toward completion. Through countless discussions and revisions of the material, weekly projects involving lots of nightly, and weekend, work on their part, real-life laboratory implementations under adverse conditions, and contagious enthusiasm, they are credited with generously driving the book project to the point of no return. My deepest appreciation to all of them for having endured the difficult times and for the "mountains" of workload involved.

The first author is indebted to Professor Vicente Feliu-Batlle of the Universidad de Castilla La Mancha (UCLM), Ciudad Real, Spain and his superb PhD students Juan Ramón Trapero, Jonathan Becedas, and Gabriela Mamani for having put the theory to a definite test with many challenging laboratory experiments that gave us the chance to publish our results in credited journals and conferences. Such an interaction was made possible thanks to Professor Feliu's administrative skills, resulting in a full sabbatical year spent in Ciudad Real. The more recent interaction with Dr. Rafael Morales, of UCLM, has proven to be most fruitful in the use of this theory in some other challenging laboratory applications.

H. Sira-Ramírez dedicates his work in this book to his friend Professor Michel Fliess, for his constant support, kind advice, and encouragement in the writing of this book and in many other academic matters.

Hebertt Sira-Ramírez

1

Introduction

One of the main obstacles related to key assumptions in many appealing feedback control theories lies in the need to perfectly know the system to be controlled. Even though mathematical models can be derived precisely for many areas of physical systems, using well-established physical laws and principles, the problems remain of gathering the precise values of the relevant system parameters (or, obtaining the information stored in the inaccessible-for-measurement states of the system) and, very importantly, dealing with unknown (i.e., non-structured) perturbations affecting the system evolution through time. These issues have been a constant concern in the feedback control systems literature and a wealth of approaches have been developed over the years to separately, or simultaneously, face some, or all, of these challenging realities involved in physical systems operation. To name but a few, system identification, adaptive control, energy methods, neural networks, and fuzzy systems have all been developed and have tried out disciplines that propose related approaches, from different viewpoints, to deal with, or circumvent, the three fundamental obstacles to make a clean control design work: parameter identification, state estimation, and robustness with respect to external perturbations.

This book deals with a new approach to the three fundamental problems associated with the final implementation of a nicely justified feedback control law. We concentrate on the ways to handle these obstacles from an algebraic viewpoint, that is one resulting from an algebraic vision of systems dynamics and control. As for the preferred theory to deal with the ideal control problem, we emphasize the fact that the methods to be presented are equally applicable to any of the existing theories. We propose examples where sliding-mode control is used, others where passivity-based control methods are preferred, and yet others where flatness and generalized proportional integral controllers are implemented. The algebraic approach is equally suitable when dynamic observers are used. The book therefore does not concentrate on, or favor, any particular feedback control theory. We choose the controller as we please. Naturally, since the theoretical basis of the proposed algorithms and techniques stems from the differential algebraic approach to systems analysis and control, we often present the background material in detail at the end of chapters, so that the mathematically inclined reader has a source for the basics being illustrated in that chapter through numerous physically oriented examples.

Algebraic Identification and Estimation Methods in Feedback Control Systems, First Edition.
Hebertt Sira-Ramírez, Carlos García-Rodríguez, Alberto Luviano-Juárez, John Alexander Cortés-Romero.
© 2014 John Wiley & Sons, Ltd. Published 2014 by John Wiley & Sons, Ltd.

1.1 Feedback Control of Dynamic Systems

The control systems presented here are designed using an algebraic estimation methodology in combination with well-known control design techniques, which are applicable to linear and nonlinear systems. Since the algebraic estimation methodology is independent of the particular controller design method being used and, furthermore, it is quite easy to understand, it will be profitable for the reader to combine this tool with his preferred controllers or with conventional control techniques. This book introduces a wide variety of application examples and detailed explanations to illustrate the use of the algebraic methodology in identification, state estimation, and disturbance estimation. However, this work is not an introductory control textbook. A basic control course is a prerequisite for a deeper understanding of these examples.

1.1.1 Feedback

A control system is one whose objective is to positively influence the performance of a given system in accordance with some specific objectives. A *control law* or *controller* is a set of rules that allows us to determine the commands to be sent to the governed plant (via an actuator) to achieve the desired evolution. These rules can be described as either *open-loop* control or *closed-loop* (feedback) control. Figure 1.1 shows both control strategies, where $y(t)$ and $u(t)$ are, respectively, the output and input of the plant, and $x(t)$ represents a finite collection of internal variables (called state variables) that completely characterize, from the present, the system behavior in the future provided the future control inputs are available. The first scheme, an *open-loop control system*, does not measure (or feed back) the output to determine the control action; its accuracy depends on its calibration, and thus it is only used when there are no perturbations and the plant is perfectly known. A *perturbation* (or *disturbance*) is an unknown signal that tends to adversely affect the output value of the plant. This undesired signal may be internal (*endogenous*) or external (*exogenous*) to the system. Figure 1.1(b) shows a *closed-loop control system*, where a sensor is used to obtain an error signal. This signal is processed by the controller to determine the action necessary to reduce the difference between the output and its desired value.

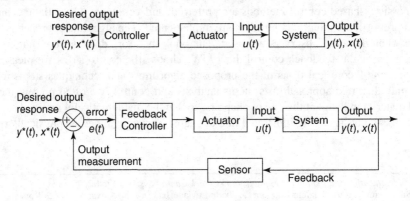

Figure 1.1   Typical control scheme

Feedback is a mechanism to command a system to evolve in a desired fashion so that the states, and outputs, exhibit a desired evolution (e.g., to track a *reference trajectory*) or stay at a prescribed equilibrium. Feedback enables the current state or the current outputs (*feedback signal*) to be measured, determining how far the behavior is from the desired state or desired outputs (i.e., to assess the *error signal* and then automatically generate a suitable *control signal* to bring the system closer and closer to the desired state). Feedback can be used to stabilize the state of a system, while also improving its performance.

Feedback control has ancient origins, as noted by Mayr (1970). Throughout history, many examples of ingenious devices, based on feedback, can be found. In ancient Greece, China, during the Middle Ages, and in the Renaissance, many examples have been recovered and explained in modern terms. These artifacts were improved and specialized to become pressure regulators, float valves, and temperature regulators. One of these devices was the fly-ball or centrifugal governor, used to control the speed of windmills; it was later adapted by James Watt, in 1788, to control steam engines.

It was only in the 1930s that a theory of feedback control was fully developed by Black and Nyquist at Bell Laboratories. They studied feedback as a means to vastly improve the amplifier performance in telephone lines. They had to face the (well-known) *closed-loop instability problem* when the feedback gain was set too high, transforming the amplifier into an oscillator.

1.1.2   Why Do We Need Feedback?

Fundamentally, feedback is necessary for the following reasons.

- Counteracting *disturbance* signals affecting the plant: A controller must reject the effects of unknown undesirable inputs and maintain the output of the plant within desired values.
- Improving system performance in the presence of model *uncertainty*: Uncertainty arises from two sources – unknown or unpredictable inputs (disturbance, noise, etc.) and uncertain or unmodeled dynamics.
- Stabilizing an *unstable plant*.

This book addresses only *additive perturbations*, which will be divided into structured and unstructured perturbations. *Structured perturbations* are generated by the initial conditions of the system and unknown exogenous inputs that can be modeled as families of time polynomials. The *unstructured perturbations* are considered as highly fluctuating, or oscillating, phenomena affecting the behavior of the system. A common unstructured perturbation is the case of a zero-mean noisy signal. Notwithstanding this, it is still possible to build an algebraic estimator that takes into account these disturbances and mitigates their effects.

The main objectives of feedback control are to ensure that the variables of interest in the system either track prescribed reference trajectories (*tracking problem*) or are maintained close to their constant set-points (*regulation problem*).

1.2   The Parameter Identification Problem

A model is a mathematical representation of the essential characteristics of an existing system. When a system model can be defined by a finite number of variables and parameters, it is called a parametric model. Examples of parametric models are the transfer function of an electrical circuit, the equations of motion of a mechanical suspension system, and so on. The whole family of functions and equations that integrate a model is called the model structure. In general, this structure can be linear or nonlinear. The models used in modern control theory are, with a few exceptions, parametric models in terms of linear and nonlinear state equations. To implement a model-based controller, it is necessary to know precisely the structure of the model of the system and its associated parameters. Therefore, if parameters are initially unknown, the process of parameter identification is quite important for the design of the control system.

1.2.1   Identifying a System

According to Eykhoff (1974), model building begins with the application of basic physical laws (Newton's laws, Maxwell's laws, Kirchhoff's laws, etc.). From these laws, a number of relations are established between variables describing the system (e.g., ordinary differential or difference equations or, sometimes, partial differential equations). If all external and internal conditions of the system are known, and if our physical knowledge about the plant is complete, then in principle the numerical value of all the parameters in those relations may be determined. Eykhoff (1974) recalls that model building consists of four steps: (a) selection of a model structure based on physical knowledge, (b) fitting of parameters to available data (identification), (c) verification and testing of the adopted model, and (d) application of control theory to the model to achieve a desired purpose. The word *process* is just another term for referring to a given system.

The identification problem is sometimes tackled via the inverse problem of system analysis; *given an input and output time history, determine the equations and parameters that describe the system behavior.* Zadeh (1956) defines *identification* as the determination, on the basis of inputs and outputs of a system, of a system model which produces an equivalent behavior to the real plant. *Parameter estimation*, in contrast, is the experimental determination of values of parameters that govern the dynamic behavior of the system, assuming that the structure of the process is well known. The method implies comparing the actual process output, contaminated with measurement noise, and the response of the hypothesized model output when both are subject to the same input signal. Because of the uncertainty introduced by noise and neglected dynamics, it is then necessary to find an adjustment of the model according to a certain *criterion.*

1.3   A Brief Survey on Parameter Identification

The problem of parameter estimation in dynamical systems, from available measurements, dates back to 300 B.C. when the motion of celestial bodies was characterized by six parameters (see Sorenson, 1980).

Estimation, based on the minimization of an error function, can be attributed to Galileo Galilei (see Favier, 1982). One of the most important early contributions in this area was the statistical view of the parameter identification problem, proposed by Gauss (1809) and known as the "inverse problem of computing the response of a system with known characteristics," and the maximum likelihood procedure introduced by Fisher (see Aldrich, 1997).

The term "identification" was coined by Zadeh (1956), as the problem of determining the input–output relation by experimental means, considering the linear class of models as a "black-box" model.

Linear estimation for stochastic systems was introduced in the works of Wiener (1949) and Kolmogorov (1941). These works placed the identification problem in the context of important engineering problems that remained open mathematical problems. An excellent survey paper on this theory can be found in Kailath (1974).

The Kalman filter (Kalman, 1960; Kalman and Bucy, 1961) represents a state-space-based algorithmic procedure geared to optimally solve the problem of state estimation in the presence of measurement noises and uncertain initial states. The approach is based on the minimization of an integral quadratic estimation error performance criterion. The great legacy of the Kalman filter has led to extensions of nonlinear systems and to the formulation and solution of some other related problems, such as the parameter identification problem. Åstrom and Bohlin (1965) introduced the concept of identifiability in the context of the maximum likelihood approach. The survey by Åstrom and Eykhoff (1971) presents a complete account of classical identification schemes.

An important reference text, from the early years of identification, is the book by Box and Jenkins (1976). This book is concerned with the identification of discrete-time stochastic systems, the forecasting of time series, and the estimation of parameters in transfer functions, with emphasis on applications. Other authors focused on the problem of model approximation and system order reduction (see Anderson *et al.*, 1978 and the references cited therein).

The book by Ljung (1987) is a most important reference in the context of linear systems identification. This book focuses on the "engineering approach to system identification," as mentioned in Gevers (2006), where numerical schemes are made to play an important rôle in the identification procedure. Other important books regarding this topic are Åstrom and Wittenmark (1995), Eykhoff (1974), Goodwin and Sin (1984), Johansson (1993), Landau (1990), Sastry and Bodson (1989), Söderström and Stoica (1989), and Sorenson (1980). A more recent authoritative survey of system identification can be found in the work of Gevers (2006).

1.4   The State Estimation Problem

The *state* of a system is a minimum set of variables (state variables) whose present values, together with the values of the input signals in the future, completely determine the future behavior of the system.

Conditions and practical ways of estimating the state are generally set according to the mathematical approach being used (dynamical systems, signal theory point of view, etc.), the system nature (stochastic, deterministic, linear, nonlinear), the amount of variables to estimate, etc. Here, it will be assumed that the system to be controlled is described by differential equations in a state-space representation.

We consider the state estimation problem as the problem of determining the states from observations of the outputs. We usually seek to test the observability of the system model with respect to its output to find an answer to this problem. This property indicates how well internal states of a system can be reconstructed by knowledge of its external outputs. The concept of observability was introduced by Rudolf E. Kalman for linear systems (Kalman, 1959, 1963).

To understand the importance of state estimators in the feedback control, we need to analyze the block diagram of a control system as shown in Figure 1.2.

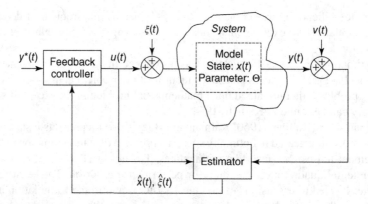

Figure 1.2   Block diagram of a control system with state and perturbation estimators

For the continuous-time case, the system model consists of a set of differential equations which describe the evolution of the state variables, $x(t)$, and a set of parameters θ. The known input signal, or control input, is denoted by $u(t)$. Sometimes, the system is also affected by an unknown external disturbance, $v(t)$. The output, $y(t)$, may frequently be contaminated by measurement noise, $\xi(t)$. The reference signal, the set point or desired output value that the control system will aim to reach, is $y^*(t)$. As Figure 1.2 shows, the block for the feedback controller requires the actual state value to produce the control signal. Since not all of the state variables are measurable, we need to build a state estimator in order to implement this feedback controller. In most practical cases, a state estimator requires an adequate noise immunity level.

In order to cancel, or at least reduce, the disturbance effects, a robust control law based on disturbance estimators can be proposed. In some cases, the disturbance estimation problem is dealt with as the estimation of an additional state.

Generally, to estimate the system state we can use two types of tools: state observers and time derivative estimators. In both approaches, the model of the system needs to satisfy a particular condition guaranteeing the feasibility of the state computation or reproduction from inputs and outputs or outputs alone. Such a property is the system observability. In the linear system case (continuous and discrete), the property may easily be assessed from the rank of a matrix called the observability matrix. Its conceptual introduction is due to Kalman.

1.4.1   Observers

The state can be generated from a dynamical system called the observer, which is based on the model of the plant, driven by measurements of the plant inputs and outputs. If the order of the observer is equal to the order of the system model, the observer is said to be a full-order observer. If the order of the observer is less than the order of the plant, then the observer is said to be a reduced-order observer. Reduced-order observers take advantage of the possibility of directly computing some state variables from the available outputs.

An observer for a linear system is readily feasible whenever the system is observable, that is, when its observability matrix has full rank. Reconstructible systems are referred to as those systems whose observability matrix is not full rank but such that the unobservable part of the system exhibits asymptotic behavior toward a certain trivial equilibrium state. Observers have been studied extensively in linear systems, via the appropriate selection of gains, showing that observers can be constructed in such a way that their states approach the true states with dynamics having prescribed stability features. If the observer gains are optimized for the noise input to the system and to the sensor(s), the observer is called a Kalman filter. If the gains are not so optimized, and the setting is deterministic, the observer is called a Luenberger observer. The concept of an observer for a dynamic process was introduced in 1966 by D. Luenberger. The generic *Luenberger observer* appeared several years after the Kalman filter.

1.4.2   Reconstructing the State via Time Derivative Estimation

Another way to estimate the state is via numerical differentiation. If a system is observable, the state-space can be reconstructed from a set of measurements of the input, the output, and a finite number of their time derivatives. However, a problem arises from estimating derivatives; a typical differentiator is greatly affected by noise. Therefore, if we choose this estimation technique, we should necessarily consider implementing an additional filtering algorithm in our estimation procedure to obtain reasonable results under noisy measurement conditions.

A systematic way to reconstruct the state-space via time derivatives is by exploiting the differential flatness property of a system. We say that a system is flat if it can be described in terms of a set of special output variables and their time derivatives up to finite order. For example, for a single-input/single-output (SISO) flat system, all state variables, including the control input, can be parameterized in terms of the flat output and a finite number of its time derivatives.

From an algebraic estimation point of view, two types of state estimators may be proposed. The first is based directly on the model of the plant, and the second is based on time-polynomial approximations of the output signals. Both schemes constitute time-derivative estimators. The flatness property of the system can be used advantageously to reconstruct the complete state. Unlike classical estimators, the algebraic estimators are robust with respect to initial conditions of the system. They have an adequate noise immunity level and include mechanisms to mitigate the effects of unknown but bounded external disturbances.

The works of Norbert Wiener on optimal filtering inspired the development of the state observation theory accomplished and completed by Rudolph Kalman, in the state-space stochastic context. A deterministic approach to the problem of estimation of inaccessible-for-measurement states, through available outputs, was developed by David G. Luenberger (1964, 1966, 1971).

Observability in nonlinear systems is developed in the work of Hermann and Krener (1977). Other important contributions to the theory of nonlinear observers are given by Brockett (1972) and Sussmann (1977, 1979). In relation to the differential geometric approach, observability was studied by Isidori (1989) and Nijmeijer and van der Schaft (1990). Another important observability concept, given in terms of the differential algebraic approach, was introduced by Diop and Fliess (1991a,b). In this approach, a system variable is said to be algebraically observable if it can be represented in terms of the input, the output, and a finite number of their time derivatives. In this particular sense, observer design is reduced to the design of a numerical differentiator (Diop *et al.*, 1993, 2000). In the realm of signal differentiators, many interesting approaches have been proposed, ranging from Luenberger-like observers to sliding-mode-based differentiators (Levant, 1998). High-gain observers (Doyle and Stein,

1979; Khalil, 1999) have also become popular in recent times, in both theory and applications, as evidenced by the survey of Khalil (2008).

1.5   Algebraic Methods in Control Theory: Differences from Existing Methodologies

This identification methodology has been used to estimate parameters (Trapero-Arenas *et al.*, 2008), states (Barbot *et al.*, 2007), derivatives of signals (Villagra *et al.*, 2008), polynomial disturbances, and also to detect faults (Fliess *et al.*, 2004). In general, the algebraic technique allows us to deal with three fundamental obstacles in controller design tasks: parameter identification, state estimation, and robustness with respect to additive perturbations (such as constant, ramp, and parabolic, i.e., classical, perturbations).

This technique is applicable to linear and nonlinear systems, and is easy to understand. For the linear case, there are two general (equivalent) approaches: the time-domain approach and the frequency-domain (operational calculus) approach. In the first case, only time differentiations of expressions, multiplications by positive (suitable) powers of the time variable, and iterated integrations (with advantageous use of the integration by parts formula) are sufficient to obtain linear expressions in the parameters, clean from the influence of initial conditions and classical perturbation inputs. The second approach only uses tools such as Laplace transforms, derivation with respect to the complex variable s (also called "algebraic derivations"), and multiplication by suitable negative powers of the complex variable s. Hence, the algebraic approach is fully compatible with the concepts and tools of classic control. For the nonlinear case, one may only resort to the time-domain approach. The fundamental difference from the linear case is the need to obtain convolutions of time powers with nonlinear expressions of the measured input and output variables or, sometimes, with nonlinear expressions of state variables. For nonlinear systems where the measured outputs are naturally the outputs of the plant, the differential flatness property of the system is quite helpful. Unlike the linear case, beside the input and the output, sometimes it is necessary to also know one or several components of the state vector. Since this methodology is independent of the preferred control design technique, it can be adapted to the needs of any other model-based control law. Their design turns out to be easier, and much faster, than many adaptive identification schemes, (Gensior *et al.*, 2008).

The algebraic method enables the robust design of online parameter identifiers, and also of state estimators in the presence of noise and time-polynomial disturbances. Algebraic estimators do not need statistical knowledge of the noises corrupting the data (see, e.g., Fliess *et al.*, (2003, 2008) for linear and nonlinear diagnoses). The basic *invariant filtering* used in this work, for the treatment of noise effects, is based on the noise-attenuating properties of the integration operation. Traditional low-pass filtering and even nonlinear filtering is suitably merged with the algebraic parameter and state-estimation procedures.

Unlike the asymptotic observers, the algebraic estimators do not rely on asymptotic convergence arguments and Lyapunov stability theory. The identifications are nearly instantaneous in nature. In García-Rodríguez *et al.* (2009), the sentinel parameter was proposed as an alternative to experimentally decide the moment at which the estimated parameters have converged to acceptable values. Another approach, based on the condition number involved in the identifier equation system, is given in Trapero-Arenas *et al.* (2008). This simplifies the controller design and enables online tuning of the certainty-equivalence controllers.

The algebraic identification method has already been compared with standard recursive identification algorithms, showing that the unknown parameters are obtained in a substantially shorter time period (Garrido and Concha 2013). Given that the algebraic parameter identification is quite fast, the estimated values are accurate enough that they can readily be used by the controller in order to accomplish a given task. It may be shown that, thanks to the almost instantaneous nature of the algebraic identification procedure, it does not require the classical "persistency of excitation condition" characteristic of slow parameter convergence in traditional adaptive control schemes. This is replaced by an algebraic consistency condition which avoids singularities over small open intervals of time (see Fliess and Sira-Ramírez, 2003). Contrary to the persistency of excitation, which is known to fundamentally interfere with desired control objectives and has to be sustained typically over rather long periods of time, the algebraic requirements only need to be valid during the small time interval needed to compute the parameter via a *static formula*. After the parameters are determined, the identification process may be permanently stopped if the parameters are known to be perfectly constant. Otherwise, one may let the process be reinitialized at a later time, provided the degradation in performance can be attributed to ongoing parameter variations in time.

Algebraic parameter estimators have been used successfully in the control of DC-to-DC power converters (Gensior *et al.*, 2008); Linares Flores *et al.*, 2011) and, more recently, a most interesting model-free control design technique based on the algebraic approach to systems control has been proposed by Michel *et al.* (2010).

Generally speaking, there are still many aspects that require development within the algebraic methodology for parameter and state estimation. These are:

- a higher noise immunity;
- to achieve a relaxation of the conditions of model-based design to cope with dynamic uncertainties;
- an extension to estimation in time-varying systems;
- a more complete extension into the realm of nonlinear systems;
- advantageous combinations with other estimation techniques;
- new application areas and extensions to new classes of systems (hybrid, networked, etc.).

All these topics are attractive options for further investigation, given their inherent simplicity and formidable power.

1.6 Outline of the Book

This book is essentially a tutorial. Each chapter has a number of examples, which set out to illustrate and explain the use of the theory or, in some instances, to motivate it. The examples, although mostly physically oriented, are therefore quite elementary, quite low dimensional, and, frankly, rather simple. We believe that with this option at hand, the reader can readily grasp the essential features of the approach and try the techniques in their own, more complex, applications. If something fails, there is always an exposition of, or reference to, the theory to find the inherent limitations of the approach or the overlooked and unsatisfied properties of that particular example. Our experience indicates that the proposed theory always yields a correct and satisfactory answer.

Chapter 2 deals with the parameter identification problem in the context of linear systems. Whether state or input–output representations, we begin with systems undergoing structured perturbations (i.e., perturbations generated by homogeneous, linear, time-invariant systems whose initial states and parameters cannot be obtained). To this class of perturbations belong the so-called classical perturbations: constant perturbations, ramps, polynomial and sinusoidal perturbations, etc. Efforts will be geared to obtain, or identify, the system or plant parameters rather than the perturbation-defining parameters. The objective we keep in mind is that of being able to implement a feedback controller where such parameter information is lacking and deemed to be crucial. In simple terms, we adopt the certainty-equivalence control approach. We design the feedback controller as if the unknown set of parameters is at our disposal but, initially, with an arbitrary value being used in its place. The identification of the actual value of the parameter is to be carried out online, rather fast, and in the presence of noise. Once the parameter is identified with accuracy, its value is immediately replaced in the designed controller. Algebraic calculation of parameters from measured inputs and outputs must be capable of dealing with unavoidable measurement noises. In this chapter we introduce a technique known as invariant filtering. Instead of pre-filtering the involved signals, we post-filter the involved expressions from where the calculation is possible. This is shown to enhance the signal-to-noise ratio and provide accurate parameter estimations in the presence of significant measurement noise.

Chapter 3 is devoted to the algebraic approach to parameter identification in a class of nonlinear systems where the vector of unknown parameters is weakly linearly identifiable. Contrary to the case of linear systems, it is not always possible to obtain explicit formulae for the components of the parameter vector which depend only on inputs and outputs, or even nonlinear functions of inputs and outputs alone. In general, the state vector components are needed to be able to compute the unknown parameters. This important limitation will be lifted in later chapters by means of online, non-asymptotic algebraic state estimation methods. In this chapter we concentrate mainly on demonstrating that the algebraic approach to parameter estimation can equally be extended to nonlinear systems by centering the approach around the time domain. Several examples of parameter identification in nonlinear systems are presented. Also, the extension of the certainty-equivalence control method for the nonlinear case is treated, providing a method for fast adaptive control of systems of a physical nature. An important class of nonlinear systems examples where the current parameter-identification techniques fail is the class of switched systems undergoing sliding motions. The high-frequency nature of the control input, under sliding conditions, naturally clouds the possibilities of accurate parameter estimation in traditional identification schemes. In our algebraic approach, however, such an inconvenient feature of the control signal is effectively dealt with by means of invariant filtering. In this chapter, we present several examples of a certainty-equivalence sliding-mode control approach to systems with unknown parameters. We should stress and clarify the following point: customarily, in sliding-mode control schemes for switched systems, a knowledge of system parameters may be unnecessary to achieve sliding motion and an asymptotic satisfaction of the control objectives. This is particularly so in state-space representations and, perhaps more specifically, in linear time-invariant systems. However, when no system states are available, it is still possible to resort to an output sliding-mode feedback control scheme that exploits knowledge of the system and implements an average output feedback controller scheme through a $\Sigma - \Delta$ modulation circuit. Such an average feedback control law requires fast identification of the unknown parameters while the measured switch position function, acting as the only measurable input signal of the system, is a high-frequency bang–bang signal.

Chapter 4 deals with the discrete-time counterpart of the online, continuous, linear and non-linear parametric identification approach presented in Chapters 2 and 3. The extension of the algebraic approach for parameter identification to this ubiquitous class of systems is also based on the module-theoretic vision of discrete-time linear dynamics, which has become classic. As in the continuous-time case, we achieve closed-loop identification in a relatively small time interval involving few samples and with no need for statistics. Several physically oriented case studies of linear and nonlinear nature, mono-variable and multi-variable, are discussed in detail along with their corresponding computer simulations.

Chapter 5 shifts our attention to the problem of state estimation in linear systems from the algebraic viewpoint. Algebraic manipulations, either in the time domain or in the frequency domain, of the system equations allow for explicit formulae to be developed for the efficient computation of the states of a system. In this chapter, we show how states can be computed continuously, in an online fashion, after a rather short time interval. Several examples of these manipulations are presented, along with simulation results. One may quickly realize that the possibilities for simultaneous, online, state and parameter estimations are indeed certain, although perhaps limited to some particularly nice linear systems. However, by resorting to suitable piecewise polynomial approximations, simultaneous state and parameter estimations can be pushed to the realm of nonlinear systems. This chapter dwells on several case studies that illustrate the powerful possibilities of the algebraic approach in the combined state and parameter identification task. Clearly, a clean algebraic approach to solve both problems simultaneously may easily be hampered by the richness of system nonlinearities and their possible combinations with unknown parameters. For this reason, an altogether fresh view of the state-estimation problem, for perturbed input–output models, was undertaken for the natural class of differentially flat systems with available flat outputs. The end of the chapter presents applications of the algebraic method to the handling of the popular synchronization problem in chaotic systems. This problem essentially involves the development of efficient state estimations for nonlinear systems. The algebraic approach is shown to fit the problem of state estimation suitably in chaotic systems. Several well-studied chaotic systems are used as illustrative instances in the application of the algebraic state-estimation problem via differentiations of measurable outputs. As a byproduct, the problem of encrypted message detection, and remote message identification, is readily solved using the developed viewpoint.

Every differentially flat system is an observable system from the vector of flat outputs. For this reason, the state-estimation problem for flat systems is intimately related to the problem of computing the successive time derivatives of the flat outputs. In general, however, the state-estimation problem of a nonlinear dynamical system is tied to time differentiations of the available outputs and of the input signals in a sufficiently large number. We propose, in Chapter 6, a non-asymptotic algebraic procedure for the approximate, piecewise local, estimation of a finite number of time derivatives of a general time signal that may be corrupted by measurement noise. The method is based on results from differential algebra and furnishes some general formulae for the time derivatives of any measurable signal. Naturally, the state-estimation method proposed may be combined with the notion of differential flatness to complete a feedback loop, with desirable dynamics, based on the flat output feedback. In this respect, we no longer regard the nonlinear model of the output variables as a nonlinear function of the state, but as a linear, time-invariant, homogeneous model describing a finite-order, local, polynomial approximation to the realization of the output signal as a time signal.

Lastly, Chapter 7 is devoted to presenting a series of alternatives in the application of the algebraic method for parameter and state estimation in the context of feedback control. Several options are explored, such as the possibility of using bounded exponential functions instead of powers of time variables in the parameter identification for linear systems.

The appendices at the end of the book contain background material, and tutorial introductions, in the algebraic methods of control systems. In particular, linear control systems.

References

Aldrich, J. (1997) R.A. Fisher and the making of maximum likelihood 1912–1922. *Statistical Science* **12**(3), 162–176.

Anderson, B. Moore, J. and Hawkes, R. (1978) Model approximations via prediction error identification. *Automatica* **14**, 615–622.

Åstrom, K. and Bohlin, T. (1965) etc. Numerical identification of dynamic systems from normal operating records. *Proceedings of the IFAC Symposium of Self-Adaptive Systems*, pp. 96–111, Teddington, UK.

Åstrom, K. and Eykhoff, P. (1971) System identification–a survey. *Automatica* **7**, 123–162.

Åstrom, K. and Wittenmark, B. (1995) *Adaptive Control*, 2nd edn. Addison-Wesley, New York.

Barbot, JP. Fliess, M. and Floquet, T. (2007) An algebraic framework for the design of nonlinear observers with unknown inputs. *Proceedings of the 46th IEEE Conference on Decision and Control*, pp. 384–389, New Orleans, LA.

Box, GE. and Jenkins, GM. (1976) *Time Series Analysis, Forecasting and Control*, Series in Time Series Analysis and Digital Processing. Holden-Day, New york.

Brockett, R. (1972) System theory on group manifolds and coset spaces. *SIAM Journal of control.* **10**, 265–284.

Diop, S. and Fliess, M. (1991a) Nonlinear observability, identifiability and persistent trajectories. *Proceedings of the 36th IEEE Conference on Decision and Control*, pp. 714–719, Brighton, UK.

Diop, S. and Fliess, M. (1991b) On nonlinear observability. *Proceedings of the European Control Conference*, pp. 152–157, Hermès, Paris.

Diop, S. Grizzle, J. and Chaplais, F. (2000) On numerical differentiation algorithms for nonlinear estimation. *Proceedings of the 39th IEEE Conference on Decision and Control*, pp. 1133–1138, Sydney, Australia.

Diop, S. Grizzle, J. Morral, PE. and Stefanoupoulou, AG. (1993) Interpolation and numerical differentiation for observer design. *Proceedings of the American Control Conference*, pp. 1329–1333, Evanston, IL.

Doyle, J. and Stein, G. (1979) Robustness with observers. *IEEE Transactions on Automatic Control* **AC-24**(4), 607–611.

Eykhoff, P. (1974) *System Identification. Parameter and State Estimation*. Wiley-Interscience, London, chapter 8.

Favier, G. (1982) *Filtrage, modélisation et identification des systèmes linéaires, stochastiques à temps discret*. CNRS.

Fliess, M. and Sira-Ramírez, H. (2003) An algebraic framework for linear identification. *ESAIM, Control, Optimization and Calculus of Variations* **9**(1), 151–168.

Fliess, M. Join, C. and Sira-Ramírez, H. (2004) Robust residual generation for linear fault diagnosis: An algebraic setting with examples. *International Journal of Control* **77**(14), 1223–1242.

Fliess, M. Join, C. and Sira-Ramírez, H. (2008) Non-linear estimation is easy. *International Journal of Modelling, Identification and Control* **4**(1), 12–27.

Fliess, M. Mboup, M. Mounier, H. and Sira-Ramírez, H. (2003) *Algebraic Methods in Flatness, Signal Processing and State Estimation*. Editiorial Lagares, Mexico, chapter 1.

García-Rodríguez, C. Cortés-Romero, J. and Sira-Ramírez, H. (2009) Algebraic identification and discontinuous control for trajectory tracking in a perturbed 1-dof suspension system. *IEEE Transactions on Industrial Electronics* **56**(9), 3665–3674.

Garrido, R. and Concha, A. (2013) An algebraic recursive method for parameter identification of a servo model. *IEEE/ASME Transactions on Mechatronics* **18**(5), 1572–1580.

Gauss, KF. (1809) *Theory of the Motion of the Heavenly Bodies Moving About the Sun in Conic Sections*. Theoria Motus. Translated by Charles Henry. Davis, Dover Publications, Inc., New York, 1847.

Gensior, A. Weber, J. Rudolph, J. and Guldner, H. (2008) Algebraic parameter identification and asymptotic estimation of the load of a boost converter. *IEEE Transactions on Industrial Electronics* **55**(9), 3352–3360.

Gevers, M. (2006) A personal view of the development of system identification, a 30-year journey through an exiting field. *IEEE Control Systems Magazine* **26**(6), 93–105.

Goodwin, G. and Sin, K. (1984) *Adaptive Filtering, Prediction and Control*. Prentice-Hall, Englewood Cliffs, NJ.

Hermann, R. and Krener, A. (1977) Nonlinear controlability and observability. *Transactions on Automatic Control* **AC-22**, 728–740.

Isidori, A. (1989) *Nonlinear Control Systems*. Springer-Verlag, New York.

Johansson, R. (1993) *System Modelling and Identification*. Prentice-Hall, Englewood Cliffs, NJ.

Kailath, T. (1974) A view of three decades of linear filtering theory. *Transactions of the IEEE* **IT-20**(2), 146–181.

Kalman, R. (1959) On the general theory of control systems. *IRE Transactions on Automatic Control* **4**(3), 110–110.

Kalman, R. (1960) A new approach to linear filtering and prediction. *Transactions of the ASME. Series D, Journal of Basic Engineering* **82**, 35–45.

Kalman, R. and Bucy, R. (1961) New results in linear filtering. *Transactions of the ASME. Series D, Journal of Basic Engineering* **83**, 95–108.

Kalman, RE. (1963) Mathematical description of linear dynamical systems. *Journal of the Society for Industrial and Applied Mathematics, Series A: Control*.

Khalil, HK. (1999) *New Directions in Nonlinear Observer Design*, vol. 244 of Lecture Notes in Control and Information Sciences. Springer-Verlag, Berlin, pp. 249–268.

Khalil, HK. (2008) High-gain observers in nonlinear feedback control. *Proceedings of the International Conference on Control, Automation and Systems*, pp. 47–57, Seoul, Korea.

Kolmogorov, A. (1941) Interpolation and extrapolation of stationary random sequences. *Bulletin of the Academy of Sciences of URSS, Series Math*. Translated by W. Doyle, and I. Selin, RAND Corporation Report RM-3090-PR, Santa Monica, CA, (1962).

Landau, I. (1990) *System Identification and Control Design*. Prentice-Hall, Englewood Cliffs, NJ.

Levant, A. (1998) Robust exact differentiation via sliding mode technique. *Automatica* **34**(3), 379–384.

Linares Flores, J. Avalos, JLB. and Espinosa, C. (2011) Passivity-based controller and online algebraic estimation of the load parameter of the dc-to-dc power converter CUK type. *Latin America Transactions, IEEE (Revista IEEE America Latina)* **9**(1), 784–791.

Ljung, L. (1987) *System Identification: Theory for the User*. Prentice-Hall, Englewood Cliffs, NJ.

Luenberger, DG. (1964) Observing the state of a linear system. *IEEE Transactions on Military Electronics* **MIL-8**, 74–80.

Luenberger, DG. (1966) Observers for multivariable systems. *IEEE Transactions on Automatic Control* **11**, 190–197.

Luenberger, DG. (1971) An introduction to observers. *IEEE Transactions on Automatic Control* **AC-16**(6), 596–602.

Mayr, O. (1970) *Origins of Feedback Control*. MIT Press, Cambridge, MA.

Michel, L. Join, C. Fliess, M. Sicard, P. and Cheriti, A. (2010) Model-free control of dc/dc converters. *2010 IEEE 12th Workshop on Control and Modeling for Power Electronics (COMPEL)*, pp. 1–8, Kyoto, Japan.

Nijmeijer, H. and van der Schaft, A. (1990) *Non Linear Dynamical Control Systems*. Springer-Verlag, Berlin.

Sastry, S. and Bodson, M. (1989) *Adaptive Control, Stability, Convergence and Robustness*. Prentice-Hall, Englewood Cliffs, NJ.

Söderström, P. and Stoica, P. (1989) *System Identification*. Prentice-Hall, Englewood Cliffs, NJ.

Sorenson, HW. (1980) *Parameter Estimation, Principles and Problems*, vol. 9 of Control and Systems Theory. Marcel Dekker, New York.

Sussmann, H. (1977) Existence and uniqueness of minimal realizations of nonlinear systems. *Mathematical Systems Theory* **10**, 263–284.

Sussmann, H. (1979) Single input observability of continuous time systems. *Mathematical Systems Theory* **12**, 371–393.

Trapero-Arenas, J. Mboup, M. Pereira-Gonzalez, E. and Feliu, V. (2008) On-line frequency and damping estimation in a single-link flexible manipulator based on algebraic identification. *16th Mediterranean Conference on Control and Automation, 2008*, pp. 338–343, Ajaccio Corsica, France.

Villagra, J. d'Andrea Novel, B. Fliess, M. and Mounier, H. (2008) Estimation of longitudinal and lateral vehicle velocities: An algebraic approach. *Proceedings of the American Control Conference, 2008*, pp. 3941–3946, Seattle, WA.

Wiener, N. (1949) *The Extrapolation, Interpolation and Smoothing of Stationary Time Series*. John Wiley & Sons, Inc., New York.

Zadeh, L. (1956) On the identification problem. *IRE Transactions on Circuit Theory* **3**(4), 277–281.

2

Algebraic Parameter Identification in Linear Systems

2.1 Introduction

As in every parameter-identification scheme, the algebraic identification methodology requires a rather precise knowledge of the model of the plant. The fundamental assumption is that the structure of the model of the system, as it relates to the set of differential equations describing it, is quite well known except for some of the parameters involved. The algebraic identification method is model based and aimed at obtaining an exact, static, formula for the unknown parameters. This formula is based on measurable quantities, such as inputs and outputs; sometimes inputs and measured states are used. The parameter-calculation formulae are obtained via specific algebraic manipulations carried out on the model equations. Generally speaking, these manipulations entitle: time differentiations, iterated integrations of the measured variables with suitable powers of the time variable (i.e., input and output convolutions), integrations by parts, and, possibly, low-pass filtering. These operations are known to eliminate the influence of initial conditions, and, also, the so-called classical perturbations (step inputs, ramps, etc.). The net result is a set of possibly linear relations in the unknown parameters. In some instances, one obtains a set of relations involving nonlinear functions of such parameters. It is assumed that an accurate calculation of the parameters may be devised immediately from these relations.

In classical adaptive identification techniques, a set of parameters can be calculated provided the input belongs to the class of "persistently exciting." In contrast, algebraic identification does not require such persistency of excitation. The method does not depend on asymptotic convergence analysis or Lyapunov stability theory. In fact, the identification process via algebraic methods is quite fast (i.e., almost instantaneous) in nature. The classical "persistency of excitation condition" characteristic of slow parameter convergence in traditional adaptive control schemes is replaced by an algebraic "consistency condition" which avoids singularities over small open intervals of time (see Fliess and Sira-Ramírez, 2003). The control inputs to be used do not interfere with the desired control objectives, and the counterpart of the algebraic requirements is that the consistency condition is valid only during a small time interval; the one that is just required to compute the parameter via the static formula.

Algebraic Identification and Estimation Methods in Feedback Control Systems, First Edition.
Hebertt Sira-Ramírez, Carlos García-Rodríguez, Alberto Luviano-Juárez, John Alexander Cortés-Romero.
© 2014 John Wiley & Sons, Ltd. Published 2014 by John Wiley & Sons, Ltd.

Although the calculation formulae for the parameters turn out to be singular at time $t = 0$, in the lapse of a quite small time interval, say $[0, +\epsilon]$ for $\epsilon \in \mathbb{R} > 0$, they yield the right parameter value provided the signal-to-noise ratio is sufficiently high. An *invariant filtering* procedure is usually invoked to improve this ratio in noisy cases (see Fliess *et al.*, 2008; García-Rodríguez *et al.*, 2009).

2.1.1   The Parameter-Estimation Problem in Linear Systems

This chapter begins our exposition on algebraic identification methods in feedback control design. We start with the parameter identification problem in the specific context of linear time-invariant systems. Whether we are facing a state or an input–output representation of the system, we first restrict ourselves to the case of systems undergoing structured perturbations (i.e., perturbations generated by homogeneous, linear, time-invariant systems whose initial states and parameters are not known). The most elementary set of perturbation inputs belonging to this class is constituted by the so called classical perturbation inputs. Specifically, we address constant perturbations, ramp perturbations, and, in general, polynomial perturbations. The other most common types of perturbations also belonging to this class are the sinusoidal perturbations, exponentially fading perturbations, and so on. We focus on identifying the system parameters in an online fashion. The main objective in identifying the unknown plant parameters is to specify precisely the required feedback controller. We therefore assume that such parameter information is lacking but crucially needed in the feedback controller specification. Naturally, there are instances in which we do not need to know the system parameters for the specification of the controller, as is the case in many classical feedback designs. For instance, the great majority of existing industrial proportional integral derivative (PID) controllers do not need *a priori* knowledge about the plant parameters. In those cases, extensive tuning is required or more complex automatic adaptation algorithms are advised. We do not advocate either of the two previous possibilities. We therefore center our developments around certainty-equivalence controllers designed for systems whose models are well known except for, possibly, some of their constant parameters. The feedback controllers are derived as if the system parameters were perfectly known. Using arbitrary values, or well-educated guesses of such parameters, we produce an incipient input–output evolution of the system with the help of the "wrong" controller. If the inputs and outputs are properly processed, we can obtain, in a very short amount of time, the precise values of the system parameters. Once the parameters are identified with accuracy, their values are immediately replaced in the designed certainty-equivalence feedback controller, thus imposing the "right" controller. Naturally, the identification of the actual values of the parameters is to be carried out online, in a very short amount of time, and in the presence of plant and measurement noises. The basic assumption is that the parameters are, indeed, constant throughout time. At the end of the chapter, we relax this assumption and allow the parameters to be piecewise constant (i.e., they change to new constant values at unpredictable instants in time). This requires an updating strategy for our parameter computations. We address this strategy as the *resetting* of the identifier. The resetting can be carried out either automatically or periodically.

The theoretical basis for the online algebraic identification method used in the several examples presented here is found in Fliess and Sira-Ramírez (2003). Finally, the addition of an extra known artificial state as a "sentinel" parameter is proposed in Section 2.3 in order to have a criterion for achieving sufficiently precise parameter estimation in the presence of noise. Two major approaches exist for the algebraic identification process: a time-domain approach and a frequency-domain approach. The second is essentially centered around linear time-invariant systems. Both approaches are entirely equivalent.

This chapter considers a series of examples aimed at familiarizing the reader with the details of the algebraic parameter-identification procedure. The variants and complements of the method are also presented via examples. For instance, the operational calculus-based (frequency-domain) alternative is introduced through simple-enough examples dealing with a visual servoing problem and the balancing of a plane rotor. Some extensions of the method into systems with nonlinearities are also presented in terms of easy-to-understand application examples. This is the case of systems exhibiting Coulomb frictions or requiring the computation of a nonlinear function of some parameter for control purposes (as in the linear motor example). Switched control systems of linear average nature can also use the algebraic identification procedure for fast parameter determination, as demonstrated by a power electronics example. Piecewise linear systems are not foreign to the benefits of online identification, as shown in the variable gain motor illustrative example. A complement of the algebraic identification procedure is represented by invariant filtering and by the introduction of a sentinel parameter. This is represented by a known parameter that is algebraically identified in order to have an objective criterion for stopping, or disconnecting, the identification process. This issue is examined in the context of a sliding-mode-controlled suspension system example at the end of the chapter.

2.2 Introductory Examples

2.2.1 Dragging an Unknown Mass in Open Loop

Consider the problem of dragging an unknown mass along a straight line over a frictionless horizontal surface . The model of the system is, from Newton's second law, given by

$$m\ddot{x} = u(t) \tag{2.1}$$

where x is the mass displacement, perfectly measured from some reference point or origin of coordinates, precisely marked on the line, and labeled 0. The quantity $u(t)$ is the applied force, which is assumed to be non-identically zero over any open interval of time. To make the problem simple, assume that $u(t)$ is a known, nonzero, open-loop control force at our disposal. The entire purpose of applying such a force to the mass is to gather some input–output information so that we can identify the unknown mass parameter m. The mass is initially placed at a distance x_0 from the origin, that is, $x(t_0) = x_0$ with an initial velocity $\dot{x}(t_0) = \dot{x}_0$.

The model itself suggests that a relation for m may be obtained by integrating the equation twice, from the initial instant t_0, and using the fact that m is constant. We obtain

$$m[x(t) - x_0 - (t - t_0)\dot{x}_0] = \int_{t_0}^{t} \int_{t_0}^{\sigma} u(\lambda)d\lambda d\sigma \tag{2.2}$$

It is clear from this last relation that a new piece of information becomes necessary: $\dot{x}(t_0) = \dot{x}_0$, the initial velocity. If the experiment is started with the mass at rest, we know we can set $\dot{x}(t_0) = \dot{x}_0 = 0$, but then the measurement of the initial position x_0 becomes important. Somehow it is troubling that our estimation of the mass should depend on *where* we start the experiment along the line. We reach the conclusion that our mass-identification process should not depend on any of the initial conditions. To achieve this independence, multiply the system equation by $(t - t_0)^2$. We obtain

$$m(t - t_0)^2\ddot{x} = (t - t_0)^2 u(t) \tag{2.3}$$

Integrating once, between t_0 and t, the last expression and using the integration by parts formula in the left-hand side we obtain

$$m\left[(t - t_0)^2\dot{x}(t) - 2\int_{t_0}^{t}(\sigma - t_0)\dot{x}(\sigma)d\sigma\right] = \int_{t_0}^{t}(\sigma - t_0)^2 u(\sigma)d\sigma \tag{2.4}$$

In other words,

$$m\left[(t - t_0)^2\dot{x}(t) - 2(t - t_0)x(t) + 2\int_{t_0}^{t}x(\sigma)d\sigma\right] = \int_{t_0}^{t}(\sigma - t_0)^2 u(\sigma)d\sigma \tag{2.5}$$

We succeeded in making the expression independent of the initial conditions, but now we need to know the entire velocity history of the mass. We integrate once more to finally obtain:

$$m\left[(t - t_0)^2 x(t) - 4\int_{t_0}^{t}(\sigma - t_0)x(\sigma)d\sigma + 2\int_{t_0}^{t}\int_{t_0}^{\sigma}x(\lambda)d\lambda d\sigma\right]$$
$$= \int_{t_0}^{t}\int_{t_0}^{\sigma}(\lambda - t_0)^2 u(\lambda)d\lambda d\sigma \tag{2.6}$$

This expression has the advantage of being completely independent of the initial conditions and it only requires the measurement of the input and output to compute either m or its inverse $1/m$. We then have the following calculation formula for $1/m$:

$$\frac{1}{m} = \frac{n(t)}{d(t)}$$

$$n(t) = (t - t_0)^2 x(t) - 4\int_{t_0}^{t}(\sigma - t_0)x(\sigma)d\sigma + 2\int_{t_0}^{t}\int_{t_0}^{\sigma}x(\lambda)d\lambda d\sigma$$

$$d(t) = \int_{t_0}^{t}\int_{t_0}^{\sigma}(\lambda - t_0)^2 u(\lambda)d\lambda d\sigma \tag{2.7}$$

There is, nevertheless, a problem with the previous formula; at time $t = t_0$, both the numerator and the denominator are zero. The quotient is clearly undetermined. We must, therefore, begin to evaluate the formula not at time t_0 but at a later time, say $t_0 + \epsilon$ with $\epsilon > 0$ and small. We propose then the following estimation process for $1/m$, denoted now as $1/m_e$ to emphasize that it is an estimated quantity:

$$\frac{1}{m_e} = \begin{cases} \text{arbitrary} & \text{for } t \in [t_0, t_0 + \epsilon) \\ \\ \frac{n(t)}{d(t)} & \text{for } t > t_0 + \epsilon \end{cases} \tag{2.8}$$

The evaluation of the quotient is, of course, valid as long as the denominator does not go through zero. Note that the denominator cannot be identically zero on any open interval of time. For, suppose this to be true, then $u(t)$ is also identically zero on such an interval, which contradicts our assumption about the test signal $u(t)$.

In order to implement the calculations, and given that time integrations are needed to synthesize the numerator and the denominator expressions, we would like to give to these two quantities the character of an output of a certain dynamic system involving differential equations. We propose then the following linear time-varying "filters":

$$
\begin{cases}
n(t) = & (t - t_0)^2 x(t) + z_1 \\
\dot{z}_1 = & -4(t - t_0)x(t) + z_2 \\
\dot{z}_2 = & 2x(t)
\end{cases}
\qquad
\begin{cases}
d(t) = & \eta_1 \\
\dot{\eta}_1 = & \eta_2 \\
\dot{\eta}_2 = & (t - t_0)^2 u(t)
\end{cases}
\tag{2.9}
$$

with $z_1(t_0) = z_2(t_0) = 0$ and $\eta_1(t_0) = \eta_2(t_0) = 0$.

Figure 2.1 depicts the involved signals, the numerator $n(t)$, the denominator $d(t)$, the input u, the output $y = x$, and the estimate of $1/m$. The wrong guess for the parameter value during the time interval $[0, \epsilon)$ was taken to be $1/m_e = 0.5\,\text{kg}^{-1}$, as can be seen from the figure. We have set, in this case, $\epsilon = 0.01$ s, but yet a smaller real value could have been used. Also, we have let $u(t) = 1$ N for all t. The actual value of the mass was set to $m = 1$ kg.

Several distinctive features emerge from the simulations of this rather simple example: (1) The estimation of the mass parameter can be reliably achieved in a quite short amount of time that only depends on the arithmetic processor precision in being able to carry out the quotient of two very small quantities; namely, the numerator and denominator signals. (2) The test input signal $u(t)$ being used does not exhibit the classical " persistency of excitation" requirement. (3) The estimator of the inverse mass parameter is comprised of unstable signals in both the numerator and the denominator.

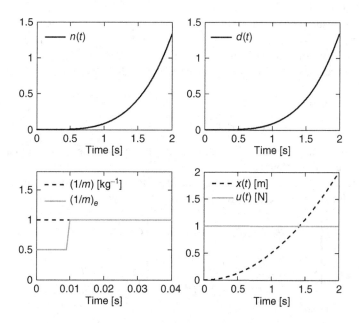

Figure 2.1 Identification of the inverse mass parameter

Regarding the first observation above, we should remark that the accurate precision with which we have obtained the mass parameter is not at all surprising, due to the fact that the formula used is as exact as the model and, very importantly, because we have not included any measurement noise in our simulations. This last feature may compromise not only the precision of the computation but, also, the fast character of the identification. The second feature, of not needing a persistently exciting signal, is certainly an unchallenged advantage. The last negative feature regarding our internally unstable scheme may be overcome in a simple manner by prescribing the need to, at least temporarily, "switch off" the estimator right after the precise parameter estimation is obtained. The noise-related aspects seem to be essential. We propose below a possible approach.

The expression

$$\frac{1}{m} = \frac{n(t)}{d(t)} \tag{2.10}$$

is evidently equivalent to the following one:

$$n(t) = \left[\frac{1}{m}\right] d(t) \tag{2.11}$$

Clearly, the numerator and denominator signals are exponentially bounded, although unstable, signals. Let $G(s)$ denote the rational transfer function of a stable low-pass filter. Abusing the notation, we can write

$$G(s)n(t) = \left[\frac{1}{m}\right] G(s)d(t) \tag{2.12}$$

In other words:

$$\frac{1}{m} = \frac{G(s)n(t)}{G(s)d(t)} \tag{2.13}$$

Whenever the signals $d(t)$ and $n(t)$ are unaffected by additive noise, the previous expressions show that the value of the inverse mass is invariant with respect to low-pass filtering. If, on the contrary, the measurement of the output signal of the system feeding $n(t)$, or that of the input signal feeding $d(t)$, is affected by noise, the invariant filtering has the effect of bettering the signal-to-noise ratio in both the numerator and denominator signals. The quotient, therefore, enjoys greater accuracy. Note that $G(s)$ may be constituted of a pure integration chain of certain finite length.

Consider then the perturbed system

$$m\ddot{x} = u(t) + \zeta(t), \qquad y(t) = x(t) + \xi(t) \tag{2.14}$$

where $\xi(t)$ and $\zeta(t)$ are zero-mean, rapidly varying noise processes. $\zeta(t)$ is a plant noise input and $\xi(t)$ is a measurement noise.

Corresponding to the inverse mass-parameter estimation, we propose then the following time-varying filters, with second-order integration low-pass-filtered outputs:

$$\begin{cases} n(t) &= z_1 \\ \dot{z}_1 &= z_2 \\ \dot{z}_2 &= (t-t_0)^2 y(t) + z_3 \\ \dot{z}_3 &= -4(t-t_0)y(t) + z_4 \\ \dot{z}_4 &= 2y(t) \end{cases} \qquad \begin{cases} d(t) &= \eta_1 \\ \dot{\eta}_1 &= \eta_2 \\ \dot{\eta}_2 &= \eta_3 \\ \dot{\eta}_3 &= \eta_4 \\ \dot{\eta}_4 &= (t-t_0)^2 u(t) \end{cases} \tag{2.15}$$

Here $z_1(t_0) = z_2(t_0) = z_3(t_0) = z_4(t_0) = 0$ and $\eta_1(t_0) = \eta_2(t_0) = \cdots = \eta_4(t_0) = 0$.

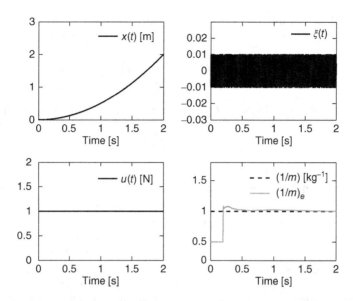

Figure 2.2 Identification of the inverse mass parameter under noise measurements and using invariant filtering

We set $\xi(t)$ and $\zeta(t)$ to be computer-generated random processes, respectively consisting of piecewise constant random variables uniformly distributed in the intervals $[-0.01, 0.01]$ m and $[-0.5, 0.5]$ N of the real line.

Figure 2.2 depicts the outcome of the invariant filtering modification of our previously proposed parameter-estimation scheme for the unknown dragged mass. We note that a larger ϵ parameter was used in this instance ($\epsilon = 0.2$ s) to allow for a reliable quotient yielding the inverse mass, once the signal-to-noise ratio becomes important in the numerator.

2.2.1.1 Controlling an Unknown Mass

The procedure outlined in the previous example does not really distinguish between open-loop and closed-loop schemes. The values of u intervene, in either case, in exactly the same manner in the calculation formula for the unknown mass. To illustrate this fact, we now pose the problem of controlling the unknown mass along a prespecified trajectory. We also include a Coulomb friction perturbation term in order to show how our identification algorithm naturally blends with existing robust feedback-control strategies.

Consider then the system with unknown mass

$$m\ddot{x} = u(t) - \mu \, sign(\dot{x})$$

$$y = x \tag{2.16}$$

where μ is the unknown constant amplitude of the Coulomb friction force. Note that while the mass is in motion, the Coulomb friction term is piecewise constant. Close to the equilibrium ($\dot{x} \approx 0$), this term exhibits a typical "bang–bang" behavior. If the mass parameter m were completely known, we would find that the following generalized proportional integral (GPI) controller (see Fliess *et al.* 2002 for details), written in classical form, yields the accurate tracking of a given trajectory $x^*(t)$ in spite of the unknown Coulomb friction perturbation:

$$u = m \left[\ddot{x}^*(t) - \left[\frac{k_2 s^2 + k_1 s + k_0}{s(s + k_3)} \right] (x - x^*(t)) \right] \tag{2.17}$$

where the set of coefficients $\{k_2, \cdots, k_0\}$ is chosen so that the closed-loop characteristic polynomial

$$p(s) = s^4 + k_3 s^3 + k_2 s^2 + k_1 s + k_0 \tag{2.18}$$

exhibits all its roost in the left half of the complex plane.

The performance of the GPI controller, when the mass is known, is found to be satisfactory as judged from Figure 2.3. Here we set the task of smoothly dragging the mass from the initial rest position to a final rest position 5 m away in 5 seconds.

The identification of the unknown mass parameter is now faced with the effects of the unknown perturbation term coming from the Coulomb friction phenomenon. Since this perturbation adopts the form of a piecewise constant perturbation while the object is moving, we may proceed as if we were dealing with a constant perturbation. The idea is that if this perturbation remains constant for a small period of time, then we should be able to accurately estimate the parameter within that period. But, clearly, for this we also have to make the identification process independent of the friction force amplitude μ, whatever its sign. In contrast, when the mass velocity is in the vicinity of zero, the Coulomb friction term exhibits a rapidly varying bang–bang behavior. This is tantamount to a high-frequency noise which, as far as the parameter-estimation process is concerned, will be averaged out to zero by the iterated integrations being considered.

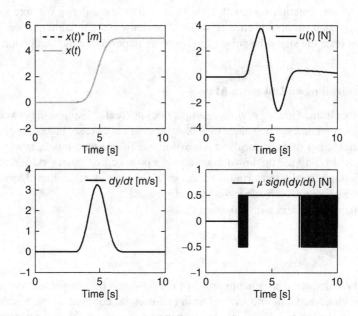

Figure 2.3 Performance of GPI controller for trajectory tracking under Coulomb friction perturbation

Taking one time derivative of the model, under the assumption that the Coulomb term is temporarily yielding a constant force, we obtain

$$mx^{(3)} = \dot{u}$$

$$y = x \tag{2.19}$$

Thus, in this case, we multiply out the modified system model by $(t - t_0)^3$ and integrate by parts three times both members of the expression. We obtain[1]

$$\frac{1}{m} = \frac{6\left(\int_{t_0}^{(3)} y\right) - 18\left(\int_{t_0}^{(2)}(t - t_0)y\right) + 9\left(\int_{t_0}(t - t_0)^2 y\right) - (t - t_0)^3 y}{\left(\int_{t_0}^{(3)}(t - t_0)^2 u\right) - \left(\int_{t_0}^{(2)}(t - t_0)^3 u\right)} \tag{2.20}$$

In order to overcome possible high-frequency noise effects present in the measurement of the position variable, one further integrates two more times the numerator and denominator expressions. We get the following formula for the estimation of the unknown mass in the presence of measurement noises and friction perturbation inputs:

$$\frac{1}{m_e} = \begin{cases} \text{arbitrary} & \text{for } t \in [t_0, t_0 + \epsilon) \\ \frac{n(t)}{d(t)} & \text{for } t \geq t_0 + \epsilon \end{cases}$$

$$\begin{cases} n(t) = \zeta_1 \\ \dot{\zeta}_1 = \zeta_2 \\ \dot{\zeta}_2 = \zeta_3 - (t - t_0)^3 y(t) \\ \dot{\zeta}_3 = \zeta_4 + 9(t - t_0)^2 y(t) \\ \dot{\zeta}_4 = \zeta_5 - 18(t - t_0)y(t) \\ \dot{\zeta}_5 = 6y(t) \end{cases} \qquad \begin{cases} d(t) = \eta_1 \\ \dot{\eta}_1 = \eta_2 \\ \dot{\eta}_2 = \eta_3 \\ \dot{\eta}_3 = \eta_4 \\ \dot{\eta}_4 = \eta_5 - (t - t_0)^3 u(t) \\ \dot{\eta}_5 = 3(t - t_0)^2 u(t) \end{cases} \tag{2.21}$$

where $y(t)$ is the measured mass position signal, assumed to undergo unavoidable noise processes characterized by their high-frequency or rapidly varying nature: $y(t) = x(t) + \xi(t)$.

The GPI trajectory-tracking controller plays the role of a *certainty-equivalence* controller (Åstrom and Eykhoff, 1971). As such, it should now be specified in an adaptive manner as follows:

$$u = m_e \left[\ddot{y}^*(t) - \left[\frac{k_2 s^2 + k_1 s + k_0}{s(s + k_3)} \right] (y - y^*(t)) \right] \tag{2.22}$$

where $y^*(t) = x^*(t)$ is the desired position-reference trajectory and m_e is the estimated mass.

Figure 2.4 depicts the simulations of a closed-loop controlled trajectory for an unknown mass with unknown Coulomb friction terms affecting the mass dynamics. The feedback controller is a certainty-equivalence adaptive GPI controller based on fast online identification of the mass parameter. Clearly, the parameter-identification formula advises not to start the mass at a rest position where the control input is identically zero. A small-state perturbation from the equilibrium position triggers a control action that is sufficient to identify the parameter in $\epsilon = 0.1$ s.

---

[1] We use here the following notation: $(\int_{t_0}^{(n)} \phi(t))$, which represents the iterated integral $\int_{t_0}^{t} \int_{t_0}^{\sigma_1} \cdots \int_{t_0}^{\sigma_{n-1}} \phi(\sigma_n) d\sigma_n \cdots d\sigma_1$ with $(\int_{t_0} \phi(t)) = (\int_{t_0}^{(1)} \phi(t)) = \int_{t_0}^{t} \phi(\sigma) d\sigma$. Whenever $t_0 = 0$, we use $(\int^{(n)} \phi(t))$.

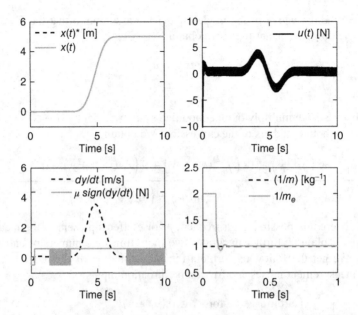

Figure 2.4 GPI-controlled trajectory tracking for uncertain mass under Coulomb friction perturbation and measurement noise with online fast identification

In the meantime, a "wrong" value of the mass is used in the GPI controller. The position measurement $y = x + \xi(t)$ is carried out under noisy conditions. We used a computer-generated noise for the signal $\xi(t)$ consisting of a random process synthesized on the basis of a piecewise constant random variable uniformly distributed in the interval $[-0.5a, 0.5a]$ with $a = 10^{-2}$ m.

2.2.2 A Perturbed First-Order System

We now turn our attention to those cases where more than one parameter is present in the system and the need arises for a closed-loop identification. As before, regarding the control part, we resort to the certainty-equivalent control method. As for how to handle several parameters, we will have to generate as many algebraic equations as there may be unknown parameters.

Consider the following linear, parameter-uncertain, perturbed, first-order system:

$$\dot{y} = ay + bu + \kappa + \xi(t) \tag{2.23}$$

where a, b, and κ are unknown constants and $\xi(t)$ is a zero-mean noise process. We assume, only temporarily, that the output signal y, can be measured precisely without further noisy perturbations.

We would like to specify a feedback control law such that the following problem finds a solution: *Devise a feedback control law that forces the output signal y, to follow a given reference trajectory $y^*(t)$, in spite of the lack of knowledge about the plant parameters, a and b, the uncertainty about the constant perturbation, κ, and the presence of the zero-mean, rapidly varying, plant perturbation noise.*

2.2.2.1 A Certainty-Equivalence Controller

In order to start the quest for a solution to the posed problem, we ask ourselves what type of controller we need if the parameters a and b are perfectly known and, moreover, if there exists no plant perturbation noise (i.e., if $\xi(t) = 0$ for all t). We let the system still be affected by the constant perturbation κ. We thus consider the noise-free dynamics, with perfectly known a and b, and abusively keep the same notation for the output y. We assume throughout that $t_0 = 0$:

$$\dot{y} = ay + bu + \kappa \tag{2.24}$$

Under these circumstances, we would certainly use a classical proportional integral controller given its robustness with respect to constant, unknown, perturbations:

$$u = \frac{1}{b}\left[\dot{y}^*(t) - ay - k_1(y - y^*(t)) - k_0 \int_0^t \left(y(\sigma) - y^*(\sigma)\right) d\sigma\right] \tag{2.25}$$

Defining $e := y - y^*(t)$, the tracking error, we have that the closed-loop tracking error satisfies

$$\dot{e} + k_1 e + k_0 \int_0^t e(\sigma)d\sigma - \kappa = 0 \tag{2.26}$$

Define an auxiliary state $z(t)$ as

$$z := \int_0^t e(\sigma)d\sigma - \frac{\kappa}{k_0} \tag{2.27}$$

Clearly, $z(0) = -\kappa/k_0$. The resulting closed-loop error dynamics may then be written, in matrix form, as

$$\frac{d}{dt}\begin{bmatrix} e \\ z \end{bmatrix} = \begin{bmatrix} -k_1 & -k_0 \\ 1 & 0 \end{bmatrix}\begin{bmatrix} e \\ z \end{bmatrix} \tag{2.28}$$

The characteristic polynomial of the resulting closed-loop system is given by

$$p(s) = \det\begin{bmatrix} s + k_1 & k_0 \\ -1 & s \end{bmatrix} = s^2 + k_1 s + k_0 \tag{2.29}$$

For suitable choices of the design parameters k_1 and k_0 so that the roots of $p(s)$ are all strictly located in the left portion of the complex plane, the system responses are known to exponentially and asymptotically converge toward zero. The tracking task is asymptotically accomplished.

We must therefore concentrate our efforts on obtaining the right values of a and b.

2.2.2.2 Parameter Identification

Suppose $\xi(t)$ is identically zero. We try to generate a linear system of equations for the unknown parameters a and b. This system should be independent of the plant initial condition, independent of the constant perturbation κ and, moreover, it should rely only on knowledge of the signals y and u (i.e., the input and output). This last consideration is obviously much more relevant in higher-dimensional systems, but it is convenient that we become accustomed to constraining ourselves to the fact that only inputs and outputs will be available (in some other cases not even inputs will be available).

We perform, to get rid of initial conditions and constant perturbations, the following seemingly unrelated steps:

- differentiation of the plant equations (once),
- multiplication by t^2, the square value of time,
- iterated integration of the resulting expressions a sufficient number of times—in this case, we shall integrate only twice.

We then start with the noise-free dynamics. Differentiation yields

$$\ddot{y} = a\dot{y} + b\dot{u} \tag{2.30}$$

while multiplication by t^2 results in

$$t^2\ddot{y} = at^2\dot{y} + bt^2\dot{u} \tag{2.31}$$

Integrating twice the resulting expression, we have

$$\int^{(2)} t^2\ddot{y} = \int t^2\dot{y} - 2\int^{(2)} t\dot{y} = t^2 y - 4\int ty + 2\int^{(2)} y$$

$$\int^{(2)} t^2\dot{y} = \int t^2 y - 2\int^{(2)} ty$$

$$\int^{(2)} t^2\dot{u} = \int t^2 u d\sigma - 2\int^{(2)} tu \tag{2.32}$$

We then obtain:

$$\left[\int t^2 y - 2\int^{(2)} ty\right] a + \left[\int t^2 u - 2\int^{(2)} tu\right] b = t^2 y(t) - 4\int ty + 2\int y \tag{2.33}$$

Integrating once more, we obtain a linear system for the constant parameters a and b. We arrive at a linear time-varying equation of the form

$$P(t, y, u)\begin{bmatrix} a \\ b \end{bmatrix} = q(t, y) \tag{2.34}$$

with the matrix $P(t, y, u)$ and the vector $q(t, y)$ being given by

$$P(t, y, u) = \begin{bmatrix} p_{11}(t, y) & p_{12}(t, u) \\ p_{21}(t, y) & p_{22}(t, u) \end{bmatrix}$$

where

$$p_{11}(t, y) = \int t^2 y - 2\int^{(2)} ty$$

$$p_{12}(t, u) = \int t^2 u - 2\int^{(2)} tu$$

$$p_{21}(t, y) = \int^{(2)} t^2 y - 2\int^{(3)} ty$$

$$p_{22}(t, u) = \int^{(2)} t^2 u - 2 \int^{(3)} tu$$

$$q(t, y) = \begin{bmatrix} t^2 y - 4 \int ty + 2 \int^{(2)} y \\ \int t^2 y - 4 \int^{(2)} ty + 2 \int^{(3)} y \end{bmatrix}$$

We find that the matrix $P(t, y, u)$ and the vector $q(t, y)$ are singular at time $t = 0$. Nevertheless, it is easy to verify that the matrix $P(t, y, u)$ is indeed invertible at a small time $t = \epsilon > 0$.

Under noise-free circumstances, we may then compute a and b *exactly* at time $t = \epsilon > 0$ regardless of the constant perturbation input κ and, moreover, for any initial condition on the plant dynamics output, or state, y.

2.2.2.3 Effects of Noise

In the previous identification scheme, the effect of system noise is to produce an additive term on $q(t)$ of the form

$$q'(t, \xi) = \begin{bmatrix} \int^{(2)} t^2 \dot{\xi} \\ \int^{(3)} t^2 \dot{\xi} \end{bmatrix} \tag{2.35}$$

Under relatively mild assumptions on the noise, like zero-mean, high-frequency noise of deterministic nature, the integration and convolution operations render the noise contributions negligible during intervals of the form $[0, \epsilon]$. Low-pass filtering attenuates even further such a contribution to the parameter calculation formula.

In order to have an intuitive feeling for such a fact, recall, for example, that iterated integral convolutions of a high-frequency periodic signal, like $\sin(\omega t)$, yield negligible contributions, over an ϵ length interval of the noisy signal to the identified values

$$\int_0^\epsilon \int_0^\sigma \lambda^2 \omega \sin(\omega \lambda) d\lambda d\sigma = \frac{-\omega^2 \epsilon^2 \cos(\omega \epsilon) + 6 \cos(\omega \epsilon) + 4 \omega \epsilon \sin(\omega \epsilon)}{\omega^3} \tag{2.36}$$

$$\int_0^\epsilon \int_0^\sigma \int_0^\lambda \rho^2 \omega \sin(\omega \rho) d\rho d\lambda d\sigma =$$

$$- \frac{(-\omega^2 \epsilon^2 \cos(\omega \epsilon) + 12[\cos(\omega \epsilon) - 1] + 6 \omega \epsilon \sin(\omega \epsilon))}{\omega^4} \tag{2.37}$$

Clearly, the higher the value of ω, the smaller the contribution of the noise.

In general, for non-differentiable noise inputs the contribution of such a perturbation is of the form

$$q'(t, \xi) = \begin{bmatrix} \left(\int^{(2)} t^2 \dot{\xi} \right) \\ \left(\int^{(3)} t^2 \dot{\xi} \right) \end{bmatrix} = \begin{bmatrix} \int t^2 \xi - 2 \left(\int^{(2)} t\xi \right) \\ \left(\int^{(2)} t^2 \xi - 2 \int^{(3)} t\xi \right) \end{bmatrix} \tag{2.38}$$

which is free of time derivatives of the noisy signal.

Rapidly varying signals, in the sense of a non-standard analysis, can only be approximated. Thus, our last digression establishes the required validity for the invariant filtering that we propose next.

A rigorous approach to the evaluation of noise effects in the algebraic approach to parameter estimation was given in Fliess (2006).

2.2.2.4   Invariant Filtering

Consider the following general expression yielding the online estimation of the unknown constant parameters summarized in the r-dimensional vector Θ:

$$P(t)\Theta = q(t) + q'(t) \tag{2.39}$$

where $P(t)$ is an invertible square matrix of dimensions $r \times r$ and $q(t) \in \mathbb{R}^{r \times 1}$ is a nonzero vector involving iterated integrals of inputs and outputs. We have assumed that the set of parameters can be obtained from a linear, though time-varying, set of algebraic equations.

Let $G(s)$ denote a scalar rational transfer function of the complex variable s, viewed as an operator, representing a low-pass filter. In fact, $G(s)$ may well be a scalar transfer function representing a high order of iterated integrations. Note that if I_r denotes the $r \times r$ identity matrix, the quantity $G(s)I_r$ can be expressed as $\mathrm{diag}G(s)$, that is, as an $r \times r$ diagonal matrix containing the same scalar transfer function of the low-pass filter on each entry of the diagonal matrix.

We have, with an abuse of notation which combines operational calculus or frequency-domain, notation with time-domain quantities:

$$[\mathrm{diag}G(s)P(t)]\Theta = \mathrm{diag}G(s)q(t) + \mathrm{diag}G(s)q'(t) \tag{2.40}$$

If $q'(t)$ is constituted by rapidly varying functions, the time-varying entries of the vector $\mathrm{diag}G(s)q'(t)$ are small relative to the signals in the vector $\mathrm{diag}G(s)q(t)$. The signal-to-noise ratio is clearly enhanced by the proposed low-pass-invariant filtering. As a consequence, the resulting computation of the parameter vector Θ is found to be more precise and less affected by noise compared with the case in which the filtering is not exercised.

2.2.2.5   Feedback Controller

A certainty-equivalence controller of the following form is proposed:

$$u = \frac{1}{\hat{b}}\left[\dot{y}^*(t) - \hat{a}x - k_1(y - y^*(t)) - k_0\int_0^t \left(y(\sigma) - y^*(\sigma)\right)d\sigma\right] \tag{2.41}$$

with

$$\begin{bmatrix}\hat{a}\\\hat{b}\end{bmatrix} = \begin{cases}\text{arbitrary,} \quad \hat{b}\neq 0 \quad \text{for} \quad t\in[0,\epsilon)\\[2ex] P^{-1}(t)q(t) \qquad\quad \text{for} \quad t\in[\epsilon,+\infty)\end{cases} \tag{2.42}$$

Any sudden, unforseen, change of the parameters may be handled by reinitialization of the integrations in the calculations.

2.2.2.6   Simulation Results

Figure 2.5 depicts the fast adaptation system response in a rest-to-rest trajectory-tracking task. As can be seen, the determination of the system parameters happens quite fast, in approximately 4×10^{-3} s. The absence of measurement and plant noises certainly makes the

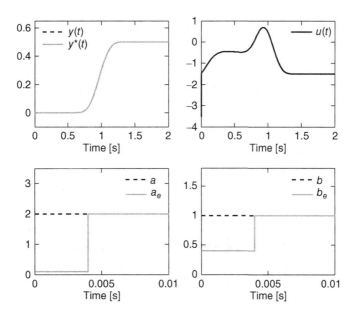

Figure 2.5 System response, parameter determination, and control input

algebraic estimation task quite precise and rather fast. The integral action on the proposed certainty-equivalence controller annihilates the effects of the unknown constant perturbation input, while our estimation technique is shown to be totally independent of the constant perturbation input amplitude.

To carry out our previously proposed algebraic parameter-estimation approach to fast adaptive control, we considered the following intimately related perturbed system:

$$\dot{x} = ax + bu + k + \eta(t), \qquad y(t) = x(t) + \xi(t) \tag{2.43}$$

where $\eta(t)$ and $\xi(t)$ are zero-mean computer-generated noises consisting of a sequence of piecewise constant random variables uniformly distributed in the interval $[-0.5R, 0.5R]$.

For the case of measurement noises, the same computational algorithm was used but now including an invariant filtering strategy. We low-pass filtered both members of each of the algebraic equations derived before for the online calculation of the parameters. A second-order integration was used in each case. In the simulation shown in the next figure, we have assumed a zero-mean computer-generated measurement noise of significant amplitude. The output signal, and the control input signal, do exhibit the influence of the measurement noise but the parameter estimates converge quite precisely and fast enough to the actual value of the parameters. The computation time is substantially increased in the noisy case. Nevertheless, the estimation of the unknown parameters is still quite accurate.

Figure 2.6 depicts the performance of the algebraic parameter identifier including invariant filtering along with the systems response in a rest-to-rest trajectory-tracking task and the evolution of the applied feedback control input. For the measurement noise $\xi(t)$ we have chosen R to be 0.01 and for the plant system noise η the corresponding R value was set to be 0.1.

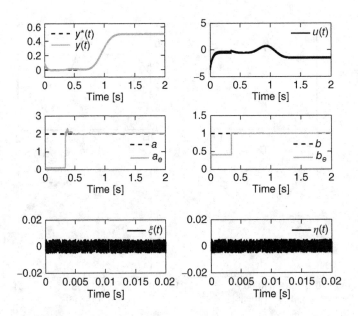

Figure 2.6 System response, parameter determination, and control input

Some systems apparently pose a nonlinear parameter identification problem. In the next example, by resorting to complex variables, the identification problem is ultimately defined on a simple linear system. Moreover, in this case, for control purposes one does not even need to identify all the unknown parameters, but a single nonlinear expression in these parameters.

2.2.3 The Visual Servoing Problem

Consider a planar, two-degree-of-freedom robot, partially controlled with a fast-feedback velocity control loop (Fliess and Sira-Ramírez, 2002). The robot dynamics, characterized by the joint position coordinates $q = [q_1, q_2]^T$, reduces then, to first-order dynamics, to

$$dq/dt' = \tau \tag{2.44}$$

where τ can be regarded as the reference velocity trajectory in the task space, to be generated from a corresponding trajectory in the visual frame coordinates (y_1, y_2):

$$y = \begin{bmatrix} y_1 \\ y_2 \end{bmatrix} \in \mathbb{R}^2 \tag{2.45}$$

The visual flow dy/dt' evolves according to the following dynamics:

$$\frac{dy}{dt'} = a\frac{f}{f - Z} \begin{bmatrix} -1 & 0 \\ 0 & 1 \end{bmatrix} \begin{bmatrix} \cos(\theta) & -\sin(\theta) \\ \sin(\theta) & \cos(\theta) \end{bmatrix} \begin{bmatrix} u_1 \\ u_2 \end{bmatrix} \tag{2.46}$$

where a is the unknown camera scaling factor. The parameter f represents the focal distance of the camera lens, which is assumed to be known. Z is the known orthogonal distance from the lens to the workspace (with $Z > f$). The parameters a, f, and Z are assumed to be constant. The angular position θ is the angle formed by the camera visual frame angular position with respect to the manipulator planar frame. The control inputs u_1, u_2 are the transformed velocity vector reference, $u = \mathcal{J}(q)\tau$.

We introduce a change of time coordinates

$$dt = -\frac{f}{f - Z} \, dt' \tag{2.47}$$

to obtain the model

$$\frac{dy}{dt} = \dot{y} = -a \begin{bmatrix} -\cos(\theta)u_1 + \sin(\theta)u_2 \\ \sin(\theta)u_1 + \cos(\theta)u_2 \end{bmatrix} \tag{2.48}$$

The visual servoing problem has been treated from the perspective of a *nonlinear* adaptive control problem. See, for instance, Hutchinson *et al.* (1996), Kelly (1996), Hsu and Aquino (1999), and Hsu *et al.* (2001).

Define the following complex-valued coordinates:

$$z = y_1 + jy_2, \qquad v = u_1 + ju_2 \tag{2.49}$$

We have

$$\dot{z} = ae^{-j\theta}[u_1 - ju_2] = -ae^{-j\theta}\overline{v} \tag{2.50}$$

Let ξ denote the complex constant $\xi = ae^{-j\theta}$. We then have

$$\dot{z} = \xi\overline{v} \tag{2.51}$$

The visual servoing problem, including a constant perturbation, is formulated as follows: *Given the uncertain first-order complex dynamics for the visual flow (where $H(t)$ is the Heaviside unit step)*

$$\dot{z} = ae^{-j\theta}\overline{v} + \kappa H(t) = \xi \, \overline{v} + \kappa H(t) \tag{2.52}$$

find a complex-valued control input v such that z asymptotically converges toward a given desired trajectory $z^*(t)$, regardless of the unknown character of the parameters a, θ (i.e., regardless of ξ), and κ.

2.2.3.1   A Certainty-Equivalence Controller

A tracking controller, robust w.r.t. unmodeled constant perturbations, which assumes perfect knowledge of the parameter ξ, is given by

$$\overline{v} = \frac{1}{\xi} \left[\dot{z}^*(t) - k_1(z - z^*(t)) - k_0 \int_0^t (z - z^*(\sigma))d\sigma \right] \tag{2.53}$$

The closed-loop tracking error evolves according to

$$\dot{e} = -k_1 e - k_0 \int_0^t e(\sigma)d\sigma \tag{2.54}$$

A remarkable feature of algebraic identification is that one may also proceed to obtain the unknown parameters by considerations in the frequency domain, and then switch to the time domain. We illustrate this feature next.

2.2.3.2 An Algebraic Identifier for ξ

Consider the visual flow dynamics $\dot{z} = \xi \bar{v} + \kappa H(t)$. In the usual operational calculus notation, we obtain

$$sZ(s) - z(0) = \xi \bar{V}(s) + \frac{\kappa}{s} \tag{2.55}$$

Multiplying by s and differentiating the expression twice with respect to s yields an expression free of $z(0)$ and κ

$$s^2 \frac{d^2 Z}{ds^2} + 4s \frac{dZ}{ds} + 2Z = \xi \left[2 \frac{d\bar{V}}{ds} + s \frac{d^2 \bar{V}}{ds^2} \right] \tag{2.56}$$

In the time domain, this expression reads

$$2z - 4 \frac{d}{dt}[tz] + \frac{d^2}{dt^2}[t^2 z] = \xi \left(-2t\bar{v} + \frac{d}{dt}[t^2 \bar{v}] \right) \tag{2.57}$$

Integrating twice the last expression yields, after rearrangement,

$$\xi = \frac{2 \int^{(2)} z - 4 \int tz + t^2 z}{-2 \int^{(2)} t\bar{v} + \int t^2 \bar{v}} \tag{2.58}$$

which is *not well defined* at the initial time $t = 0$, but which after an arbitrarily small time interval $[0, \delta)$, with $\delta > 0$, renders the exact value for the uncertain parameter ξ.

2.2.3.3 A Feedback Controller

The proposed feedback controller is summarized as follows:

$$\bar{v} = \frac{1}{\xi} \left[\dot{z}^*(t) - k_1(z - z^*(t)) - k_0 \int_0^t (z - z^*(\sigma)) d\sigma \right] \tag{2.59}$$

$$\hat{\xi} = \begin{cases} \text{arbitrary nonzero} & \text{for } t \in [0, \delta) \\ \dfrac{2 \int^{(2)} z - 4 \int tz + t^2 z}{-2 \int^{(2)} t\bar{v} + \int t^2 \bar{v}} & \text{for } t \geq \delta \end{cases} \tag{2.60}$$

2.2.3.4 Simulation Results

Figure 2.7 depicts the response of the system to a rather simple trajectory-tracking task. The parameter identification is performed quite fast and the results of such an identification are immediately replaced in the certainty-equivalence controller. No plant noise was considered for these simulations.

Figure 2.8 depicts the response of the system to the same circle trajectory-tracking task. In this instance, we added some plant noise in the linear complex dynamics in the following manner:

$$\dot{z} = ae^{-j\theta} \bar{v} + v, \qquad v = v_1(t) + jv_2(t) \tag{2.61}$$

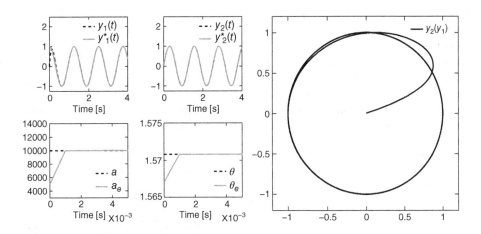

Figure 2.7   Servoing system response, parameter determination, and control input

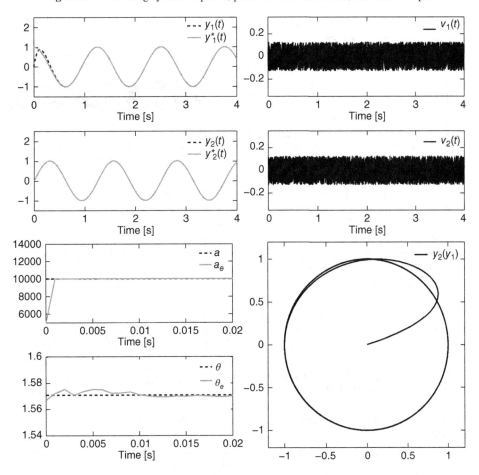

Figure 2.8   Servoing system response, parameter determination, and control input in a noisy environment

where v_1 and v_2 are computer-generated noise processes. The parameter identification is again performed quite fast and the results of such an identification are replaced in the certainty-equivalence controller.

Figure 2.9 depicts the response of the system to a slightly more complex trajectory-tracking task. Plant noise in the linear complex dynamics is present in the same manner as in the previous case.

In the next example, resorting again to complex variables significantly simplifies the apparently nonlinear identification problem to one defined on a simple linear system. However, lumped identification of a single parameter, as in the previous example, is no longer possible due to its time-varying nature.

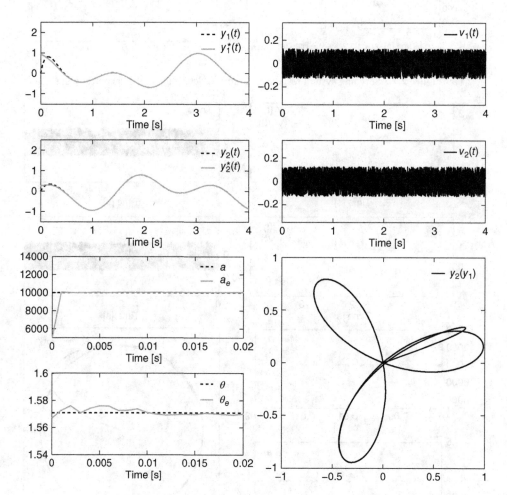

Figure 2.9 Servoing system response, parameter determination, and control input in a noisy environment

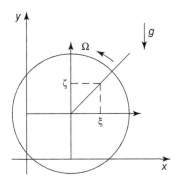

Figure 2.10 Plane rotor

2.2.4 Balancing of the Plane Rotor

Consider the following model, taken from Lum *et al.* (1998), of a plane rotor supported by a radial active magnetic bearing:

$$m\ddot{x} = \Omega^2 \xi \cos \Omega t - \Omega^2 \zeta \sin \Omega t + u_x$$
$$m\ddot{y} = \Omega^2 \xi \sin \Omega t + \Omega^2 \zeta \cos \Omega t + u_y \qquad (2.62)$$

where m is the rotor mass. The variables x, y denote cartesian coordinates of the rotor center. The variables u_x and u_y represent the magnetic forces acting as control inputs on the rotor, with u_y including gravity compensation. Ω is the unknown, constant, rotating speed and the parameters ξ and ζ are the constant, unknown, cartesian body coordinates of the rotor mass center. See Figure 2.10.

In order to simplify our treatment, we will consider complex coordinates and define

$$z = x + jy, \qquad \varphi = \xi + j\zeta = \gamma \exp(j\phi), \qquad (2.63)$$

A complex control input is defined as

$$u = u_x + ju_y, \qquad j = \sqrt{-1} \qquad (2.64)$$

We then have

$$m\ddot{z} = \Omega^2 \gamma \exp(j(\Omega t + \phi)) + u \qquad (2.65)$$

2.2.4.1 Formulation of the Problem

Assume m to be known. Given the uncertain second-order complex dynamics for the rotor center position, devise a feedback control law for u such that z → 0 regardless of the unknown, but constant, nature of the rotor speed Ω, the real eccentricity parameter γ, and the phase angle φ.

2.2.4.2 A Certainty-Equivalence Controller

Assume, just for a moment, that the constant rotating speed Ω, the eccentricity parameter γ, and the phase angle ϕ of the rotor are perfectly known. Let v be a complex auxiliary control input. The dynamic feedback controller

$$u = -\psi + v$$
$$\psi = \Omega^2 \gamma \exp(j(\Omega t + \phi)) \tag{2.66}$$

reduces the control design problem to that of regulating to zero the second-order system

$$mz^{(2)} = v \tag{2.67}$$

So, a pole placement-based controller for the system is readily given by

$$v = -m(k_1 \dot{z} + k_0 z) \tag{2.68}$$

It is not difficult to see that the characteristic polynomial of the closed-loop system is given by

$$p(s) = s^2 + k_1 s + k_0 \tag{2.69}$$

Thus, a suitable choice of the design coefficients $\{k_1, k_0\}$, in the characteristic polynomial results in a Hurwitz polynomial and a globally asymptotically stable closed-loop system. Lack of knowledge of the parameters Ω, γ, and ϕ requires an online identification scheme to complete the proposed GPI controller.

2.2.4.3 An Algebraic Identifier for Ω

Consider again,

$$mz^{(4)} + m\Omega^2 \ddot{z} = \ddot{u} + \Omega^2 u \tag{2.70}$$

The Laplace transform of the system results in

$$m(s^4 z(s) - s^3 z^{(3)}(0) + s^2 \ddot{z}(0) - s\dot{z}(0) + z(0))$$
$$+ m\Omega^2 (s^2 z(s) - s\dot{z}(0) + z(0)) = [s^2 + \Omega^2] u(s) \tag{2.71}$$

where we have explicitly assumed that $\dot{u}(0) = u(0) = 0$.

Taking *four* consecutive derivatives with respect to the complex variable s and solving for Ω^2, we obtain, independently of initial conditions,

$$\Omega^2 = \frac{a(s) + b(s)}{\delta(s)} \tag{2.72}$$

where

$$a(s) = m\left[s^4 \frac{d^4 z}{ds^4} + 16s^3 \frac{d^3 z}{ds^3} + 72s^2 \frac{d^2 z}{ds^2} + 96s \frac{dz}{ds} + 24z\right]$$

$$b(s) = -\left[s^2 \frac{d^4 u}{ds^4} + 8s \frac{d^3 u}{ds^3} + 12 \frac{d^2 u}{ds^2}\right]$$

$$\delta(s) = \frac{d^4 u}{ds^4} - m\left[s^2 \frac{d^4 z}{ds^4} + 8s \frac{d^3 z}{ds^3} + 12 \frac{d^2 z}{ds^2}\right]$$

Multiplying the numerator and denominator by s^{-4}, we obtain

$$\Omega^2 = \frac{s^{-4}a(s) + s^{-4}b(s)}{s^{-4}\delta(s)} \tag{2.73}$$

with

$$s^{-4}a(s) = m\left[\frac{d^4z}{ds^4} + 16s^{-1}\frac{d^3z}{ds^3} + 72s^{-2}\frac{d^2z}{ds^2} + 96s^{-3}\frac{dz}{ds} + 24s^{-4}z\right]$$

$$s^{-4}b(s) = -\left[s^{-2}\frac{d^4u}{ds^4} + 8s^{-3}\frac{d^3u}{ds^3} + 12s^{-4}\frac{d^2u}{ds^2}\right]$$

and

$$s^{-4}\delta(s) = s^{-4}\frac{d^4u}{ds^4} - m\left[s^{-2}\frac{d^4z}{ds^4} + 8s^{-3}\frac{d^3z}{ds^3} + 12s^{-4}\frac{d^2z}{ds^2}\right]$$

Using the well-known operational calculus correspondence which assigns, in the time domain, multiplication by $-t$ to each derivation with respect to the complex variable s, we obtain the following formula for the online fast computation of the unknown constant parameter Ω^2 which we denote by Ω_e^2 to differentiate it from the real one:

$$\Omega_e^2 = \frac{\alpha(t) + \beta(t)}{\delta(t)} \tag{2.74}$$

where

$$\alpha(t) = m\left[t^4z - 16\left(\int t^3z\right) + 72\left(\int^{(2)} t^2z\right) - 96\left(\int^{(3)} tu\right) + 24\left(\int^{(4)} z\right)\right]$$

$$\beta(t) = -\left[\left(\int^{(2)} t^4u\right) - 8\left(\int^{(3)} t^3u\right) + 12\left(\int^{(4)} t^2u\right)\right]$$

$$\delta(t) = \left(\int^{(4)} t^4u\right) - m\left[\left(\int^{(2)} t^4z\right) - 8\left(\int^{(3)} t^3z\right) + 12\left(\int^{(4)} t^2z\right)\right]$$

The formula for computing Ω_e^2 exhibits a singularity at $t = 0$ but yields, depending on the accuracy of the arithmetic processor used in the controller setup, accurate numerical results after a small interval of time, of duration say $\epsilon > 0$, has elapsed. In the next section we illustrate the convergence characteristics of the proposed parameter-identification formula.

2.2.4.4   An Algebraic Identifier for γ and ϕ

Now, Ω is assumed known such that the unknown parameters ξ and ζ can be estimated from the original model equations

$$m\ddot{x} = \xi\beta_1(t) - \zeta\beta_2(t) + u_x(t)$$
$$m\ddot{y} = \xi\beta_2(t) + \zeta\beta_1(t) + u_y(t) \tag{2.75}$$

where $\beta_1(t) = \Omega_e^2 \cos \Omega_e t$ and $\beta_2(t) = \Omega_e^2 \sin \Omega_e t$. Applying the Laplace transform to the model,

$$m\left(s^2 X(s) - sX(0) - \dot{x}(0)\right) = \xi \beta_1(s) - \zeta \beta_2(s) + U_X(s)$$

$$m\left(s^2 Y(s) - sY(0) - \dot{Y}(0)\right) = \xi \beta_2(s) + \zeta \beta_1(s) + U_Y(s) \tag{2.76}$$

Taking two consecutive derivatives with respect to the complex variable s, and multiplying the results by s^{-2}, we obtain

$$m\left(2s^{-2} X(s) + 4s^{-1}\frac{dX}{ds} + \frac{d^2 X}{ds^2}\right) = \xi\left(s^{-2}\frac{d^2\beta_1}{ds^2}\right) - \zeta\left(s^{-2}\frac{d^2\beta_2}{ds^2}\right) + s^{-2}\frac{d^2 U_X}{ds^2}$$

$$m\left(2s^{-2} Y(s) + 4s^{-1}\frac{dY}{ds} + \frac{d^2 Y}{ds^2}\right) = \xi\left(s^{-2}\frac{d^2\beta_2}{ds^2}\right) + \zeta\left(s^{-2}\frac{d^2\beta_1}{ds^2}\right) + s^{-2}\frac{d^2 U_Y}{ds^2} \tag{2.77}$$

Returning the expressions to the time domain, a system of equations can be built to calculate the unknown parameters:

$$m\left(t^2 x - 4\int tx + 2\int^{(2)} x\right) - \int^{(2)} t^2 u_x = \xi\int^{(2)} t^2\beta_1 - \zeta\int^{(2)} t^2\beta_2$$

$$m\left(t^2 y - 4\int ty + 2\int^{(2)} y\right) - \int^{(2)} t^2 u_y = \xi\int^{(2)} t^2\beta_2 + \zeta\int^{(2)} t^2\beta_1 \tag{2.78}$$

The estimated values of γ and ϕ are given by

$$\gamma_e = \sqrt{\xi_e^2 + \zeta_e^2} \quad \phi_e = \arctan\left(\frac{\zeta_e}{\xi_e}\right) \tag{2.79}$$

2.2.4.5 Simulation Results

We considered a rigid rotor of mass $M = 0.5$ kg with a nominal angular velocity of $\Omega = 100$ rad/s. The results are shown in Figures 2.11 and 2.12.

In linear motors, and also in induction motors, knowledge of the initial position (or initial angular position) may be crucial for control purposes. Here we illustrate that a lumped identification of a nonlinear expression involving the initial position may radically ease the controller design problem.

2.2.5 On the Control of the Linear Motor

Consider the following simplified model of a bi-phasic linear motor:

$$\ddot{D} = \frac{K}{m}\left(u_1 \sin\left(\frac{2\pi}{P}(x_0 + D)\right) + u_2 \cos\left(\frac{2\pi}{P}(x_0 + D)\right)\right) - \frac{f}{m}\text{sign}\left(\dot{D}\right) \tag{2.80}$$

where D is the measured displacement. The variables u_1 and u_2 stand for the control inputs, and they represent currents flowing through the spires of the motor. P is the spatial period, x_0 is the unknown initial position, and the parameters K, m, and f are, respectively, the unknown motor gain, the unknown mass load, and the unknown Coulomb friction coefficient.

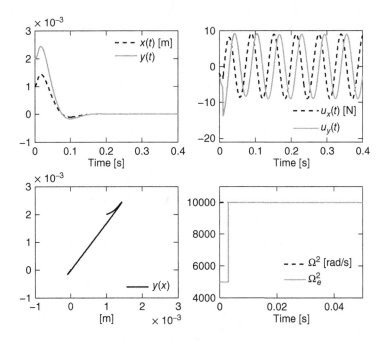

Figure 2.11   Performance of controlled rotor and performance of identification scheme

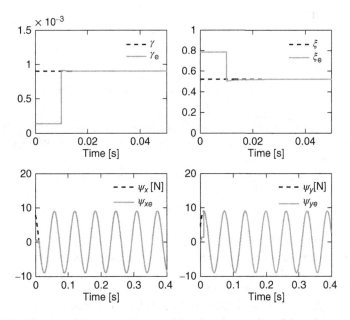

Figure 2.12   Identification of the parameters γ and ξ and reconstruction of the unknown dynamic terms $\psi_x(t)$ and $\psi_y(t)$

2.2.5.1 A Certainty-Equivalence GPI controller

We propose the following certainty-equivalence field-oriented GPI feedback controller with $\hat{x}_0$ arbitrary:

$$u_1 = I(t) \, \sin\left(\frac{2\pi}{P}(\hat{x}_0 + D)\right) \tag{2.81}$$

$$u_2 = I(t) \, \cos\left(\frac{2\pi}{P}(\hat{x}_0 + D)\right) \tag{2.82}$$

where $I(t)$ is a signal generated by a GPI controller. Defining $\hat{x}_0 = x_0 + \Delta x_0$ and using the identities

$$\sin(x_0 + \Delta x_0) = \sin(x_0)\cos(\Delta x_0) + \cos(x_0)\cos(\Delta x_0)$$

$$\cos(x_0 + \Delta x_0) = \cos(x_0)\cos(\Delta x_0) - \sin(x_0)\sin(\Delta x_0) \tag{2.83}$$

the closed-loop system takes the form

$$\ddot{D} = \gamma I(t) - \frac{f}{m}sign(\dot{D}) \tag{2.84}$$

where the unknown parameter γ is defined as

$$\gamma = \frac{K}{m}\cos\left(\frac{2\pi}{P}(\hat{x}_0 - x_0)\right) \tag{2.85}$$

So, $I(t)$ is proposed to be

$$I(t) = \beta u(t) \tag{2.86}$$

with $u(t)$ given by

$$u(t) = \ddot{D}^*(t) - \left[\frac{k_2 s^2 + k_1 s + k_0}{s(s + k_3)}\right](D - D^*(t)) \tag{2.87}$$

The design parameter β is obtained from a fast estimation procedure. Initially, for a small time interval of value ϵ, we shall let it take the particular value of 1. We shall determine the appropriate value of β after $t = \epsilon$.

 The classical GPI controller gains are chosen so that the fourth-order characteristic polynomial

$$p(s) = s^4 + k_3 s^3 + k_2 s^2 + k_1 s + k_0 \tag{2.88}$$

has all its roots in the left half of the complex plane.

 The previous controller may be proven to be robust with respect to the Coulomb friction term once all the involved unknown parameters are canceled. The rationale is as follows. Since the friction force is piecewise constant before the velocity becomes zero, or close to zero, the integral action embedded in the classical form of the GPI controller takes care of the piecewise constant perturbations. When the velocity is arbitrarily close to zero a chattering motion may appear which is compensated by the high-pass-filtering characteristic of the proposed feedback controller. In simulations, the performance is found to be quite acceptable.

2.2.5.2 An Online Identification Scheme

From the closed-loop system 2.84, during a small interval of time of the form $[0, \epsilon]$ we shall implement a fast estimation procedure for γ yielding an accurate estimate $\hat{\gamma}$, within the interval of time $[0, \epsilon)$. Once $\hat{\gamma}$ is found to converge to a constant value, we will set β in the proposed controller so as to cancel the parameter γ. That is,

$$\beta = \begin{cases} 1 & \text{for} \quad t \in [0, \epsilon) \\ \\ 1/\hat{\gamma} & \text{for} \quad t \in [\epsilon, +\infty) \end{cases} \tag{2.89}$$

Using algebraic parameter-estimation techniques, we compute an estimate $\hat{\gamma}$ of the uncertain parameter γ as follows. Take a time derivative of the closed-loop model 2.84, with $I(t)$ as control input (this eliminates the piecewise constant Coulomb friction term). Multiply out by t^3 the resulting expression and integrate by parts four times (this achieves independence from initial conditions and introduces some invariant filtering). Finally, solve for γ. We set

$$\hat{\gamma} = \begin{cases} \text{arbitrary} & \text{for} \quad t \in [0, \epsilon) \\ \\ \dfrac{\left(6\smallint^{(4)}D\right) - 18\left(\smallint^{(3)}tD\right) + 9\left(\smallint^{(2)}t^2D\right) - (\smallint\, t^3 D)}{3\left(\smallint^{(4)}t^2 I\right) - \smallint^{(3)}t^3 I} & \text{for} \quad t \in [\epsilon, +\infty) \end{cases} \tag{2.90}$$

2.2.5.3 Simulation Results

We used the following parameter values:

$$x_0 = 5 \times 10^{-3} \text{ m}, \qquad \hat{x}_0 = 1 \times 10^{-3} \text{ m}$$

$$P = 42 \times 10^{-3}, \quad f/m = 0.1 \text{ m/s}\char`^2, \quad K/m = 2 \text{ m/s}\char`^2$$

The design parameters for the GPI controller were set to be such that the closed-loop characteristic polynomial

$$p(s) = s^4 + k_3 s^3 + k_2 s^2 + k_1 s + k_0$$

coincided with

$$p_d(s) = (s^2 + 2\zeta\omega_n s + \omega_n^2)^2$$

with $\zeta = 1$, $\omega_n = 60$. The calculation interval parameter ϵ was taken to be $\epsilon = 0.002$. We devised a periodic trajectory for $D^*(t)$ built up by means of interpolating Bézier polynomials.

In order to test the robustness of the proposed scheme with respect to high-frequency measurement noises, we used a noisy measurement of the displacement in the form $D(t) + \xi(t)$, with $\xi(t)$ being an unbiased computer-generated random process. The results are shown in Figure 2.13.

In the following example, two features are emphasized: (1) Algebraic identification is equally applicable to switched controlled systems, especially when switchings are of very high frequency as in power electronics. (2) Even if individual parameters are not identifiable, controllers can still be synthesized on the basis of algebraic identification of nonlinear expressions of the unknown parameters.

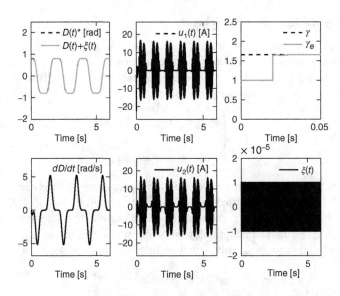

Figure 2.13   Closed-loop online identification of unknown initialization parameter for the linear motor

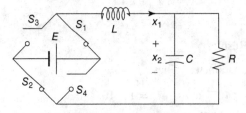

Figure 2.14   The double-bridge buck converter

2.2.6   Double-Bridge Buck Converter

Consider a double-bridge buck converter as shown in Figure 2.14. In reference to the figure, the several electronic switches take position values according to

$$\begin{cases} u = 1 & S_1 = ON, S_2 = ON, S_3 = OFF, S_4 = OFF \\ u = 0 & S_1 = OFF, S_2 = OFF, S_3 = ON, S_4 = ON \end{cases} \tag{2.91}$$

The (average) model of a double-bridge buck converter (see Sira-Ramírez *et al.*, 2002) is given by

$$L\dot{x}_1 = -x_2 + (2v - 1)E$$

$$C\dot{x}_2 = x_1 - \frac{x_2}{R}$$

$$y = x_2 \tag{2.92}$$

where x_1 is the inductor current and x_2 represents the capacitor voltage. The average control input v is assumed to take values in the closed interval: $v \in [0, 1]$ of the real line. This variable actually represents the *duty ratio* of the switch positions. All parameters $L, C, R,$ and E are assumed to be unknown constants.

The input–output relation of the converter is

$$\ddot{y} + \frac{1}{RC}\dot{y} + \frac{1}{LC}y = \mu\frac{E}{LC} \tag{2.93}$$

The three parameters of interest will be denoted respectively by $\gamma_1 = \frac{1}{RC}$, $\gamma_0 = \frac{1}{LC}$, and $\gamma = \frac{E}{LC}$. Clearly, these parameters are linearly identifiable while the parameters $L, C,$ and R are not weakly linearly identifiable, they are simply non-identifiable. Only E is identifiable. This fact is not an obstacle for controlling the system on the basis of algebraic identification.

We rewrite the linearly identifiable system as follows:

$$\ddot{y} + \gamma_1\dot{y} + \gamma_0 y = \mu\gamma \tag{2.94}$$

and we proceed to design the controller as if these parameters were all perfectly known.

2.2.6.1 Problem Formulation

It is required to design an output feedback controller, possibly of dynamic nature, which induces in the uncertain system representing the double-bridge buck converter average model an exponentially asymptotic convergence of the output signal y toward the desired reference signal $y^*(t)$.

In other words, we want

$$y \rightarrow y^*(t) \qquad \text{exponentially as time elapses}$$

2.2.6.2 A Certainty-Equivalence Controller

We propose the following certainty-equivalence generalized GPI controller which abusively combines frequency-domain quantities and time-domain quantities:

$$\mu = \mu^*(t) - \frac{1}{\gamma}\left\{\frac{[\gamma_1(\gamma_1 - k_1) + k_0 - \gamma_0]s + \gamma_0(\gamma_1 - k_1) + k_{-1}}{s + (k_1 - \gamma_1)}\right\}(y - y^*(t)) \tag{2.95}$$

2.2.7 Closed-Loop Behavior

The closed-loop behavior of the tracking error, where the parameters are perfectly known, is given by the following linear dynamics:

$$\ddot{e}_y + k_1\dot{e}_y + k_0 e_y + k_{-1}\int_0^t e_y(\sigma)d\sigma = 0 \tag{2.96}$$

where $e_y = y - y^*(t)$ is the trajectory-tracking error.

The appropriate choice of the coefficients $\{k_1, k_0, k_{-1}\}$, in the characteristic polynomial of the tracking-error dynamics, turns it into a Hurwitz polynomial with the associated asymptotically and exponentially stable nature of the origin of coordinates of the natural tracking error state-space $\{e_y = 0, \dot{e}_y = 0, \ddot{e}_y = 0\}$.

Under the assumption of perfect parameter knowledge we obtain, modulo control input saturations,

$$e_y(t) \to 0 \quad \text{exponentially}$$

The problem now becomes one of accurate determination of the unknown parameters of the system *as required by the proposed GPI controller*.

2.2.7.1 Algebraic Determination of the Unknown Parameters

Consider the average input–output model of the converter system

$$\ddot{y} + \gamma_1 \dot{y} + \gamma_0 y = \gamma \mu \tag{2.97}$$

In the notation of operational calculus, we have

$$s^2 y(s) - s y_0 - \dot{y}_0 + \gamma_1 (s y(s) - y_0) + \gamma_0 y(s) = \gamma \mu(s) \tag{2.98}$$

Taking the derivative twice with respect to s we obtain

$$(s^2 + \gamma_1 s + \gamma_0)\frac{d^2 y}{ds^2} + (4s + 2\gamma_1)\frac{dy}{ds} + 2y = \gamma \frac{d^2 \mu}{ds^2} \tag{2.99}$$

The last expression can be rewritten as follows:

$$\left[s\frac{d^2 y}{ds^2} + 2\frac{dy}{ds}\right]\gamma_1 + \left[\frac{d^2 y}{ds^2}\right]\gamma_0 - \left[\frac{d^2 \mu}{ds^2}\right]\gamma = -s^2\frac{d^2 y}{ds^2} - 4s\frac{dy}{ds} - 2y \tag{2.100}$$

Multiplying out by a sufficient power of s^{-1}, say by s^{-4}, so that an invariant filtering effect is obtained, we also get rid of possible derivations in the time domain. We obtain

$$\left[s^{-3}\frac{d^2 y}{ds^2} + 2s^{-4}\frac{dy}{ds}\right]\gamma_1 + \left[s^{-4}\frac{d^2 y}{ds^2}\right]\gamma_0 - \left[s^{-4}\frac{d^2 \mu}{ds^2}\right]\gamma =$$

$$- \left[s^{-2}\frac{d^2 y}{ds^2} + 4s^{-3}\frac{dy}{ds} + 2s^{-4}y\right] \tag{2.101}$$

Reverting the previous expression to the time domain, we obtain a linear equation with time-varying coefficients in three unknowns $\{\gamma_1, \gamma_0, \gamma\}$. We write such an equation as

$$p_{11}(t)\gamma_1 + p_{12}(t)\gamma_0 + p_{13}(t)\gamma = q_1(t) \tag{2.102}$$

We create a system of three equations in three unknowns by simply adjoining to the previous equation its first integral and its iterated integral, that is,

$$p_{11}(t)\gamma_1 + p_{12}(t)\gamma_0 + p_{13}(t)\gamma = q_1(t)$$

$$\left(\int p_{11}(t)\right)\gamma_1 + \left(\int p_{12}(t)\right)\gamma_0 + \left(\int p_{13}(t)\right)\gamma = \left(\int q_1(t)\right) \qquad (2.103)$$

$$\left(\int^{(2)} p_{11}(t)\right)\gamma_1 + \left(\int^{(2)} p_{12}(t)\right)\gamma_0 + \left(\int^{(2)} p_{13}(t)\right)\gamma = \left(\int^{(2)} q_1(t)\right)$$

This linear system of equations allows us to determine γ_1, γ_0, and γ for $t \geq \epsilon$, with ϵ being a very small positive real number.

2.2.7.2 Simulation Results

We considered the average model of a double-bridge buck converter with the following (unknown) parameters:

$$R = 39.52\ \Omega, \qquad L = 1\ \text{mH}, \qquad C = 1\ \mu\text{F}, \qquad E = 30\ \text{V}$$

It is desired that the average output voltage signal will track a rest-to-rest trajectory starting at 21.0 V and landing at 9.0 V in approximately 0.25 ms. The tracking maneuver is to start at $t_{init} = 0.25$ ms. and end at $t_f = 0.5$ ms. It was assumed that the output voltage could be measured through an additive noise process simulated with a computer-generated sequence of random variables uniformly distributed in the interval $A[-0.5, 0.5]$ with the factor A taken to be $A = 0.1$.

Figure 2.15 depicts the simulated closed-loop performance of the GPI-controlled double-bridge boost converter along with the performance of the proposed algebraic parameter estimator. The value of ϵ used to avoid the singularity of the formulae at time $t = 0$ was taken to be 50 μs. Once identified, the value of the parameters is immediately substituted in the GPI feedback control law.

We tested our fast adaptive estimation algorithm now using a switched control signal for the control input whose average coincides with the previous control input. This is achieved using a double-sided Sigma–Delta modulator described by the following discontinuous dynamics:

$$\dot{z} = \mu - u, \qquad u = \frac{1}{2}[sign(\mu) + sign(z)] \qquad (2.104)$$

The actual control input signal u being used in the identification algorithm is now a high-frequency signal actively switching and taking values in the discrete set $\{-1, 0, +1\}$. As can be inferred from Figure 2.16, the algebraic identifier works perfectly well with this input. In this instance, the maneuver enabled a trajectory-tracking task with the same time duration constraints as before to take the output voltage from an initial equilibrium of 21 V toward a final equilibrium of -9 V.

Figure 2.16 shows the closed-loop response for the trajectory-tracking task of the switched input model as well as the precision and rapidity of the unknown parameter-estimation process. The output voltage signal is also assumed to be measured through additive noisy–means. A sample of the noise process is also depicted in the figure. The actual bang–bang control input is shown along with the nominal value of the average control input.

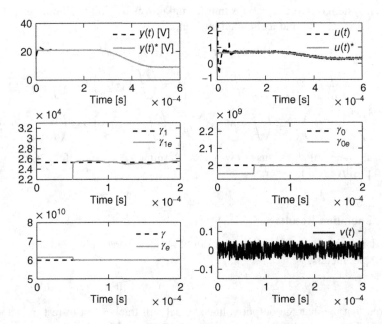

Figure 2.15   Closed-loop average converter response with online identification of all linearly identifiable system parameters

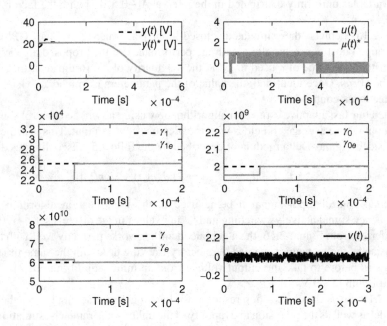

Figure 2.16   Closed-loop switched converter response with online identification of all linearly identifiable system parameters

2.2.8   Control of an Unknown Variable Gain Motor

The fast online features of the algebraic approach to parameter estimation result in new feedback control schemes for handling traditionally difficult problems involving "hard nonlinearities." Here we address the problem of controlling a simplified model of a DC motor with an input voltage that must go through a piecewise linear gain characteristic, or through a variable gain.

Consider the simplified model of a DC motor

$$\ddot{y} = -a\dot{y} + u$$

$$u = K_r v + \varphi \tag{2.105}$$

where y denotes the position of the motor shaft and u is the input voltage to the motor arising from a saturating amplifier whose gain parameter K_r is completely unknown and known to change values in a step-like fashion according to the magnitude of v. The piecewise constant parameter φ, which makes the characteristic curve $u - v$ continuous, is also unknown. Typically, one has a variable gain characteristic as shown in Figure 2.17. We assume that the parameter a is perfectly known.

The system is then described by the second-order differential equation

$$\ddot{y} = -a\dot{y} + K_r v + \varphi \tag{2.106}$$

If K_r were perfectly known, the following GPI controller, in classical form, would stabilize the system trajectory around a desired reference trajectory, specified by $y^*(t)$, in spite of the

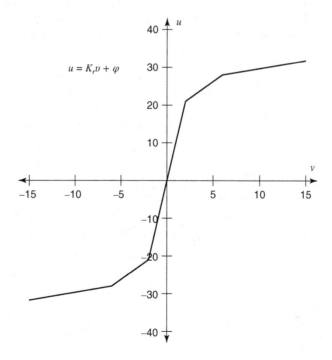

Figure 2.17   Typical variable gain characteristic for motor input voltage

unknown value of φ which should be treated as a piecewise constant unknown perturbation. In fact, the integral action of the certainty-equivalent controller cancels the effect of such a varying bias on the closed-loop response of the system:

$$v = \frac{1}{K_r} \left\{ v^*(t) - \left[\frac{k_2 s^2 + k_1 s + k_0}{s(s + k_3)} \right] (y - y^*(t)) \right\} \tag{2.107}$$

where

$$v^*(t) = \frac{1}{K_r} [\ddot{y}^*(t) + a\dot{y}^*(t)] \tag{2.108}$$

and we have ignored the parameter φ in the feedforward term as is customarily done with unknown constant perturbation inputs that do not have a nominal character.

The controller gains may be chosen so that the closed-loop characteristic polynomial, given by

$$p(s) = s^4 + (a + k_3)s^3 + (ak_3 + k_2)s^2 + k_1 s + k_0$$

coincides with a desired characteristic polynomial of the form

$$p_d(s) = s^4 + \gamma_3 s^3 + \gamma_2 s^2 + \gamma_1 s + \gamma_0$$

with coefficients γ such that $p_d(s)$ has all its roots in the left half of the complex plane.

What makes this problem particularly difficult in some other adaptive contexts is that, during transients and start-ups, the varying value of gain K_r can visit several of its possible unknown slopes (which may be orders of magnitude different), spending very little time at each value. A fast, reliable, online adaptation scheme is therefore required. Figure 2.18 depicts the typical situation of controlling such a variable gain system with a fixed, arbitrary gain in the controller. Notice the tracking performance degradation and the rapidly varying nature of K_r.

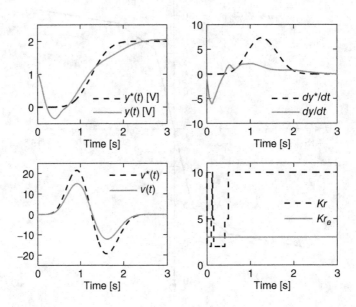

Figure 2.18 Typical system response to a GPI controller with integral action and a fixed gain

To proceed using the algebraic method, we first observe that the unknown constant parameter φ can be eliminated by differentiating both sides of the system equation. We get

$$y^{(3)} = -a\ddot{y} + K_r \dot{v} \tag{2.109}$$

Our estimation strategy for online computing K_r will consist of resetting, very often, the computations involved in the formulae found for K_r. We will do so at each resetting instant t_i. Using the customary algebraic manipulations with provisions for invariant filtering through a second-order integration, we obtain

$$K_r = \frac{(\int_{t_i}^{(5)} 6(1-at)y) - (\int_{t_i}^{(4)} (18t - 6at^2)y) + (\int_{t_i}^{(3)} (9t^2 - at^3)y) - (\int_{t_i}^{(2)} t^3 y)}{3(\int_{t_i}^{(5)} t^2 v) - (\int_{t_i}^{(4)} t^3 v)} \tag{2.110}$$

At each resetting a small amount of time ϵ is allowed for the computation to be accurately performed. We used the following online adaptive GPI controller:

$$v = \frac{1}{K_{re}} \left\{ v^*(t) - \left[\frac{k_2 s^2 + k_1 s + k_0}{s(s + k_3)} \right] (y - y^*(t)) \right\} \tag{2.111}$$

where

$$v^*(t) = \frac{1}{K_{re}} [\ddot{y}^*(t) + a\dot{y}^*(t)] \tag{2.112}$$

that is, we exercised fast adaptation on both the feedforward part of the controller and the gain of the classical controller. The estimate of K_r is expressed as

$$K_{re} = \begin{cases} \text{arbitrary} & \text{for } t \in [t_i, t_i + \epsilon) \\ \dfrac{n(t)}{d(t)} & \text{for } t \in [t_i + \epsilon, t_{i+1}) \end{cases}$$

$$\begin{aligned}
n(t) &= \xi_1 & d(t) &= \eta_1 \\
\dot{\xi}_1 &= \xi_2 & \dot{\eta}_1 &= \eta_2 \\
\dot{\xi}_2 &= -(t - t_i)^3 y(t) + \xi_3 & \dot{\eta}_2 &= \eta_3 \\
\dot{\xi}_3 &= [9 - a(t - t_i)](t - t_i)^2 y(t) + \xi_4 & \dot{\eta}_3 &= \eta_4 \\
\dot{\xi}_4 &= -[18 - 6a(t - t_i)](t - t_i)y(t) + \xi_5 & \dot{\eta}_4 &= -(t - t_i)^3 v(t) + \eta_5 \\
\dot{\xi}_5 &= 6(1 - a * (t - t_i))y(t) & \dot{\eta}_5 &= 3(t - t_i)^2 v(t)
\end{aligned} \tag{2.113}$$

Figure 2.19 depicts the performance of the proposed GPI controller, with integral action and adaptive gain, in a trajectory-tracking task. The algebraic parameter estimator was reset every 0.1 s with a waiting time $\epsilon = 0.01$ s. The estimated gain was immediately replaced in the feedback controller.

2.2.8.1 A Periodic Probing and Parameter Updating Strategy

Suppose now that the uncertain constant parameters are subject to sudden, unmodeled step variations. A periodic "probing and updating" strategy will produce, upon resetting of the algebraic identifier integrations, the current actual values of the changing parameters. Figure 2.20 depicts the scheme of the proposed strategy.

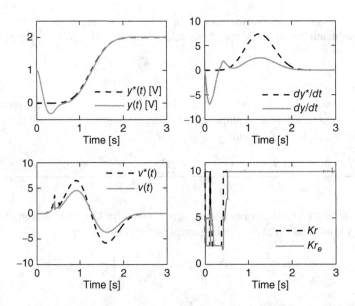

Figure 2.19 GPI controller with integral action and fast online algebraic gain estimation and adaptation

2.2.9 Identifying Classical Controller Parameters

The purpose of this section is to illustrate the use of an algebraic online identification scheme for determining PI controller parameters and PD controller parameters acting on sufficiently simple systems.

It is customary that the parameters of a PI or those of a PD controller need eventually to be evaluated while the system is performing in a closed loop. The parameter-identification scheme must then be carried out in an "online" fashion. Our purpose here is to provide a simple computational tool for evaluating such controller parameters without "disconnecting" the controller from the process.

2.2.9.1 Identification of PI Controller Parameters

Consider the following elementary perturbed linear system:

$$\dot{x} = u + k\,\mathbf{1}(t) + \xi(t)$$

$$y = x + v(t) \tag{2.114}$$

where k is an uncertain constant parameter, $\mathbf{1}(t)$ is the unit-step heaviside function, and the signals $\xi(t)$ and $v(t)$ are stochastic piecewise constant noises uniformly distributed on symmetric intervals of the real line. This class of random processes is usually synthesized with the help of the function $rect(t)$, which is commonly found in most computer simulation packages.

A PI controller safeguards the performance of the system regarding the constant-step perturbation input, $k\mathbf{1}(t)$, while regulating the system output toward a desired constant value Y.

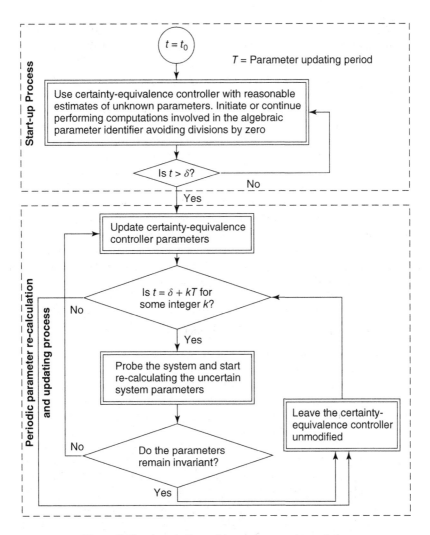

Figure 2.20 A periodic probing and parameter updating

Such a controller is given by

$$u = -k_2(y - Y) - k_1 \int_0^t [y(\sigma) - Y]d\sigma \qquad (2.115)$$

The parameters k_1 and k_2 are usually specified as

$$k_2 = 2\zeta\omega_n, \qquad k_1 = \omega_n^2$$

where ζ is the desired *damping ratio* and ω_n is the desired *natural frequency* of the closed-loop system.

Using the output stabilization error signal $e(t) = y(t) - Y$ and the control input signal $u(t)$, an identifier for the PI controller parameters k_2 and k_1 can immediately be proposed.

Such an identifier must start running at an arbitrary time $t_e > 0$, and obtain accurate results in a quite small time interval $[t_e, t_e + \delta]$. The form of the identifier is as follows:

$$k_{2e} = \frac{\left[-u \int_{t_e}^{(2)} (y - Y) + \int_{t_e} u \int_{t_e} (y - Y) \right]}{(y - Y) \int_{t_e}^{(2)} (y - Y) - \left(\int_{t_e} (y - Y) \right)^2} \tag{2.116}$$

$$k_{1e} = \frac{u \int_{t_e} (y - Y) d\sigma - (y - Y) \int_{t_e}^{t} u}{(y - Y) \int_{t_e}^{(2)} (y - Y) - \left(\int_{t_e} (y - Y) \right)^2} \tag{2.117}$$

Figures were obtained with $\zeta = 0.8$ and $\omega_n = 1$, with system and output measurement noises: $\xi = 0.5(rect(t) - 0.5)$ and $v = 0.1(rect(t) - 0.5)$, respectively. The evaluation time was set to be $t_e = 0$ and the set-point $Y = 1$. The perturbation parameter amplitude was set to be $k = 0.2$.

Figure 2.21 depicts the efficiency of the controller parameter identifier from which the value of the damping ratio ζ and the natural frequency can be determined to be 0.8 and 1, respectively in a significantly small amount of time (approximately $\delta = 0.01$ s).

2.2.9.2 Identification of PD Controller Parameters

Consider the following second-order linear system:

$$\ddot{x} = u + \xi(t)$$
$$y = x(t) + v(t) \tag{2.118}$$

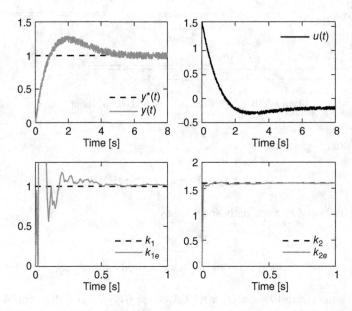

Figure 2.21   System closed-loop performance and PI controller parameter identification

where κ is an uncertain constant parameter and the signals $\xi(t)$ and $v(t)$ are, as before, stochastic piecewise constant noises uniformly distributed on symmetric intervals of the real line.

A PD controller regulating the system output trajectories towards a constant reference set-point Y is given by

$$u = -k_2\dot{y} - k_1(y - Y) \tag{2.119}$$

where k_2 is of the form $k_2 = 2\zeta\omega_n$ and $k_1 = \omega_n^2$ represents the desired damping factor and natural frequency-independent term of the closed-loop system characteristic polynomial.

Using the output stabilization error signal $e(t) = y(t) - Y$ and the control input signal $u(t)$, an identifier for the PD controller parameters k_2 and k_1 can be proposed in a similar manner to the previous case. Such an identifier is to be started at an arbitrary evaluation time $t_e > 0$ and be capable of obtaining accurate results about the value of the parameters in a quite small time interval of the form $[t_e, t_e + \delta]$. An identifier for the controller parameters, independent of the initial condition on y, is readily found to be

$$k_{2e} = \frac{n_2(t)}{r(t)}, \qquad k_{1e} = \frac{n_1(t)}{r(t)} \tag{2.120}$$

$$n_1(t) = \left(\int_{t_e} ty - \int_{t_e}^{(2)} y \right) \left(\int_{t_e} tu \right) - \left(ty - \int_{t_e} y \right) \left(\int_{t_e}^{(2)} tu \right)$$

$$n_2(t) = \left(\int_{t_e} t(y - Y) \right) \left(\int_{t_e}^{(2)} tu \right) - \left(\int_{t_e}^{(2)} t(y - Y) \right) \left(\int_{t_e} tu \right)$$

$$r(t) = \left(ty - \int_{t_e} y \right) \left(\int_{t_e}^{(2)} t(y - Y) \right) - \left(\int_{t_e} ty - \int_{t_e}^{(2)} y \right) \left(\int_{t_e} t(y - Y) \right)$$

Figures 2.21 and 2.22 depict the efficiency of the controller parameter identifier from which the value of the damping ratio ζ and the natural frequency can be determined approximately to be 0.8 and 1, respectively.

Figures were obtained with $\zeta = 0.8$ and $\omega_n = 1$ (i.e., $k_2 = 1.6, k_1 = 1$); the system and output measurement noises were set to be $\xi = 0.1(rect(t) - 0.5)$ and $v = 0.001(rect(t) - 0.5)$, respectively. The parameter evaluation instant t_e was set to be $t_e = 0$ and the desired set-point $Y = 1$.

2.3   A Case Study Introducing a "Sentinel" Criterion

It is common to find electromechanical systems in interaction with mechanical vibrations, whose purpose is to attenuate or increase those vibrations. An example of this is suspension systems in vehicles, shakers, or vibration generators, or the electrostatic actuators in hard disk drives. The suspension system can be classified into passive, semi-active, and active suspension systems, according to its ability to add, or extract, energy from the overall system. Our work is centered around an active suspension system given that the control task requires tracking a reference trajectory with exactitude and robustness.

A one-degree-of-freedom (1-DOF) suspension system is a key element for the analysis, study, and control of the class of vibrating systems. A 1-DOF suspension system can be represented in a simple way by a mass-damping-stiffness configuration, which is a linear

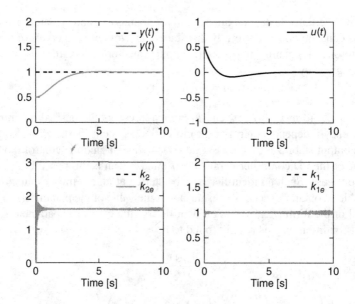

Figure 2.22   System closed-loop performance and PD controller parameter identification

approximation to the actual physical plant. It is possible to use a high-fidelity mathematical model which captures the realistic dynamic behavior of an active suspension and identify the model parameters using a Kalman filter, Tan and Bradshaw, 1997). Considering a simple suspension model, under noisy measurement conditions, perturbed by a constant disturbance, we design an output feedback controller for an accurate trajectory-tracking task. The parameters of this linear model are assumed uncertain, so that the online algebraic identification process can be implemented. The algebraic parameter-estimation process is improved here with an automated disconnection algorithm. Owing to design requirements, it is necessary to implement a control law based only on output measurements which renders a satisfactory trajectory-tracking performance.

2.3.1   A Suspension System Model

Consider a perturbed 1-DOF suspension model whose simplified diagram is shown in Figure 2.23 (García-Rodríguez *et al.*, 2009):

$$m\ddot{x} + c\dot{x} + kx = F + \kappa + \eta(t)$$

$$y(t) = x(t) + \xi(t) \tag{2.121}$$

where m stands for the mass of the suspension system, c denotes a viscous friction coefficient, k is a linear stiffness coefficient, x represents the displacement of the mass, F represents the applied input force, and κ is an unknown constant disturbance. The signals $\eta(t)$ and $\xi(t)$ are zero-mean, high-frequency noisy processes.

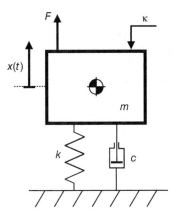

Figure 2.23   Perturbed 1-DOF suspension system

2.3.1.1   Design of the GPI Control Law

Setting the initial conditions to zero, the transfer function of the unperturbed system is given by

$$G(s) = \frac{\frac{1}{m}}{s^2 + \frac{c}{m}s + \frac{k}{m}} \tag{2.122}$$

The tracking-error dynamics is readily found to be given by $e_y(s) = G(s)e_{u_{av}}(s)$, with $e_y = y - y^*(t)$ and $e_{u_{av}} = u_{av} - u_{av}^*(t)$, where $u = F$, and $y^*(t)$ and $u_{av}^*(t)$ are the desired nominal trajectories for the output and control, respectively. In order to ameliorate the effect of constant disturbances, the tracking trajectory control law is designed using a classical compensator with an additional integral control action. Such control option is entirely equivalent to the GPI control scheme based on integral reconstruction. We propose the following trajectory-tracking controller:

$$e_{u_{av}}(s) = C(s)e_y(s) = -m\left(\frac{k_2 s^2 + k_1 s + k_0}{s\left(s + k_3\right)}\right)e_y(s) \tag{2.123}$$

This control law leads to the following closed-loop tracking-error characteristic polynomial:

$$(s^4 + \lambda_3 s^3 + \lambda_2 s^2 + \lambda_1 s + \lambda_0)e_y(s) = 0 \tag{2.124}$$

Setting the coefficients λ_i, $i = 0, \dots, 3$ according to a Hurwitz polynomial, the tracking error converges to zero in an asymptotically exponential fashion.

Consider now the system affected by a disturbance $\delta(t)$ which involves the parameter uncertainty and the constant perturbation κ satisfying $\|\delta^{(j)}\| \le \mu_j$, $j = 0, 1, 2 < \infty$. So, the polynomial which governs the closed-loop tracking-error dynamics for the perturbed system is

$$(s^4 + \lambda_3 s^3 + \lambda_2 s^2 + \lambda_1 s + \lambda_0) \; e_y(s) = s(s + k_3)\delta(s)$$

where $\lambda_0 = k_0$, $\lambda_1 = \left(k_1 + \frac{k}{m}k_3\right)$, $\lambda_2 = \left(k_2 + \frac{k}{m} + \frac{c}{m}k_3\right)$, $\lambda_3 = \left(k_3 + \frac{c}{m}\right)$. An upper bound on $e_y(t)$ is given by

$$0 \le \limsup \|e_y(t)\| \le \frac{\mu_2 + k_3\mu_1}{\lambda_0} \tag{2.125}$$

With a proper choice of the set of design coefficients k_0, k_1, k_2, k_3, the error tracking can be made to converge in the vicinity of the origin bounded by a small disk where its radius is inversely proportional to λ_0. The coefficients can be chosen to obtain a closed-loop characteristic polynomial of Hurwitz type, such as: $(s^2 + 2\zeta\omega_n s + \omega_n^2)^2$. Taking into account that ζ and ω_n are positive quantities, the expressions for the controller gains are

$$k_3 = 4\zeta\omega_n - \frac{c}{m}$$

$$k_2 = (2\omega_n^2 + 4\zeta^2\omega_n^2) - \frac{k}{m} - \frac{c}{m}k_3$$

$$k_1 = 4\zeta\omega_n^3 - \frac{k}{m}k_3$$

$$k_0 = \omega_n^4 \tag{2.126}$$

Notice that, in this case, $\lambda_0 = \omega_n^4$. In order to obtain the best possible values for the design parameter, there are optimal control strategies of linear systems that are well established in the literature. All of them are based on minimization of a certain cost function or performance index. Two of the most classical ones that can properly be used for the underlying optimization problem are the quadratic optimal regulator and the integral of time multiplied by absolute error (ITAE) optimal system.

Rewriting (2.123) in the time domain, the implementation of a GPI controller is achieved using (2.127), (2.128), and the compensation term (2.129):

$$u_{av} = u_{av}^* - m\left[k_2\left(y - y^*\right) + (k_1 - k_2 k_3)\xi_2. + k_0\xi_1\right] \tag{2.127}$$

$$\dot{\xi}_1 = \xi_2$$

$$\dot{\xi}_2 = (y - y^*) - k_3\xi_2 \tag{2.128}$$

$$u_{av}^* = m\ddot{y}^* + c\dot{y}^* + ky^* \tag{2.129}$$

2.3.1.2 $\Sigma - \Delta$ Modulator

The implementation of the control input signal in electric actuators is generally accomplished via pulse-width modulation techniques in order to regulate the average power supplied to the plant. In our case,

$$u = \frac{W}{2}[sign(u_{av}) + sign(z)]$$

$$\dot{z} = u_{av} - u \tag{2.130}$$

2.3.1.3   Parameter Identification

We now apply the frequency-domain algebraic identification methodology to (2.121), free from the signals $\eta(t)$ and $\xi(t)$. Differentiating once with respect to time in order to eliminate the constant disturbance,

$$mx^{(3)} + c\ddot{x} + k\dot{x} = \dot{u}$$

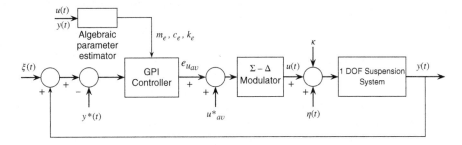

Figure 2.24   Control scheme of the perturbed 1-DOF suspension system

and applying the Laplace transform operator we have

$$m[s^3X(s) - s^2x(0) - s\dot{x}(0) - \ddot{x}(0)] + c[s^2X(s) - sx(0) - \dot{x}(0)]$$

$$+k[sX(s) - x(0)] = [sU(s) - u(0)] \qquad (2.131)$$

Take three derivatives of (2.131) with respect to the complex variable s to get rid of the initial conditions:

$$m\left[6X(s) + 18s\frac{dX(s)}{ds} + 9s^2\frac{d^2X(s)}{ds^2} + s^3\frac{d^3X(s)}{ds^3}\right] + c\left[6\frac{dX(s)}{ds} + 6s\frac{d^2X(s)}{ds^2}\right]$$

$$+s^2\frac{d^3X(s)}{ds^3}\right] + k\left[3\frac{d^2X(s)}{ds^2} + s\frac{d^3X(s)}{ds^3}\right] = 3\frac{d^2U(s)}{ds^2} + s\frac{d^3U(s)}{ds^3} \qquad (2.132)$$

Multiplying both sides by s^{-3} to avoid differentiations with respect to time and returning to the time domain one obtains

$$mv_1(t) + cv_2(t) + kv_3(t) = \gamma(t) \qquad (2.133)$$

where

$$v_1(t) = 6\int^{(3)} x - 18\int^{(2)} tx + 9\int t^2x - t^3x$$

$$v_2(t) = -6\int^{(3)} tx + 6\int^{(2)} t^2x - \int t^3x$$

$$v_3(t) = 3\int^{(3)} t^2x - \int^{(2)} t^3x$$

$$\gamma(t) = 3\int^{(3)} t^2u - \int^{(2)} t^3u$$

We can generate a linear system of equations by successive integration with respect to time of (2.133), that is,

$$D(t)\theta = H(t) \qquad (2.134)$$

where $\theta = [m \; c \; k]^T$, and H and D are, respectively, a vector and a matrix whose entries are defined by

$$d_{ij} = \int^{(i-1)} v_j(t)$$

$$h_i = \int^{(i-1)} \gamma(t) \quad i,j = 1,2,3$$

The unknown parameters are estimated by solving this system of equations. So, the system is said to be identifiable if, and only if, $\det(D) \neq 0$. The input signal u used in this algebraic approach does not necessarily exhibit the classical persistency of excitation requirement; however, the zero crossing of $\det(D)$ may produce a local loss of identifiability. This drawback can be eliminated using invariant nonlinear filtering, which will be explained in a later section.

2.3.1.4 Algebraic Identifier with a "Sentinel"

A modified version of the previous algorithm is now presented. This procedure uses a sentinel parameter like an indicator of the estimated parameters' convergence. This parameter is introduced through the multiplication of (2.132) by factors that hold invariant the equation. The proposed factors are $(s+1)$ and $(s+b_x)$, where b_x is a new additional parameter to be considered. This is

$$(s+1)\left(m\left[6X(s) + 18s\frac{dX(s)}{ds} + 9s^2\frac{d^2X(s)}{ds^2} + s^3\frac{d^3X(s)}{ds^3}\right] + c\left[6\frac{dX(s)}{ds} + 6s\frac{d^2X(s)}{ds^2}\right]\right.$$
$$\left. + s^2\frac{d^3X(s)}{ds^3}\right] + k\left[3\frac{d^2X(s)}{ds^2} + s\frac{d^3X(s)}{ds^3}\right]\right) = (s+b_x)\left(3\frac{d^2U(s)}{ds^2} + s\frac{d^3U(s)}{ds^3}\right) \quad (2.135)$$

where b_x must be 1, such that the equation is preserved. This expression corresponds to an extended dynamic system in which it is possible to build the identifier of the system parameters plus the sentinel parameter.

In order to avoid derivatives of the input and output in the identifier, we expand (2.135) and multiply both sides by s^{-4}. Then, returning to the time domain we have

$$mp_1(t) + cp_2(t) + kp_3(t) + b_x p_4(t) = q_1(t) \quad (2.136)$$

where

$$p_1(t) = 6\int^{(4)} x - 18\int^{(3)} tx + 6\int^{(3)} x + 9\int^{(2)} t^2 x - 18\int^{(2)} tx$$
$$- \int^{(1)} t^3 x + 9\int^{(1)} t^2 x - t^3 x$$

$$p_2(t) = -6\int^{(4)} tx + 6\int^{(3)} t^2 x - 6\int^{(3)} tx - \int^{(2)} t^3 x + 6\int^{(2)} t^2 x - \int^{(1)} t^3 x$$

$$p_3(t) = 3\int^{(4)} t^2 x - \int^{(3)} t^3 x + 3\int^{(3)} t^2 x - \int^{(2)} t^3 x$$

$$p_4(t) = \int^{(4)} (t^3 \dot{u}) = -3 \int^{(4)} t^2 u + \int^{(3)} t^3 u$$

$$q_1(t) = 3 \int^{(3)} t^2 u - \int^{(2)} t^3 u$$

From (2.136), it is possible to build another linear system of equations to obtain the estimated parameters including the sentinel parameter. The procedure previously described has a time-domain equivalent. The algebraic identifier obtained in this way must be the same as that obtained by the frequency-domain method.

2.3.1.5   An Automatic Disconnection Scheme

There exists a well-known feature in the algebraic identification technique; the identifier inner states grow indefinitely because they represent iterated integrals of time-polynomial factors, as seen in (2.136). This situation may produce an overflow in the mathematical processing unit when a long time has passed. Given that the algebraic parameter identification is quite fast, the estimated values are accurate enough in a short time, so it is possible to stop the identification process after a small period of time. We will name this procedure: the *disconnection* procedure.

In this context, disconnection means that the estimation process has reached acceptable estimated values for all parameters and these can be used by the controller in order to accomplish a given control task. After that, the identification process may be permanently stopped, or reinitialized at a later time, if the degradation in performance can be attributed to parameter variations.

An automatic disconnection of the parameter-estimation subsystem can be implemented based on the monitoring of the introduced sentinel parameter and simple optimization criteria. Here, we present two criteria to accomplish the disconnection task. These are applied in a simultaneous way:

$$AE(k) : |b_x - 1|$$

(absolute error)

$$IADE(k) : \int_{kT}^{(k+1)T} \left| \frac{d(b_x - 1)}{dt} \right| dt = \int_{kT}^{(k+1)T} \left| \frac{db_x}{dt} \right| dt$$

(integral of absolute derivative of error)

In order to ensure the quality of estimation, a periodic evaluation criterion may be applied. If strong restrictions are assumed for these indices, the identification subsystem will take a longer time to yield the estimated values. Two policies can be taken for the management of the disconnection; the first one is simple and for many cases is enough. A single disconnection, governed by a small constant tolerance, is performed assuming that the obtained parameters are sufficiently accurate for the correct operation of the designed control system. The second policy considers a greater initial tolerance, allowing us to have a first estimated value. Subsequently, the tolerances are strained and more accurate estimated values are obtained in each evaluation. The process terminates when the sentinel stays within a minimal specified tolerance. This disconnection scheme is advisable for systems where the estimation is slow and it is desirable to feed the control with parameters that at least guarantee an acceptable trajectory tracking while the parameters reach more precise values.

2.3.1.6   The Sentinel and the Estimated Parameters' Convergence

The online algebraic identifiers are formulae that were obtained from the system model with algebraic manipulations that yield, in an exact way, the constant parameters of the system. Thus, in the algebraic method framework it is not possible to refer to the classical asymptotic stability and asymptotic convergence arguments (see Gensior *et al.*, 2008). In Trapero-Arenas *et al.* (2008), the time required to obtain the estimates is explored based on the condition number of the matrix involved in the identifier equations system.

At the beginning of the estimation, wrong estimated values are obtained due to the signal-to-noise ratio (SNR) effects. A survey has been written (see Trapero *et al.*, 2007) concerning the minimum-time response estimation with respect to sample time and SNR, where the effectiveness of invariant linear filters is shown.

According to the construction philosophy of the sentinel identifier, as shown in Section 2.3.1, its structure is obtained from the system dynamics and it is similar to the remaining parameter identifiers. So, the sentinel parameter emulates, in a suitable way, the behavior of the remaining system parameters.

2.3.1.7   An Alternative Invariant Nonlinear Filtering

The basic invariant filtering used for the noise treatment in this work is integration. As shown before, one way to generate a system of simultaneous equations is by successive integration of the original identifier equation. This integration provides an additional basic filtering effect on the equations. Moreover, it is possible to make additional integrations over all the equations to improve the SNR.

In spite of this filtering, the terms that belong to the identifier denominator may still cross the singular value of zero, causing a locally indeterminate quotient and in some cases an unreliable identification process. In order to overcome this problem, we use an extra invariant filtering approach. This invariant nonlinear filtering is subject to three assumptions: first, the parameters are locally constant; second, the parameter signs are known; and third, we are only interested in calculating their magnitudes.

Consider the following identifier for a certain positive parameter $a > 0$:

$$a = \frac{n(t)}{d(t)}$$

If we take the integral of the absolute value of the functions $n(t)$ and $d(t)$, the fraction holds invariant:

$$a = \frac{\int_0^t |n(\tau)| d\tau}{\int_0^t |d(\tau)| d\tau}$$

This new identifier avoids crossings by zero and induces a softer behavior. We may perform additional integrations for the improvement of the underlying.

2.3.1.8   Simulation Results

The complete diagram of the control system of this suspension is shown in Figure 2.24. The nominal parameters of the plant for the numerical simulations are $m = 2.1\,\mathrm{kg}$, $c = 4.5\,\mathrm{N\,s\,m^{-1}}$, $k = 195\,\mathrm{N\,m^{-1}}$, controller gains $\omega = 10$, $\zeta = 0.85$, and for the $\Sigma - \Delta$ modulator, $W = 4\mathrm{N}$. The initial conditions for the suspension system were $x(0) = 0.0025$ m, $\dot{x}(0) = -0.01$ m s^{-1} and for the estimated parameters, $m_e(0) = 1.6$ kg, $c_e(0) = 10$ N s m^{-1}, $k_e(0) = 350$ N m$^-1$. The control task is to regulate the position of the mass and track a smooth trajectory in spite of having a constant input disturbance $\mu = -0.5$N; noise signals $\eta(t) = 0.001(rand(1, t) - 0.5)$ and $\xi(t) = 0.0003(rand(1, t) - 0.5)$ are presented, respectively, in the input and output signal measurements, where $rand(1, t)$ is a function which generates computer random numbers between 0 and 1. For a desired rest-to-rest maneuver, a smooth desired trajectory for the incremental output is proposed using an interpolating Bézier polynomial of the form:

$$\vartheta^*(t) = \begin{cases} \hbar_0 & \text{for } t < t_0 \\ \hbar_0 + (\hbar_f - \hbar_0)\phi(t, t_0, t_f) & \forall t \in [t_0, t_f] \\ \hbar_f & \text{for } t > t_f \end{cases}$$

$$\phi(t, t_0, t_f) = \left(\frac{t - t_0}{t_f - t_0}\right)^r \sum_{i=0}^{r} (-1)^i \beta_i \left(\frac{t - t_0}{t_f - t_0}\right)^i$$

where the initial and final values of the output are denoted by $\hbar_0$ and $\hbar_f$, respectively. The polynomial function of time $\phi(t, t_0, t_f)$ exhibits a sufficient number of zero derivatives at times t_0 and t_f while also satisfying $\phi(t_0, t_0, t_f) = 0$ and $\phi(t_f, t_0, t_f) = 1$. We propose $r = 8$, in order to obtain a smooth trajectory. Therefore, in this case, the interpolation function ϕ and its derivatives are

$$\phi^*(t, t_0, t_f) = \chi^8(\beta_0 - \beta_1\chi + \beta_2\chi^2 - \cdots + \beta_8\chi^8)$$

$$\dot{\phi}(t, t_0, t_f)^* = \chi^7(8\beta_0 - 9\beta_1\chi + 10\beta_2\chi^2 - \cdots + 16\beta_8\chi^8)$$

$$\ddot{\phi}^*(t, t_0, t_f) = \chi^6(56\beta_0 - 72\beta_1\chi + \cdots + 240\beta_8\chi^8)$$

with $\chi = \left(\frac{t-t_0}{t_f-t_0}\right)$, $\beta_0 = 12\ 870$, $\beta_1 = 91\ 520$, $\beta_2 = 288\ 288$, $\beta_3 = 524\ 160$, $\beta_4 = 600\ 600$, $\beta_5 = 443\ 520$, $\beta_6 = 205\ 920$, $\beta_7 = 54\ 912$, and $\beta_8 = 6435$. The trajectory to be followed is defined as

$$y^* = \begin{cases} \vartheta_1^*(t) & t < 3.5 \\ \vartheta_2^*(t) & t \geq 3.5 \end{cases} \tag{2.137}$$

where

	$\hbar_0$	$\hbar_1$	t_0	t_f
$\vartheta_1^*(t)$	0.005	0.015	2	3
$\vartheta_2^*(t)$	0.015	-0.008	4	5

Figure 2.25 shows the quality of the trajectory tracking when the average control signal u_{av} is implemented by means of a $\Sigma - \Delta$ modulator to obtain a discontinuous control signal $u(t)$. Notice that u_{av} is always bounded within the range $[-W, W]$, which produces a suitable modulation process. This figure also shows the robustness of the control strategy to the presence of constant disturbances and noise. The designed controller can be robust to a certain degree for a reasonable parametric uncertainty, as shown in Figure 2.26, where the controller uses the uncertain preset parameters $m_e(0)$, $c_e(0)$, $k_e(0)$ during the entire simulation time.

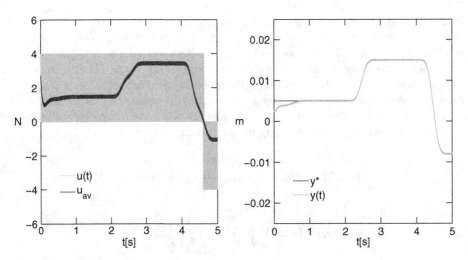

Figure 2.25   System behavior when the control law is implemented by $\Sigma - \Delta$ modulation using nominal parameters

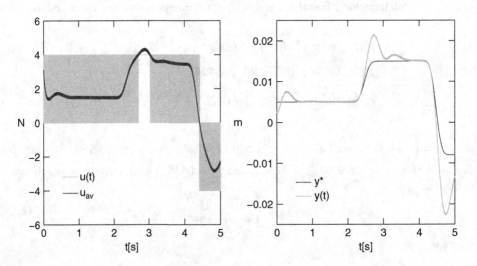

Figure 2.26   System behavior using uncertain parameters in the control

Now the identifier (2.136) is implemented in order to attenuate the effects of this uncertainty. As shown in Figure 2.27, the performance is very similar to the case illustrated in Figure 2.25. Notice, from the algebraic identification process shown in Figure 2.28, the comparison between the original outputs from the algebraic estimator (m_e, c_e, k_e), in a noisy environment, and the output signals generated by the disconnection module (m_{ed}, c_{ed}, k_{ed}).

This module based on the behavior of b_x, denoted by s in the figure, supplies aperiodically new estimated parameter values for the controller according to the second disconnection policy described in Section 2.3.1. The first estimate supplied to the controller is clearly visible in the graph of the control. Since, on average, the input is the same when we use $\Sigma - \Delta$ modulation,

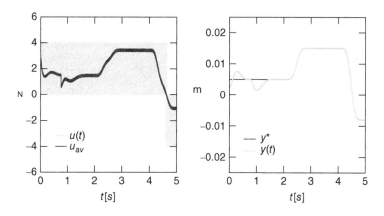

Figure 2.27 Performance of the controller with estimated parameters

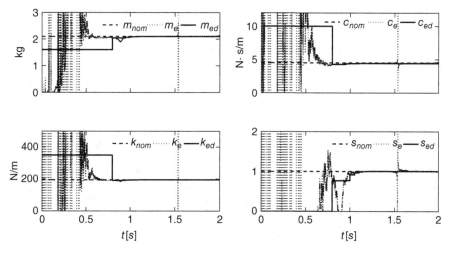

Figure 2.28 Identification of parameters via algebraic estimator and disconnection module without invariant filtering

the identification task and the tracking behavior are similar to the continuous case. In this case, the evaluation of the convergence of the sentinel parameter has allowed us to identify the convergence of the others.

At the beginning of the estimation process, some large values are obtained because of the ill-conditioning of the algebraic identifier equations. The connection and disconnection subsystem ensures that the controller does not process these wrong values. The scale of Figures 2.28 and 2.29 is adjusted to view a small region around the real values. Figure 2.29 shows the application of the invariant filtering to improve the response of the algebraic estimator under conditions of additive noise. There are marked differences in the signal shape.

Notice that, from a classic identification point of view, the trajectory (2.137) does not represent a suitable *excitation* geared to identifying parameters, due to its lack of *persistence*.

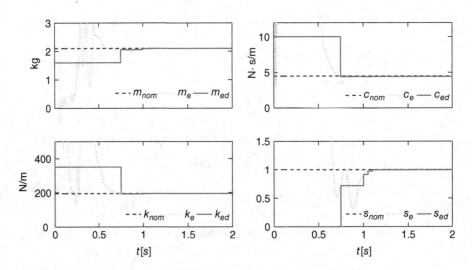

Figure 2.29 Identification of parameters using connection module and invariant filtering

Despite this, we get excellent results through the algebraic parameter-estimation methodology. The estimation is achieved, quite fast, while the system undergoes its transient response. These incipient mass motions produce enough information to quickly identify, in an accurate fashion, the unknown parameters.

2.3.1.9 Comparison with a Sliding Mode Identification Approach

In order to compare the efficiency of the proposed identification algorithm and the robustness of the tracking strategy, other identification techniques will be tested using a new irregular persistent reference trajectory that is sufficiently smooth and bounded. We propose the following trajectory be followed:

$$y^*(t) = 2\rho_1 e^{-\sin^2 0.5t} \sin t \cos t$$

where ρ_1 is governed by the following time-scaled Van der Pol dynamics:

$$\dot{\rho}_1 = 5\rho_2$$
$$\dot{\rho}_2 = 5(1.1(1 - \rho_1^2)\rho_2 - \rho_1)$$

Sliding-mode techniques with twisting algorithms can be applied for parameter estimation (see, e.g., Butt and Bhatti, 2008; Yan and Edwards, 2008). The aim of this scheme is to estimate the elements of a linear regressor through a preset uncertain parameter-based observer. The parametric uncertainty is estimated by solving this regressor via recursive least squares.

Considering the unperturbed version of (2.121) and introducing the variables $\zeta_1 = x$ and $\zeta_2 = \dot{x}$, one state-space representation of the suspension system is given by

$$\dot{\zeta}_1 = \zeta_2$$
$$\dot{\zeta}_2 = a_0 u - a_1 \zeta_2 - a_2 \zeta_1$$

with $a_0 = \frac{1}{m}$, $a_1 = \frac{c}{m}$, $a_2 = \frac{k}{m}$. So, the super-twisting observer for this system is defined as (Davila *et al.*, 2006)

$$\dot{\hat{\zeta}}_1 = \hat{\zeta}_2 + \alpha_2 |\tilde{\zeta}_1|^{1/2} sign(\tilde{\zeta}_1)$$
$$\dot{\hat{\zeta}}_2 = \bar{a}_0 u - \bar{a}_1 \hat{\zeta}_2 - \bar{a}_2 \zeta_1 + \alpha_1 sign(\tilde{\zeta}_1) \qquad (2.138)$$
$$\tilde{\zeta}_1 = \zeta_1 - \hat{\zeta}_1$$

where $\bar{a}_0$, $\bar{a}_1$, and $\bar{a}_2$ are the preset uncertain parameters. Thus, it is possible to propose a linear regressor from the estimation-error dynamics involving the parametric uncertainty as

$$\bar{z}_{eq} = \Delta_\theta \varphi \qquad (2.139)$$

where $\Delta_\theta = [a_0 - \bar{a}_0 \quad -a_1 + \bar{a}_1 \quad -a_2 + \bar{a}_2]$, $\varphi = [u \; \hat{\zeta}_2 \; y]^T$, and $\bar{z}_{eq}$ is the filtered version of the estimated term $\hat{z}_{eq} = \alpha_1 sign(\tilde{\zeta}_1)$. The coefficients α_1 and α_2 are selected as proposed by Davila *et al.* (2006), and (2.139) is solved by a recursive least-squares method. The behavior of (2.138) and the tracking of the new trajectory with the same noise characteristics using the GPI controller are shown in Figure 2.30. This simulation shows the robustness of the GPI controller to develop good tracking of an irregular trajectory. The graphs in Figure 2.31 correspond to a comparison between observer least-squares estimates m_{smls}, c_{smls}, k_{smls} and algebraic method estimates m_{ed}, c_{ed}, k_{ed}. Although this scheme is based on the rapid estimation of z_{eq}, the least-squares algorithm convergence is affected significantly by the filtering quality of z_{eq} and the persistent excitation of the system. In this simulation, the continuous control signal was used to obtain an acceptable result of the estimated values. Another important characteristic of an algebraic identifier is that it can even be used along with the discontinuous signal generated by the $\Sigma - \Delta$ modulator and not only with the continuous input control signal. This

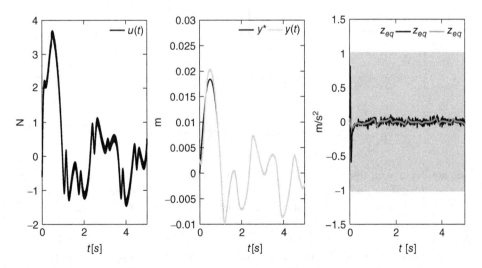

Figure 2.30 Tracking of the new trajectory and super-twisting observer performance. Here, z_{eq} denotes the system uncertainty, $\hat{z}_{eq}$ is the uncertainty estimated by the super-twisting observer, and $\bar{z}_{eq}$ is a filtered version of $\hat{z}_{eq}$

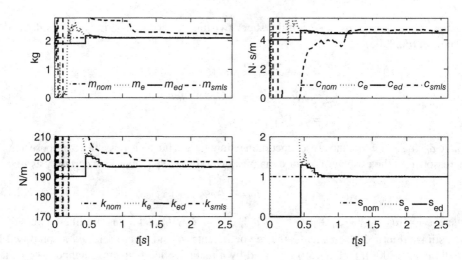

Figure 2.31 Comparisons with a sliding-mode least-squares methodology

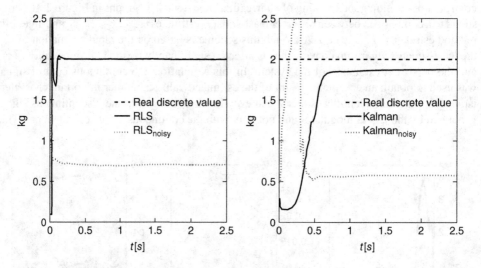

Figure 2.32 Comparisons with standard estimators

does not occur with the super-twisting algorithm. This suggests that the scheme based on algebraic identifiers and a GPI controller is a competitive control technique in real practical implementations. Figure 2.32 shows the behavior of the estimation corresponding to two classical estimators: Kalman filter and recursive least squares (RLS); both simulations were set under the same conditions of the last simulations. Additionally, a noise-free simulation is made for reference purposes. As is clear from the figure, the estimated values are far from the real values in the noisy case. The cause of these estimation errors is well established in the literature (see Ikonen and Najim, 2002), due to insufficient persistence of excitation. This common requirement is not necessary under the algebraic framework. Besides, the convergence is slower than in the case of the algebraic method and standard approaches are therefore not suitable for the studied case.

2.3.1.10   Discussion

The introduction of a sentinel parameter provides a useful criterion for assessing the computation time of the algebraic parameter-estimation procedure. The simulations show that the disconnection policy is robust with respect to the noise and takes place in an appropriate period of time. The possibility of performing algebraic manipulations that include other types of sentinel parameters −in algebraic identification schemes −by application of licit algebraic operations over the system set of differential equations is quite open for contributions. It is advisable that the operations somehow minimize the complexity of the resulting linear system yielding the parameters and performing supplementary filtering. We assumed an additive noise of zero-mean value and we do not use stochastic analysis in order to deal with this noise. The integration, with respect to time, of the terms involved in the identification equations improves the signal-to-noise ratio. These successive integrations produce an averaging and smoothing effect over the high-frequency noise signal. The nonlinear invariant filtering plays an important role and eliminates singularities, due to crossings by zero, of the identification terms in the quotients defining the parameters and simultaneously attenuates additive noise effects, thus simplifying the disconnection task.

2.4   Remarks

- An algebraic method for fast, online, estimation of unknown parameters using only inputs and outputs has been introduced for the class of linear time-invariant systems subject to "classical" unknown perturbation inputs.
- The technique, which rests on the concept of an "algebraic derivative," is also robust with respect to zero-mean, high-frequency noises and yields promising results, as seen from digital computer-based simulations.
- Identification problems in multi-variable linear time-invariant, continuous, and discrete systems can also be treated similarly. The technique seems to work well in some nonlinear systems.
- Other reported applications of the methodology with experimental results are the identification of flexible manipulators (Becedas et al., 2009), DC motor control (Mamani et al., 2009, Sira-Ramirez et al., 2007), mechanical systems (Becedas et al., 2007, García-Rodríguez et al., 2009), sinusoidal signal parameter identification (Trapero et al., 2007, Trapero-Arenas et al., 2008) among others.
- Implications of similar algebraic methods exist for solutions of problems related to signal processing (Mboup et al., 2009).

References

Åstrom, K. and Eykhoff, P. (1971) System identification–a survey. *Automatica* **7**, 123–162.

Becedas, J., Mamani, G., Feliu-Batlle, V. and Sira-Ramírez, H. (2007) Algebraic identification method for mass–spring–damper system. *International Conference on Modeling, Simulation and Control (ICMSC07)*, San Francisco, CA.

Becedas, J., Trapero, J., Feliu, V. and Sira-Ramírez, H. (2009) Adaptive controller for single-link flexible manipulators based on algebraic identification and generalized proportional integral control. *IEEE Transactions on Systems, Man, and Cybernetics, Part B: Cybernetics*, **39**(3), 735–751.

Butt, Q. and Bhatti, A. (2008) Estimation of gasoline-engine parameters using higher order sliding mode. *IEEE Transactions on Industrial Electronics* **55**(11), 3891–3898.

Davila, J., Fridman, L. and Poznyak, A. (2006) Observation and identification of mechanical systems via second order sliding modes. *International Workshop on Variable Structure Systems, 2006 (VSS'06)*, pp. 232–237.

Fliess, M. (2006) Analyse non standard du bruit. *C. R., Math., Academie des Sciences de Paris* **342**(10), 797–802.

Fliess, M. and Sira-Ramírez, H. (2002) On the non-calibrated visual based control of planar manipulators: An on line algebraic identification approach. *IEEE-SMC-2002*, Tunisia.

Fliess, M. and Sira-Ramírez, H. (2003) An algebraic framework for linear identification. *ESAIM, Control, Optimization and Calculus of Variations* **9**(1), 151–168.

Fliess, M., Join, C. and Sira-Ramírez, H. (2008) Non-linear estimation is easy. *International Journal of Modelling, Identification and Control* **4**(1), 12–27.

Fliess, M., Marquez, R., Delaleau, E. and Sira-Ramírez, H. (2002) Correcteurs proportionnels-intègraux généralisés. *ESAIM, Control, Optimization and Calculus of Variations* **7**(2), 23–41.

García-Rodríguez, C., Cortés-Romero, J. and Sira-Ramírez, H. (2009) Algebraic identification and discontinuous control for trajectory tracking in a perturbed 1-dof suspension system. *IEEE Transactions on Industrial Electronics* **56**(9), 3665–3674.

Gensior, A., Weber, J., Rudolph, J. and Guldner, H. (2008) Algebraic parameter identification and asymptotic estimation of the load of a boost converter. *IEEE Transactions on Industrial Electronics* **55**(9), 3352–3360.

Hsu, L. and Aquino, P. (1999) Adaptive visual tracking with uncertain manipulator dynamics and uncalibrated camera. *Proceedings of the 38th IEEE Conference on Decision and Control, 1999*, vol. 2, pp. 1248–1253, Phoenix, AZ.

Hsu, L., Lopes Zachi, A. and Lizarralde, F. (2001) Adaptive visual tracking for motions on smooth surfaces *Proceedings of the 40th IEEE Conference on Decision and Control, 2001*, vol. 3, pp. 2430–2435, Orlando, FL.

Hutchinson, S., Hager, G. and Corke, P. (1996) A tutorial on visual servo control. *IEEE Transactions on Robotics and Automation* **12**(5), 651–670.

Ikonen, E. and Najim, K. (2002) *Advanced Process Identification and Control*. Marcel Dekker, New York.

Kelly, R. (1996) Robust asymptotically stable visual servoing of planar robots. *IEEE Transactions on Robotics and Automation* **12**(5), 759–766.

Lum, K., Coppola, V. and Bernstein, D. (1998) Adaptive virtual autobalancing for a rigid rotor with unknown mass imbalance supported by magnetic bearings. *Journal of Vibration and Acoustics* **120**(2), 557–570.

Mamani, G., Becedas, J., Feliu, V. and Sira-Ramírez, H. (2009) *Advances in Computational Algorithms and Data Analysis*, vol. 14 of Lecture Notes in Electrical Engineering. Springer, Berlin, pp. 381–393.

Mboup, M., Join, C. and Fliess, M. (2009) Numerical differentiation with annihilators in noisy environment. *Numerical Algorithms* **50**, 439–467.

Sira-Ramírez, H., Barrios-Cruz, E. and Marquez, R. (2007) Fast adaptive trajectory tracking control of a completely uncertain dc motor via output feedback. *46th IEEE Conference on Decision and Control, 2007*, pp. 4197–4202, New Orleans, LO.

Sira-Ramírez, H., Fossas, E. and Fliess, M. (2002) On the control of an uncertain, double bridge buck converter: An algebraic parameter identification approach. *Proceedings of the 41st IEEE Conference on Decision and Control*, pp. 2462–2467, Las Vegas, NV.

Tan, HS. and Bradshaw, T. (1997) Model identification of an automotive hydraulic active suspension system. *Proceedings of the American Control Conference, 1997.*

Trapero-Arenas, J., Mboup, M., Pereira-Gonzalez, E. and Feliu, V. (2008) On-line frequency and damping estimation in a single-link flexible manipulator based on algebraic identification. *16th Mediterranean Conference on Control and Automation, 2008*, pp. 338–343, Albuquerque, NM.

Trapero, JR., Sira-Ramírez, H. and Feliu Batlle, V. (2007) An algebraic frequency estimator for a biased and noisy sinusoidal signal. *Signal Processing* **87**(6), 1188–1201.

Yan, XG. and Edwards, C. (2008) Adaptive sliding-mode-observer-based fault reconstruction for nonlinear systems with parametric uncertainties. *IEEE Transactions on Industrial Electronics* **55**(11), 4029–4036.

3

Algebraic Parameter Identification in Nonlinear Systems

3.1 Introduction

In this chapter, we present some nonlinear systems examples, of a physical nature, where it is necessary to carry out an online algebraic based parameter identification. The fundamental operations related to the algebraic derivative, devised for linear systems, can be used in a purely time domain interpretation within the realm of nonlinear systems. This poses no particular difficulty for the nonlinear case, except for the fact that the convolutions found in the linear system case are obtained here, generally speaking, in terms of nonlinear functions of the state variables. Naturally, the identification is greatly facilitated in those cases where the system unknown parameter set is linearly, or weakly linearly, identifiable. Contrary to the case of linear systems, here we assume that the whole set of state vector components is measurable; a restriction that will later be lifted thanks to the availability of algebraic methods for fast state estimation in nonlinear systems. Section 3.2.1 starts with an elementary uncertain nonlinear system control example. Here we emphasize that since the Laplace transformation is not suitable for identifying the system parameters in a nonlinear system, then a direct time-domain algebraic manipulation of the system equations yields the required setting that allows for fast identification. We also stress that, in the presence of high frequency measurement noise, the invariant filtering may be carried out in two different ways. In Section 3.2.2 we deal with a block dragging problem where all important system parameters are unknown constants. The parameter identification problem is solved here in the context of a nontrivial trajectory tracking problem for the block position. This shows how the algebraic parameter-estimation technique suitably combines with the flatness-based controller design method. Section 3.2.3 deals with the control of the rigid-body problem, when all the principal moments of inertia are unknown constants. We devote some special consideration to a particularly important switched-control-system case in which the parameter identification must be carried out under sliding-mode control conditions. This is particularly important due to the validity of the method in the presence of associated chattering responses and high-frequency bang–bang-type control inputs. The use of invariant filtering proves to be rather successful in the accurate identifica-

---

Algebraic Identification and Estimation Methods in Feedback Control Systems, First Edition.
Hebertt Sira-Ramírez, Carlos García-Rodríguez, Alberto Luviano-Juárez, John Alexander Cortés-Romero.
© 2014 John Wiley & Sons, Ltd. Published 2014 by John Wiley & Sons, Ltd.

tion of the system parameters needed for the average certainty-equivalence feedback controller design.

The control of an uncertain inverted pendulum coupled to a DC motor is presented in Section 3.2.5. An online, fast adaptive feedback controller is proposed under the restriction that the only measurable output signal is the motor angular position θ. Section 3.2.6 deals with a convey crane, where the algebraic estimator is obtained from the nonlinear model and the feedback controller is synthesized from the linearized system around an equilibrium point and by use of its flatness property. A classical example is presented in Section 3.2.7, where a simplified magnetic levitation model is consider. In Section 3.3, an alternative procedure for linear system equation generation from a regressor is developed. Some advantages are reported from the numerical point of view. Entering into the field of chaotic systems, a simple and illustrative example from the dynamics of the Genesio–Tesi chaotic system is described in Section 3.3.1 and the Ueda oscillator is tackled in Section 3.3.2. Associated with A.C. electric machines, an identification and control of an uncertain brush-less DC motor is presented in section 3.3.3 and an algebraic parameter identification for induction motors is described in detail in 3.3.5. A mechatronic application based on the inertia wheel pendulum is developed by a parameter identification combined with self-tuned control in section 3.3.4.

3.2 Algebraic Parameter Identification for Nonlinear Systems

In the last chapter, the class of dynamic input–output systems analyzed was of the form

$$\frac{d^n y(t)}{dt^n} + a_{n-1}\frac{d^{n-1}y(t)}{dt^{n-1}} + \dots + a_1\frac{dy(t)}{dt} + a_0 y(t) = b_m\frac{d^m u(t)}{dt^m} +$$

$$b_{m-1}\frac{d^{m-1}u(t)}{dt^{m-1}} + \dots + b_1\frac{du(t)}{dt} + b_0 u(t) + \xi(t) \tag{3.1}$$

where $y(t)$ stands for the measurable output, $u(t)$ is the input, a_i, b_j, $i = 1, \dots, n-1$, $j = 1, \dots, m$ are the sets of constant unknown parameters, and $\xi(t)$ represents a structured disturbance input of polynomial type. The algebraic method for parameter identification can be approached from two different, but equivalent, viewpoints:

1. Algebraic identification procedure carried out in the time domain.
2. Algebraic identification procedure carried out in the frequency domain.

Even though both schemes result in the same formulae for the parameters, the first case can be implemented in a larger class of systems while the second method is restricted to purely linear time-invariant systems. This fact can be illustrated in the following example.

Example 3.2.1 *Consider the following mass–spring system (Figure 3.1) where the spring is of "hardening type":*

$$m\ddot{y} + f_{hs}(y) = F$$

$$f_{hs}(y) := k(1 + a^2 y^2)y, \quad k, a > 0 \tag{3.2}$$

Here $f_{hs}(y)$ denotes the force exerted by the hardening spring (Khalil, 2002). It is clear that the operational calculus identification scheme cannot be applied. However it is possible to

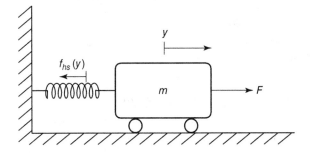

Figure 3.1 Mass hardening spring system

apply the identification procedure in the time domain. Considering that the highest order of differentiation in (3.2) is two, one multiplies the expression by t^2. Integrating twice, we have

$$\left[t^2 y - 4 \int (ty) + 2 \int\int y \right] m + \left[\int\int t^2 y \right] k + \left[\int\int t^2 y^3 \right] ka^2 = \int\int t^2 F \tag{3.3}$$

which can be expressed as

$$\left[t^2 y - 4 \int ty + 2 \int\int y \quad \int\int t^2 y \quad \int\int t^2 y^3 \right] \begin{bmatrix} m \\ k \\ ka^2 \end{bmatrix} = \int\int t^2 F \tag{3.4}$$

The last expression is a linear function with respect to the set of parameters m, k, ka^2. One may continue with the procedure of algebraic identification as performed on the class of linear systems.

3.2.0.1 Class of Systems to be Treated

The algebraic parameter-identification method can be extended to a class of continuous-time nonlinear systems whose model can be rewritten as a set of linear functions of a parameter vector, say θ. In other words, given a set of measurable outputs $y_1, y_2, \ldots, y_l$, a set of inputs $u_1, u_2, \ldots, u_m$, and a set of constant parameters $\theta = \begin{bmatrix} \theta_1 & \theta_2 & \ldots & \theta_p \end{bmatrix}$, if the nonlinear system can be expressed in the following form

$$y_1^{(n_1)}(t) = \phi_1(t, y, u)^T \theta$$

$$y_2^{(n_2)}(t) = \phi_2(t, y, u)^T \theta$$

$$\vdots \tag{3.5}$$

$$y_{l-1}^{(n_{l-1})}(t) = \phi_{l-1}(t, y, u)^T \theta$$

$$y_l^{(n_l)}(t) = \phi_l(t, y, u)^T \theta$$

the set of parameters θ is linearly identifiable in the algebraic sense.

In several cases, it is straightforward to express a wide class of nonlinear systems in the form (3.5). Notice that the set of functions ϕ_i, $i = 1, \ldots, l$ is nonlinear, but system (3.5) is linear in θ; that is, we have linearity in the parameters and not necessarily linearity in the dynamics (Middleton and Goodwin, 1990).

The next sections describe some case studies concerning nonlinear control systems which are algebraically identifiable.

3.2.1 Controlling an Uncertain Pendulum

Consider a single link as shown in Figure 3.2, with the mass concentrated on the bob which is driven by an applied torque at the pivot point. The system equations are readily obtained as

$$mR^2\ddot{\theta} = -mgR\sin(\theta) + \tau \tag{3.6}$$

where m is the unknown mass and R is the length of the massless link, here also assumed to be unknown. The parameter g is the known constant of acceleration due to gravity. The variable θ is the angular position, measured from the vertical reference line in a counterclockwise sense. The applied torque is τ, which is considered to be positive when it tries to provoke angular motion in the counterclockwise sense.

The system model can be written as

$$\ddot{\theta} = -\frac{g}{R}\sin(\theta) + \frac{1}{mR^2}\tau \tag{3.7}$$

The unknown system parameters m and R are not linearly identifiable. They are, however, weakly linearly identifiable.

Indeed, we let

$$a = \frac{g}{R}, \quad b = \frac{1}{mR^2}$$

Clearly, the parameters a and b are linearly identifiable. From the knowledge of these parameters, as identified quantities a_e and b_e, we immediately have

$$R_e = \frac{g}{a_e}, \quad m_e = \frac{a_e^2}{g^2 b_e}$$

It is desired to track a given output reference angular displacement trajectory $\theta^*(t)$ having at our disposal only angular position measurements θ.

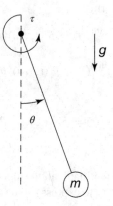

Figure 3.2 The controlled pendulum

3.2.1.1 A Classical Certainty-Equivalence Controller

A certainty-equivalence controller for the uncertain system

$$\ddot{\theta} = -a \sin(\theta) + b\tau \tag{3.8}$$

may be proposed as

$$\tau = \frac{1}{b_e}[a_e \sin(\theta) + v]$$

$$v = \ddot{\theta}^*(t) - \left[\frac{k_2 s^2 + k_1 s + k_0}{s(s + k_3)}\right](\theta - \theta^*(t)) \tag{3.9}$$

Notice that under perfect cancelation of the actual parameters with the estimated parameters, the closed-loop tracking error signal $e = \theta - \theta^*(t)$ is governed by the linear differential equation

$$e^{(4)} + k_3 e^{(3)} + k_2 \ddot{e} + k_1 \dot{e} + k_0 e = 0 \tag{3.10}$$

The design parameters $\{k_3, k_2, k_1, k_0\}$ are chosen so as to have the characteristic polynomial $p(s) = s^4 + k_3 s^3 + k_2 s^2 + k_1 s + k_0$ exhibit all its roots in the strictly open left half of the complex plane at convenient locations.

3.2.1.2 Algebraic, Online Parameter Identifier

The online computation of the unknown parameters a and b is accomplished as follows. Multiply the system equation $\ddot{\theta} = -a \sin \theta + b\tau$ by t^2. Integrate, in the interval $[0, t]$, the obtained expression with respect to time. This yields, irrespective of the initial condition for $\dot{\theta}$, the following expression:

$$t^2 \theta(t) - 2 \left[t\theta(t) - \int_0^t \theta(\sigma)d\sigma\right] = -a \int_0^t \sigma^2 \sin(\theta(\sigma))d\sigma + b \int_0^t \sigma^2 \tau(\sigma)d\sigma \tag{3.11}$$

Integrating again with respect to time we find, after some rearrangement,

$$t^2 \theta(t) - 4 \int_0^t \sigma\theta(\sigma)d\sigma + 2 \int_0^t \int_0^\sigma \theta(\sigma_1)d\sigma_1 d\sigma$$

$$= -a \int_0^t \int_0^\sigma \sigma_1^2 \sin(\theta(\sigma_1))d\sigma_1 d\sigma + b \int_0^t \int_0^\sigma \sigma_1^2 \tau(\sigma_1)d\sigma_1 d\sigma \tag{3.12}$$

The equation for the unknown parameters a and b is a linear equation of the form

$$p_{11}(t)a + p_{12}(t)b = q_1(t) \tag{3.13}$$

with

$$p_{11}(t) = -\int_0^t \int_0^\sigma \sigma_1^2 \sin(\theta(\sigma_1))d\sigma_1 d\sigma$$

$$p_{12}(t) = \int_0^t \int_0^\sigma \sigma_1^2 \tau(\sigma_1)d\sigma_1 d\sigma$$

$$q_1(t) = t^2 \theta(t) - 4 \int_0^t \sigma\theta(\sigma)d\sigma + 2 \int_0^t \int_0^\sigma \theta(\sigma_1)d\sigma_1 d\sigma$$

The generation of a second equation in the two unknowns a and b may proceed in several ways. We opt for the integration of the previously obtained linear equation. We then propose as a second linear equation the following:

$$a \int_0^t p_{11}(\sigma)d\sigma + b \int_0^t p_{12}(\sigma)d\sigma = \int_0^t q_1(\sigma)d\sigma \tag{3.14}$$

The system of equations is then of the form

$$\begin{bmatrix} p_{11}(t) & p_{12}(t) \\ \int_0^t p_{11}(\sigma)d\sigma & \int_0^t p_{12}(\sigma)d\sigma \end{bmatrix} \begin{bmatrix} a \\ b \end{bmatrix} = \begin{bmatrix} q_1(t) \\ \int_0^t q_1(\sigma)d\sigma \end{bmatrix} \tag{3.15}$$

and we have the following linear equation:

$$\begin{bmatrix} -\int_0^t \int_0^\sigma \sigma_1^2 \sin(\theta(\sigma_1))d\sigma_1 d\sigma & \int_0^t \int_0^\sigma \sigma_1^2 \tau(\sigma_1)d\sigma_1 d\sigma \\ -\int_0^t \int_0^\sigma \int_0^{\sigma_1} \sigma_2^2 \sin(\theta(\sigma_2))d\sigma_2 d\sigma_1 d\sigma & \int_0^t \int_0^\sigma \int_0^{\sigma_1} \sigma_2^2 \tau(\sigma_2)d\sigma_2 d\sigma_1 d\sigma \end{bmatrix} \begin{bmatrix} a \\ b \end{bmatrix}$$

$$= \begin{bmatrix} t^2 \theta(t) - 4\int_0^t \sigma\theta(\sigma)d\sigma + 2\int_0^t \int_0^\sigma \theta(\sigma_1)d\sigma_1 d\sigma \\ \int_0^t \sigma^2\theta(\sigma)d\sigma - 4\int_0^t \int_0^\sigma \sigma_1\theta(\sigma_1)d\sigma_1 d\sigma + 2\int_0^t \int_0^\sigma \int_0^{\sigma_1} \theta(\sigma_2)d\sigma_2 d\sigma_1 d\sigma \end{bmatrix} \tag{3.16}$$

The estimates of the parameters a and b can readily be obtained by solving the linear equation

$$\begin{bmatrix} a_e \\ b_e \end{bmatrix} = \begin{cases} \text{arbitrary} & \text{for } t \in [0, \epsilon) \\ \begin{bmatrix} \dfrac{q_1 \left(\int p_{12}\right) - p_{12}\left(\int q_1\right)}{p_{11}\left(\int p_{12}\right) - p_{12}\left(\int p_{11}\right)} \\ \dfrac{-q_1\left(\int p_{11}\right) - p_{12}\left(\int q_1\right)}{p_{11}\left(\int p_{12}\right) - p_{12}\left(\int p_{11}\right)} \end{bmatrix} & \text{for } t \geq \epsilon \end{cases} \tag{3.17}$$

Invariant filtering may be exercised on the obtained solution for the parameters by simultaneously low-pass filtering the numerator and the denominator of each component of the obtained vector of estimates, that is, filtering is carried out *after* solving the linear equations. We set

$$\begin{bmatrix} a_e \\ b_e \end{bmatrix} = \begin{bmatrix} \dfrac{G(s)\left[q_1(\int p_{12}) - p_{12}(\int q_1)\right]}{G(s)\left[p_{11}(\int p_{12}) - p_{12}(\int p_{11})\right]} \\ \dfrac{G(s)\left[-q_1(\int p_{11}) - p_{12}(\int q_1)\right]}{G(s)\left[p_{11}(\int p_{12}) - p_{12}(\int p_{11})\right]} \end{bmatrix} \tag{3.18}$$

An alternative to the proposed invariant-filtering scheme would be to filter the linear equations *before* solving them, as follows:

$$\begin{bmatrix} G(s)p_{11}(t) & G(s)p_{12}(t) \\ G(s)\left[\int_0^t p_{11}(\sigma)d\sigma\right] & G(s)\left[\int_0^t p_{12}(\sigma)d\sigma\right] \end{bmatrix} \begin{bmatrix} a \\ b \end{bmatrix} = \begin{bmatrix} G(s)q_1(t) \\ G(s)\left[\int_0^t q_1(\sigma)d\sigma\right] \end{bmatrix}$$

It is easy to check that, in this case, the invariant filtering leads to the following equation for the parameter estimates:

$$
\begin{bmatrix} a_e \\ b_e \end{bmatrix} = \begin{bmatrix} \dfrac{G^2(s)\,[q_1(\int p_{12}) - p_{12}(\int q_1)]}{G^2(s)\,[p_{11}(\int p_{12}) - p_{12}(\int p_{11})]} \\[2ex] \dfrac{G^2(s)\,[-q_1(\int p_{11}) - p_{12}(\int q_1)]}{G^2(s)\,[p_{11}(\int p_{12}) - p_{12}(\int p_{11})]} \end{bmatrix}
\tag{3.19}
$$

Such a low-pass filtering tends to enhance, in a superior manner, the signal-to-noise ratio achieved by the previous procedure.

3.2.1.3 Simulations

Figure 3.3 depicts the controlled responses of the uncertain pendulum under the action of the certainty-equivalence classical trajectory-tracking controller with nonlinearity cancelation and fast adaptation. A smooth rest-to-rest angular displacement represents the control task. The parameter identification takes place in a small amount of time, $\epsilon = 0.1$ s, in spite of a significant noisy angular measurement of the form

$$
y = \theta(t) + 0.01(rect(t) - 0.5)
$$

where $rect(t)$ is a computer-generated random process consisting of a sequence of piecewise constant random variables uniformly distributed over the closed interval $[0, 1]$.

The parameters used in the simulation were given by

$$
m = 0.5\,\text{kg}, \quad R = 1\,\text{m}, \quad g = 9.81\,\text{m/s}^2, \quad a = 9.81, \quad b = 2
$$

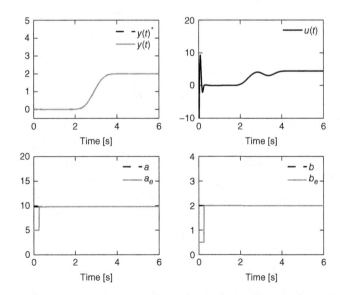

Figure 3.3   Controlled responses and parameter identification for the uncertain pendulum

The initial (wrong) values adopted for the estimates of the unknown parameters a and b were set, respectively, to $a_e = 5$ and $b_e = 0.5$. For the invariant filtering we have chosen $G(s) = 1/s^2$ with filtering after having solved the linear equations.

3.2.1.4   Robustness with Respect to the Coulomb Friction Phenomenon

The classical certainty-equivalence controller is found to be robust with respect to the simple but widely accepted "velocity sign" friction phenomenon. The reason is that, after the nonlinearity cancelation, the integral action embedded in the tracking controller helps in rejecting the piecewise constant nature of the Coulomb friction torques whenever the angular velocity is significant. Close to zero velocity, the chattering nature of the friction torques is filtered out by the closed-loop system in a low-pass manner. However, even if the feedback controller is robust with respect to Coulomb friction terms, the computation of the unknown parameters is by no means robust with respect to the unmodeled piecewise constant perturbation arising from the friction term when the velocity is nonzero. Close to zero velocity, the chattering torque input does not significantly affect the parameter computation thanks to the invariant filtering process rejecting high-frequency noise inputs. We therefore must modify the above-proposed parameter identifier suitably in order to handle friction terms.

Consider then the perturbed pendulum system

$$\ddot{\theta} = -a\sin(\theta) + bu + \mu\,sign(\dot{\theta}) \tag{3.20}$$

where μ is an unknown constant parameter. See Figure 3.4.

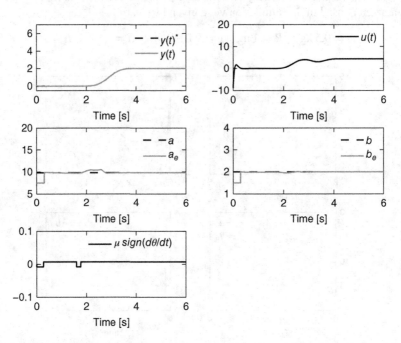

Figure 3.4 Controlled responses and parameter identification for the uncertain pendulum with Coulomb friction coefficient

For values of the angular velocity $\dot\theta$ bounded away from zero, the model exhibits a constant perturbation input of amplitude μ and angular velocity-dependent sign. We proceed to take one time derivative of the system dynamics to obtain

$$\theta^{(3)}(t) = -a\frac{d}{dt}[\sin(\theta)] + b\dot u(t) \tag{3.21}$$

Multiplying out by t^3 and integrating by parts three times, after some algebraic manipulations we obtain

$$6\left(\int^{(3)}\theta\right) - 18\left(\int^{(2)}t\theta\right) + 9\left(\int t^2\theta\right) - t^3\theta =$$

$$-a\left[3\left(\int^{(3)}t^2\sin(\theta)\right) - \left(\int^{(2)}t^3\sin(\theta)\right)\right] + b\left[3\left(\int^{(3)}t^2u\right) - \left(\int^{(2)}t^3u\right)\right] \tag{3.22}$$

Integrating once more to obtain a system of equations for the unknown parameters a and b, we now have a system of the form

$$\begin{bmatrix} p_{11}(t) & p_{12}(t) \\ \int_0^t p_{11}(\sigma)d\sigma & \int_0^t p_{12}(\sigma)d\sigma \end{bmatrix}\begin{bmatrix} a \\ b \end{bmatrix} = \begin{bmatrix} q_1(t) \\ \int_0^t q_1(\sigma)d\sigma \end{bmatrix} \tag{3.23}$$

with

$$p_{11}(t) = -\left[3\left(\int^{(3)}t^2\sin(\theta)\right) - \left(\int^{(2)}t^3\sin(\theta)\right)\right]$$

$$p_{12}(t) = \left[3\left(\int^{(3)}t^2u\right) - \left(\int^{(2)}t^3u\right)\right]$$

$$p_{21}(t) = \int_0^t p_{11}(\sigma)d\sigma$$

$$p_{22}(t) = \int_0^t p_{12}(\sigma)d\sigma$$

$$q_1(t) = 6\left(\int^{(3)}\theta\right) - 18\left(\int^{(2)}t\theta\right) + 9\left(\int t^2\theta\right) - t^3\theta$$

$$q_2(t) = \int_0^t q_1(\sigma)d\sigma$$

As before, we propose an invariant filtered estimate of the unknown vector of parameters:

$$\begin{bmatrix} a_e \\ b_e \end{bmatrix} = \begin{cases} \text{arbitrary} & \text{for } t \in [0, \epsilon) \\[2ex] \begin{bmatrix} \frac{1}{s^2}\left[q_1(\int p_{12}) - p_{12}(\int q_1)\right] \\ \frac{1}{s^2}\left[p_{11}(\int p_{12}) - p_{12}(\int p_{11})\right] \\ \frac{1}{s^2}\left[-q_1(\int p_{11}) - p_{12}(\int q_1)\right] \\ \frac{1}{s^2}\left[p_{11}(\int p_{12}) - p_{12}(\int p_{11})\right] \end{bmatrix} & \text{for } t \geq \epsilon \end{cases} \tag{3.24}$$

3.2.2 A Block-Driving Problem

Consider the problem of dragging, while orienting, a block of unknown mass on a frictionless surface through a torque T and an off-centered force F applied on an axis parallel to its main longitudinal axis located at a distance L as in Figure 3.5 (Ma *et al.*, 2002).

The model of the system is given by

$$m\ddot{x} = F\cos(\theta)$$
$$m\ddot{y} = F\sin(\theta)$$
$$J\ddot{\theta} = T + FL \tag{3.25}$$

The mass m of the object, its moment of inertia J, and the distance L are all assumed to be unknown. The control inputs are the force F and the torque T. We assume, at this point in time, that all system variables are perfectly measured.

3.2.2.1 Flatness-Based Controller

The system is differentially flat, with flat output given by the coordinates of the mass center

$$y_1 = x, \quad y_2 = y$$

Indeed, all system variables may be differentially parameterized in terms of the flat outputs y_1, y_2 and a finite number of their time derivatives:

$$\theta = \arctan\left(\frac{\ddot{y}_2}{\ddot{y}_1}\right), \quad \dot{\theta} = \frac{y_2^{(3)}\ddot{y}_1 - \ddot{y}_2 y_1^{(3)}}{\ddot{y}_1^2 + \ddot{y}_2^2}$$

$$x = y_1, \quad \dot{x} = \dot{x}_1, \quad y = y_2, \quad \dot{y} = \dot{y}_2$$

$$F = m\sqrt{\ddot{y}_1^2 + \ddot{y}_2^2}, \quad T = J\ddot{\theta} - mL\sqrt{\ddot{y}_1^2 + \ddot{y}_2^2}$$

$$\ddot{\theta} = \frac{\left(y_2^{(4)}\ddot{y}_1 - \ddot{y}_2 y_1^{(4)}\right)}{\ddot{y}_1^2 + \ddot{y}_2^2} - 2\frac{\left(y_2^{(3)}\ddot{y}_1 - \ddot{y}_2 y_1^{(3)}\right)\left[\ddot{y}_2 y_2^{(3)} + \ddot{y}_1 y_1^{(3)}\right]}{(\ddot{y}_1^2 + \ddot{y}_2^2)^2} \tag{3.26}$$

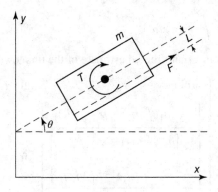

Figure 3.5   A block driven by a torque and an off-centered force

Since the inputs-to-flat-outputs relation is not invertible, we introduce, as an auxiliary control input, the second time derivative of the force F:

$$\ddot{F} = m \left(\frac{\left[\left(y_1^{(3)} \right)^2 + \ddot{y}_1 y_1^{(4)} + \left(y_2^{(3)} \right)^2 + \ddot{y}_2 y_2^{(4)} \right]}{(\ddot{y}_1^2 + \ddot{y}_2^2)^{\frac{1}{2}}} \right.$$

$$\left. - \frac{\left(\ddot{y}_1 y_1^{(3)} + \ddot{y}_2 y_2^{(3)} \right) \left(\ddot{y}_1 y_1^{(3)} + \ddot{y}_2 y_2^{(3)} \right)}{(\ddot{y}_1^2 + \ddot{y}_2^2)^{\frac{3}{2}}} \right) \tag{3.27}$$

This control input extension now yields an invertible control-input-to-flat-output relation of the form

$$\begin{bmatrix} \ddot{F} \\ T \end{bmatrix} = \begin{bmatrix} m & 0 \\ 0 & J \end{bmatrix} \begin{bmatrix} \dfrac{\ddot{y}_1}{\sqrt{\ddot{y}_1^2 + \ddot{y}_2^2}} & \dfrac{\ddot{y}_2}{\sqrt{\ddot{y}_1^2 + \ddot{y}_2^2}} \\ -\dfrac{\ddot{y}_2}{\ddot{y}_1^2 + \ddot{y}_2^2} & \dfrac{\ddot{y}_1}{\ddot{y}_1^2 + \ddot{y}_2^2} \end{bmatrix} \begin{bmatrix} y_1^{(4)} \\ y_2^{(4)} \end{bmatrix}$$

$$+ \begin{bmatrix} \phi_1(y_1, y_2, \ldots, y_1^{(3)}, y_2^{(3)}) \\ \phi_2(y_1, y_2, \ldots, y_1^{(3)}, y_2^{(3)}) \end{bmatrix} \tag{3.28}$$

The system is then found to be equivalent to the following set of linear decoupled fourth-order systems:

$$y_1^{(4)} = v_1, \quad y_2^{(4)} = v_2 \tag{3.29}$$

where v_1 and v_2 play the rôle of auxiliary control inputs to be specified in accordance with the trajectory-tracking control objectives.

A dynamic trajectory-tracking controller, including integral action, would be given by

$$\begin{bmatrix} \ddot{F} \\ T \end{bmatrix} = \begin{bmatrix} m & 0 \\ 0 & J \end{bmatrix} \begin{bmatrix} \dfrac{\ddot{y}_1}{\sqrt{\ddot{y}_1^2 + \ddot{y}_2^2}} & \dfrac{\ddot{y}_2}{\sqrt{\ddot{y}_1^2 + \ddot{y}_2^2}} \\ -\dfrac{\ddot{y}_2}{\ddot{y}_1^2 + \ddot{y}_2^2} & \dfrac{\ddot{y}_1}{\ddot{y}_1^2 + \ddot{y}_2^2} \end{bmatrix} \begin{bmatrix} v_1 \\ v_2 \end{bmatrix}$$

$$+ \begin{bmatrix} \phi_1(y_1, y_2, \ldots, y_1^{(3)}, y_2^{(3)}) \\ \phi_2(y_1, y_2, \ldots, y_1^{(3)}, y_2^{(3)}) \end{bmatrix} \tag{3.30}$$

with

$$v_1 = [y_1^*(t)]^{(4)} - \gamma_4(y_1^{(3)} - [y_1^*(t)]^{(3)}) - \gamma_3(\ddot{y}_1 - \ddot{y}_1^*(t)) - \gamma_2(\dot{y}_1 - \dot{y}_1^*(t))$$

$$- \gamma_1(y_1 - y_1^*(t)) - \gamma_0 \int_0^t (y_1 - y_1^*(\sigma)) d\sigma$$

$$v_2 = [y_2^*(t)]^{(4)} - \kappa_4(y_2^{(3)} - [y_2^*(t)]^{(3)}) - \kappa_3(\ddot{y}_2 - \ddot{y}_2^*(t)) - \kappa_2(\dot{y}_2 - \dot{y}_2^*(t))$$

$$- \kappa_1(y_2 - y_2^*(t)) - \kappa_0 \int_0^t (y_2 - y_2^*(\sigma)) d\sigma \tag{3.31}$$

All the higher-order time derivatives in the controller may be placed in terms of the states of the system. Note that $\dot{F}$ is an available state of the dynamic controller:

$$y_1 = x, \quad \dot{y}_1 = \dot{x}, \quad \dot{x}_1 = \frac{1}{m}F\cos\theta, \quad y_1^{(3)} = \frac{1}{m}[\dot{F}\cos\theta - F\dot{\theta}\sin\theta]$$

$$y_2 = y, \quad \dot{y}_2 = \dot{y}, \quad \ddot{y}_2 = \frac{1}{m}F\sin\theta, \quad y_2^{(3)} = \frac{1}{m}[\dot{F}\sin\theta + F\dot{\theta}\cos\theta] \tag{3.32}$$

In this particular instance, the homogeneity property of the differential parameterizations of $\ddot{F}$ and T avoids the need to divide by the unknown mass parameter m.

3.2.2.2 Simulations

In the simulations to be shown, we simply verify that the proposed flatness-based controller renders an excellent closed-loop response for the system. We assume then, for the moment, that all the parameters are perfectly known. For the next simulations we consider the following perturbed model:

$$m\ddot{x} = F\cos(\theta) + \xi_1$$

$$m\ddot{y} = F\sin(\theta) + \xi_2$$

$$J\ddot{\theta} = T + FL \tag{3.33}$$

with ξ_1 and ξ_2 being constant, unknown perturbations (Figure 3.6).

Suppose it is desired to follow an n-leaved rose in the x, y plane. We used the following trajectory given in parametric form:

$$x^*(t) = R\cos(n\omega t)\cos(\omega t), \quad y^*(t) = R\cos(n\omega t)\sin(\omega t) \tag{3.34}$$

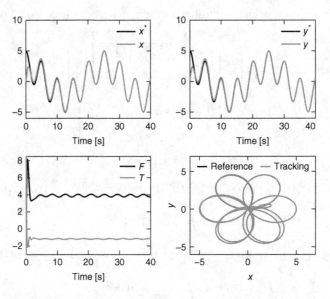

Figure 3.6　Trajectory-tracking performance of flatness-based controller for perturbed system

The poles of the closed-loop, decoupled, fifth-order linear systems for x and y were chosen to be given by the roots of the following desired characteristic polynomial:

$$p(s) = (s^2 + 2\xi\omega_n s + \omega_n^2)^2(s + p) \tag{3.35}$$

with $\xi = 0.8$, $\omega_n = 1.5$, $p = 2$ in both cases.

We adopted the following set of parameters for the simulations:

$$R = 5\,\text{m}, \quad \omega = 0.5\,\text{rad/s}, \quad J = 0.2\,\text{N m s}^2$$

$$L = 0.3\,\text{m}, \quad m = 1\,\text{kg}$$

We set the unknown constant perturbations to be $\xi_1 = \xi_2 = 0.5$ N.

3.2.2.3 Parameter Identification

The previously proposed flatness-based controller (3.30)–(3.31) is taken to be a *certainty-equivalence controller* for the given system when the parameters are unknown. We proceed to online compute the unknown system parameters, m, J, and L as follows.

We differentiate and multiply out the first equation by t^3

$$m t^3 x^{(3)} = t^3 \frac{d}{dt}[F\cos\theta] \tag{3.36}$$

This gets rid of the possible dependence of the parameter-identification process on the unknown constant perturbations and sets the stage to also become totally independent of the initial conditions.

Integrating by parts each term in (3.36) has the effect of lowering the order of the required time derivatives of the state variables. Indeed, integrating once the left and right-hand sides of equation (3.36) leads to

$$m\int_0^t \sigma^3 x^{(3)} d\sigma = m\left[t^3\ddot{x} - 3\int_0^t \sigma^2\ddot{x}d\sigma\right] = m\left[t^3\ddot{x} - 3\left(t^2\dot{x} - 2\int_0^t \sigma x d\sigma\right)\right]$$

$$\int_0^t \sigma^3 \frac{d}{d\sigma}[F\cos\theta]d\sigma = t^3[F(t)\cos\theta(t)] - 3\int_0^t \sigma^2 F(\sigma)\cos\theta(\sigma)d\sigma$$

Integrating twice more and solving for the unknown constant m results in

$$m_e = \left[\frac{3(\int^{(3)} t^2 F\cos(\theta)) - (\int^{(2)} t^3 F\cos(\theta))}{6(\int^{(3)} x) - 18(\int^{(2)} tx) + 9(\int t^2 x) - t^3 x}\right]$$

$$= \left[\frac{3(\int^{(3)} t^2 F\sin(\theta)) - (\int^{(2)} t^3 F\sin(\theta))}{6(\int^{(3)} y) - 18(\int^{(2)} ty) + 9(\int t^2 y) - t^3 y}\right] \tag{3.37}$$

Carrying out a similar program on the linear rotational dynamics, represented by the last equation in expression (3.33), we obtain

$$\begin{bmatrix} p_{11}(t) & p_{12}(t) \\ p_{21}(t) & p_{22}(t) \end{bmatrix} \begin{bmatrix} J_e \\ L_e \end{bmatrix}$$

$$= \begin{bmatrix} 3(\int^{(3)} t^2 T) - (\int^{(2)} t^3 T) \\ 3(\int^{(4)} t^2 T) - (\int^{(3)} t^3 T) \end{bmatrix} \tag{3.38}$$

where

$$p_{11}(t) = 6\left(\int^{(3)} \theta\right) - 18\left(\int^{(2)} t\theta\right) + 9\left(\int t^2\theta\right) - t^3\theta$$

$$p_{12}(t) = -3\left(\int^{(3)} t^2 F\right) + \left(\int^{(2)} t^3 F\right)$$

$$p_{21}(t) = 6\left(\int^{(4)} \theta\right) - 18\left(\int^{(3)} t\theta\right) + 9\left(\int^{(2)} t^2\theta\right) - \left(\int t^3\theta\right)$$

$$p_{22}(t) = -3\left(\int^{(4)} t^2 F\right) + \left(\int^{(3)} t^3 F\right)$$

3.2.2.4 Simulations

Figure 3.7 depicts the performance of the controlled trajectories of the uncertain perturbed multi-variable block system with online identification of all the unknown parameters.
Figure 3.8 depicts the details of the accuracy and fast features of the proposed online algebraic parameter-identification process.

3.2.3 The Fully Actuated Rigid Body

Consider now the dynamic model known as the *Euler equations* of a rigid body fixed at its center of mass in free space (Fliess *et al.*, 2008):

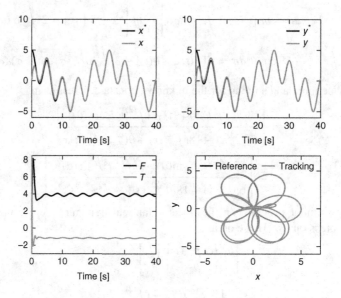

Figure 3.7   Controlled trajectories of perturbed uncertain block

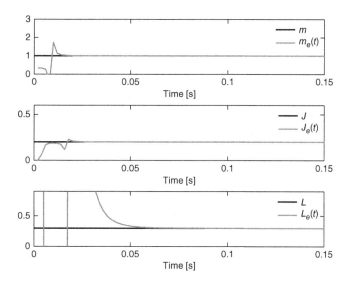

Figure 3.8 Online parameter identification

$$I_1\dot{\omega}_1 = (I_2 - I_3)\omega_2\omega_3 + u_1$$

$$I_2\dot{\omega}_2 = (I_3 - I_1)\omega_3\omega_1 + u_2$$

$$I_3\dot{\omega}_3 = (I_1 - I_2)\omega_1\omega_2 + u_3 \tag{3.39}$$

where I_1, I_2, I_3 represent the moments of inertia around the principal axes of the body (Figure 3.9). The variables ω_1, ω_2, and ω_3 are the (measurable) angular velocities of the principal axes. The control inputs are labeled u_1, u_2, u_3. They are the applied control-input torques.

3.2.3.1 A Certainty-Equivalence Controller

The system (3.39) is *flat*, with the three flat outputs being the angular velocities, ω_1, ω_2, and ω_3. Under the assumption of perfect knowledge of the moments of inertia I_1, I_2, and I_3, a stabilizing, or de-tumbling, multi-variable feedback strategy is given by the following prescription of a control law which includes integral compensation terms counteracting a possible, unknown, constant-moment perturbation vector:

$$u_1 = -(I_2 - I_3)\omega_2\omega_3 + I_1\left(-\lambda_{11}\omega_1 - \lambda_{01}\int_0^t \omega_1(\sigma)d\sigma\right)$$

$$u_2 = -(I_3 - I_1)\omega_3\omega_1 + I_2\left(-\lambda_{12}\omega_2 - \lambda_{02}\int_0^t \omega_2(\sigma)d\sigma\right)$$

$$u_3 = -(I_1 - I_2)\omega_1\omega_2 + I_3\left(-\lambda_{13}\omega_3 - \lambda_{03}\int_0^t \omega_3(\sigma)d\sigma\right) \tag{3.40}$$

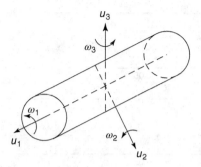

Figure 3.9 The fully actuated rigid body

The closed-loop system evolves in accordance with the following set of linear decoupled dynamics:

$$\dot{\omega}_1 = -\lambda_{11}\omega_1 - \lambda_{01} \int_0^t \omega_1(\sigma)d\sigma$$

$$\dot{\omega}_2 = -\lambda_{12}\omega_2 - \lambda_{02} \int_0^t \omega_2(\sigma)d\sigma$$

$$\dot{\omega}_2 = -\lambda_{13}\omega_3 - \lambda_{03} \int_0^t \omega_3(\sigma)d\sigma \qquad (3.41)$$

which can be made to have the origin as an asymptotically and exponentially stable equilibrium point under suitable choice of the controller design parameters $\lambda_{1i}, \lambda_{0,i}, i = 1, 2, 3$.

The performance of the proposed feedback controller, addressed as the *certainty-equivalence controller*, is depicted in Figure 3.10.

The numerical values used in the simulations for the moments of inertia, and for the design parameters, were set to be

$$I_1 = 1\,\text{N m s}^2, \quad I_2 = 0.5\,\text{N m s}^2, \quad I_3 = 0.2\,\text{N m s}^2$$

$$\lambda_{1i} = 2\zeta_i\omega_{ni}, \quad \lambda_{0i} = \omega_{ni}^2, \quad \zeta_i = 0.707, \quad \omega_{ni} = 0.5, \quad i = 1, 2, 3$$

3.2.3.2 Linear Identifiability of the Perturbed Rigid-Body Dynamics

The fundamental problem with the proposed feedback control law is that the system parameters, represented by the moments of inertia, are *not* known, except for the fact that they are constant.

We now concentrate our efforts on devising an online parameter calculation scheme which is based on an algebraic identification approach previously proposed. We first prove, in accordance with our established definitions, that the system is *linearly identifiable* even if it exhibits the external influence of a constant unknown moment vector.

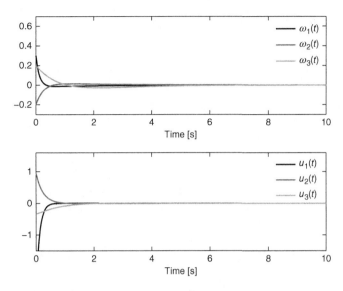

Figure 3.10 Closed-loop response of controlled rigid body with perfect knowledge of the system parameters

Consider the following perturbed rigid-body dynamics:

$$I_1\dot{\omega}_1 = (I_2 - I_3)\omega_2\omega_3 + u_1 + \xi_1$$
$$I_2\dot{\omega}_2 = (I_3 - I_1)\omega_3\omega_1 + u_2 + \xi_2$$
$$I_3\dot{\omega}_3 = (I_1 - I_2)\omega_1\omega_2 + u_3 + \xi_3 \tag{3.42}$$

where the unknown exogenous perturbation inputs ξ_1, ξ_2, ξ_3 are assumed to be *constant*.

We proceed as follows: differentiate first the set of system equations (3.42) in order to eliminate the presence of the constant perturbations. This yields

$$I_1\ddot{\omega}_1 = (I_2 - I_3)\frac{d}{dt}(\omega_2\omega_3) + \dot{u}_1$$

$$I_2\ddot{\omega}_2 = (I_3 - I_1)\frac{d}{dt}(\omega_3\omega_1) + \dot{u}_2$$

$$I_3\ddot{\omega}_3 = (I_1 - I_2)\frac{d}{dt}(\omega_1\omega_2) + \dot{u}_3 \tag{3.43}$$

Multiply out the previous expressions by the quantity t^2 and integrate *twice* the resulting expression, which is clearly independent of the values of the initial conditions. Next, use *integration by parts* wherever the sub-integral quantities exhibit a time derivative of either the control inputs or of nonlinear functions (products) of the state components.

This procedure yields a set of expressions which are independent of the initial conditions and also devoid of any time derivatives. We obtain the following *linear system* of equations for the computation of the principal moments of inertia:

$$P(t) \begin{bmatrix} I_1 \\ I_2 \\ I_3 \end{bmatrix} = \begin{bmatrix} q_1(t) \\ q_2(t) \\ q_3(t) \end{bmatrix} = q(t) \tag{3.44}$$

where

$$p_{11}(t) = t^2 \omega_1 - 4 \left(\int t\omega_1 \right) + 2 \left(\int^{(2)} \omega_1 \right)$$

$$p_{12}(t) = - \left(\int t^2 \omega_2 \omega_3 \right) + 2 \left(\int^{(2)} t\omega_2 \omega_3 \right)$$

$$p_{13}(t) = -p_{12}(t)$$

$$q_1(t) = \left(\int t^2 u_1 \right) - 2 \left(\int^{(2)} t u_1 \right) \tag{3.45}$$

$$p_{21}(t) = \left(\int t^2 \omega_3 \omega_1 \right) - 2 \left(\int^{(2)} t\omega_3 \omega_1 \right)$$

$$p_{22}(t) = t^2 \omega_2 - 4 \left(\int t\omega_2 \right) + 2 \left(\int^{(2)} \omega_2 \right)$$

$$p_{23}(t) = -p_{21}(t)$$

$$q_2(t) = \left(\int t^2 u_2 \right) - 2 \left(\int^{(2)} t u_2 \right)$$

$$p_{31}(t) = - \left(\int t^2 \omega_1 \omega_2 \right) + 2 \left(\int^{(2)} t\omega_1 \omega_2 \right)$$

$$p_{32}(t) = -p_{31}(t)$$

$$p_{33}(t) = t^2 \omega_3 - 4 \left(\int t\omega_3 \right) + 2 \left(\int^{(2)} \omega_3 \right)$$

$$q_3(t) = \left(\int t^2 u_3 \right) - 2 \left(\int^{(2)} t u_3 \right) \tag{3.46}$$

As in the linear-system parameter-identification case, the matrix $P(t) = (p_{ij}(t))$ is *not* invertible at time $t = 0$, but it is certainly invertible after an arbitrarily small time $t = \epsilon > 0$ for any nontrivial input–output trajectory. A *differential algebraic* justification of this fact follows similar lines to those encountered in Fliess and Sira-Ramírez (2003), and also in Fliess and Sira-Ramírez (2002).

In the proposed certainty-equivalence feedback controller (3.40) we utilize, during the calculation interval $[0, \epsilon]$, arbitrary numerical values for the moments of inertia I_1, I_2, and I_3 (Figure 3.11). After time $t = \epsilon$, we proceed to substitute in the controller expression the computed inertia parameters, which are now denoted by I_{1e}, I_{2e}, and I_{3e}. We have thus proposed

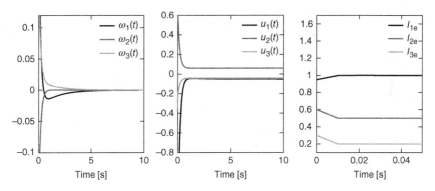

Figure 3.11 Closed-loop response of parameter-uncertain rigid body subject to external constant-perturbation moments

$$u_1 = -(I_{2e} - I_{3e})\omega_2\omega_3 + I_{1e}\left(-\lambda_{11}\omega_1 - \lambda_{01}\int_0^t \omega_1(\sigma)d\sigma\right)$$

$$u_2 = -(I_{3e} - I_{1e})\omega_3\omega_1 + I_{2e}\left(-\lambda_{12}\omega_2 - \lambda_{02}\int_0^t \omega_2(\sigma)d\sigma\right)$$

$$u_3 = -(I_{1e} - I_{2e})\omega_1\omega_2 + I_{3e}\left(-\lambda_{13}\omega_3 - \lambda_{03}\int_0^t \omega_3(\sigma)d\sigma\right) \tag{3.47}$$

with

$$\begin{bmatrix} I_{1e} \\ I_{2e} \\ I_{3e} \end{bmatrix} = \begin{cases} \text{arbitrary} & \text{for } t \in [0, \epsilon] \\ P^{-1}(t)q(t) & \text{for } t > \epsilon \end{cases} \tag{3.48}$$

3.2.3.3   Simulations

The arbitrary values of the moments of inertia during the starting of the controlled maneuver were taken to be $I_{1e} = 0.5$ N m s^2, $I_{2e} = 0.9$ N m s^2, and $I_{3e} = 0.6$ N m s^2. This faulty assignment lasts only during a time interval of the form $[0, \epsilon]$, with ϵ being a small real number of the order of tenths of seconds. Indeed, the parameter, ϵ, was determined by the violation of the condition $|\det P(t)| \leq 10^{-16}$, which happens approximately at 0.171 s.

The computed values of the moments of inertia coincide precisely with the values used for the simulation of the system, that is, $I_1 = 1.0$ N m s^2, $I_2 = 0.5$ N m s^2, and $I_3 = 0.2$ N m s^2.

The constant-moment disturbance-vector components were set to be $\xi_1 = 0.05$ N m, $\xi_2 = -0.06$ N m, and $\xi_3 = 0.04$ N m.

Computer-generated noises were used in the simulations. These are represented by piecewise constant random variables normally distributed at each instant of time. For the noise processes affecting the measurement, we used amplitudes of the order of 1×10^{-5} rad/s. The input perturbation noises were taken to be of amplitude 1×10^{-4} N m (Figure 3.12).

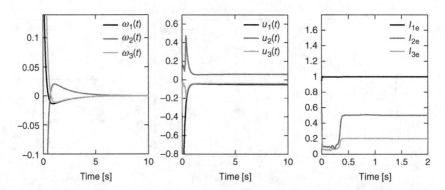

Figure 3.12   Closed-loop response of parameter-uncertain rigid body subject to external stochastic perturbation moments with constant bias and stochastic measurement noises

The constant-moment disturbances representing the bias terms in the noisy inputs were taken to be the same constant perturbations used in the previous simulation example.

3.2.4   Parameter Identification Under Sliding Motions

In this section we show, via an elementary first-order example, a particular robustness issue of our proposed algebraic parameter-estimation approach. We consider that our system is controlled by a control input that takes values in a finite set of the form $\{0, W\}$, such as may be the case for a switch position control. We base our control strategy on a continuous average design of the classical compensation network type implemented through a Sigma–Delta modulator. The average design, however, demands perfect knowledge of the system parameter. Under such circumstances a parameter estimation is required but, now the available control input is a high-frequency discontinuous signal of bang–bang type. One may justifiably worry about the possibilities of having a successful parameter estimation via the algebraic approach when such a chattering input goes through the parameter estimator. We illustrate here that *invariant filtering* adequately takes care of the situation and produces a rather precise estimate of the unknown system parameter to complete a certainty-equivalence classical controller design implemented through the $\Sigma - \Delta$ modulation alternative.

Consider the following elementary switched linear system exhibiting an unknown parameter a:

$$\dot{x} = -ax + u, \quad y = x, \quad u \in \{0, W\} \tag{3.49}$$

Suppose, just for simplicity, that it is required to regulate the state x of this uncertain system toward a fixed constant equilibrium value y^*. As is customary in this type of system, one may propose an average feedback controller design using classical control theory. A certainty-equivalence average feedback controller is given by

$$u_{av} = u^* - \left[\frac{(\gamma_1 - a\gamma_2 + a^2)s + \gamma_0}{s(s + (\gamma_2 - a))} \right] (y - y^*) \tag{3.50}$$

where u^* is the nominal average control input corresponding to the desired output equilibrium. In other words, $u^* = ay^*$ (Figure 3.13).

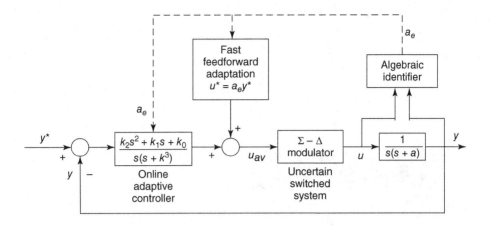

Figure 3.13 $\Sigma - \Delta$ sliding-mode-controlled uncertain system with fast adaptation

The average closed-loop stabilization error $e(t) = y - y^*$ evolves according to the linear dynamics

$$(s^2 + \gamma_2 s^2 + \gamma_1 s + \gamma_0)e(t) = 0 \tag{3.51}$$

which has the origin as an asymptotically and exponentially stable equilibrium point for a suitable set of design constants $\{\gamma_2, \gamma_1, \gamma_0\}$. Clearly, the switched nature of the control input u cannot accept the designed u_{av} as a switch command function. A $\Sigma - \Delta$ modulation implementation of u_{av} is possible which recovers all the average features of the design while providing the system with an "ON$-$OFF" binary-valued signal of amplitude W. We propose then the following modulator dynamics:

$$\dot{e}(t) = u_{av} - u, \quad u = \frac{W}{2}[1 + sign\, e(t)] \tag{3.52}$$

The certainty-equivalence controller would then be expressed as

$$u_{av} = u^* - \left[\frac{(\gamma_1 - a_e \gamma_2 + a_e^2)s + \gamma_0}{s(s + (\gamma_2 - a_e))}\right](y - y^*)$$

$$u^* = a_e y^*$$

where a_e is the fast, online algebraic estimate of the unknown parameter a obtained in a small time interval $\epsilon > 0$ on the basis of the given input$-$output dynamics using the switched input u obtained from the analog-to-binary modulator:

$$a_e = \begin{cases} \text{arbitrary for } t \in [0, \epsilon) \\ \dfrac{n(t)}{d(t)} = \dfrac{\left(\int^{(2)}[-ty + \int(y + tu)]\right)}{\int^{(3)} ty} & \text{for } t \geq \epsilon \end{cases} \tag{3.53}$$

In the identification process above, we have used a double integrator as a low-pass invariant filter affecting simultaneously the numerator and the denominator of the basic parameter identifier.

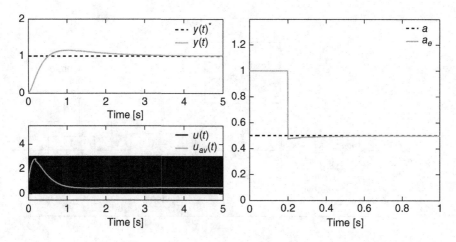

Figure 3.14   Adaptive sliding-mode-controlled uncertain system

Note that the lack of knowledge of the system parameter affects both the feedforward compensation input $a_e y^*$ as well as the controller parameters which implicitly perform an online adaptive elimination of the unknown system eigenvalue a.

Figure 3.14 depicts the performance of the adaptive feedback controller and of the fast, reliable identification process of the unknown parameter a. We set the initial arbitrary value of the parameter estimate a_e to be 1. The accurate estimation of the parameter is achieved in 0.5 time units. For the simulations presented, we have chosen

$$y^* = 1, \quad W = 3, \quad a = 0.5$$

$$\gamma_2 = 2\zeta\omega_n + p, \quad \gamma_1 = \omega_n^2 + 2\zeta\omega_n p, \quad \gamma_0 = \omega_n^2 p$$

$$\zeta = 1, \quad \omega_n = 5, \quad p = 1$$

3.2.5   Control of an Uncertain Inverted Pendulum Driven by a DC Motor

Consider the simplified model of an armature-controlled DC motor connected to a single-link pendulum whose mass is concentrated at the tip (Hernández and Sira-Ramírez, 2002), as shown in Figure 3.15. The equations of the system are given by

$$L_f \frac{di_f}{dt} = -R_f i_f + V_f$$

$$L_a \frac{di_a}{dt} = -R_a i_a - K i_f \omega + u$$

$$J_m \frac{d\omega}{dt} = -B_m \omega + K i_f i_a - \tau_L$$

$$mR^2 \frac{d\omega}{dt} = -mgR \sin(\theta) + \tau_L$$

$$\frac{d\theta}{dt} = \omega$$

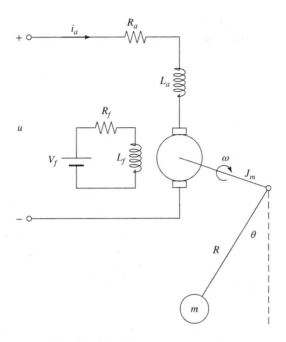

Figure 3.15 DC motor–pendular link

where i_f and i_a represent the field circuit current and the armature circuit current, and V_f is the fixed voltage applied to the field circuit. The motor shaft angular velocity is represented by ω and the pendulum angular position by θ. All resistances, inductances, lengths, and gains are assumed to be known. The mass m of the bob and the moment of inertia of the motor J_m are assumed to be unknown but constant. The variable τ_L is the load torque, represented here by the swinging pendular link. The armature circuit voltage, u, plays the role of the sole control input.

Since the field circuit produces a constant current $i_f = I_f$, we rewrite the system equations in the following reduced manner by defining $K_m = KI_f$:

$$L_a \frac{di_a}{dt} = -R_a i_a - K_m \omega + u$$

$$(mR^2 + J_m)\frac{d\omega}{dt} = -B_m \omega + K_m i_a - mgR \sin(\theta)$$

$$\frac{d\theta}{dt} = \omega$$

3.2.5.1 Problem Formulation and Main Results

Given a rest-to-rest angular position trajectory $\theta^*(t)$ for the pendular link, it is desired to have the pendulum track the prescribed reference trajectory via an online, fast adaptive feedback controller feeding back the only measurable output signal represented by the angular position θ.

We notice that the reduced system is flat, with flat output given by the angular position. Indeed, all system variables can be parameterized in terms of θ and a finite number of its time derivatives:

$$\omega = \frac{d}{dt}\theta$$

$$i_a = \frac{1}{K_m}\left[(mR^2 + J_m)\frac{d^2\theta}{dt^2} + B_m\frac{d\theta}{d\theta} + mgR\sin(\theta)\right]$$

$$u = \frac{L_a}{K_m}(mR^2 + J_m)\frac{d^3\theta}{dt^3} + \left[\frac{L_aB_m}{K_m} + \frac{R_a(mR^2 + J_m)}{K_m}\right]\frac{d^2\theta}{dt^2}$$

$$+ \left[\frac{L_a mgR\cos(\theta)}{K_m} + \frac{R_aB_m + K_m^2}{K_m}\right]\frac{d\theta}{dt} + \frac{R_a}{K_m}mgR\sin(\theta) \tag{3.54}$$

From the obtained differential parametrization (3.54) we have the following nominal value of the control input $u^*(t)$ in terms of the desired angular reference trajectory $\theta^*(t)$:

$$u^*(t) = \frac{L_a}{K_m}\Pi\frac{d^3\theta^*(t)}{dt^3} + \left[\frac{L_aB_m}{K_m} + \frac{R_a\Pi}{K_m}\right]\frac{d^2\theta^*(t)}{dt^2}$$

$$+ \left[\frac{L_a\Lambda\cos(\theta^*(t))}{K_m} + \frac{R_aB_m + K_m^2}{K_m}\right]\frac{d\theta^*(t)}{dt} + \frac{R_a\Lambda}{K_m}\sin(\theta^*(t)) \tag{3.55}$$

where we have designated $\Pi = (mR^2 + J_m)$ and $\Lambda = mgR$, the unknown system parameters.

Following Hagenmeyer and Delaleau (2003), we propose the following output time-varying certainty-equivalence feedback controller with feedforward compensation:

$$u = u^*(t) + \left(\frac{L_a}{K}\Pi\right)v$$

$$v = -\left[\frac{\gamma_3 s^3 + \gamma_2 s^2 + \gamma_1 s + \gamma_0}{s(s^2 + \gamma_5 s + \gamma_4)}\right](\theta - \theta^*(t)) \tag{3.56}$$

The closed-loop dynamics is given by

$$\left(\frac{d^3\theta}{dt^3} - \frac{d^3\theta^*(t)}{dt^3}\right) + \left[\frac{\gamma_3 s^3 + \gamma_2 s^2 + \gamma_1 s^1 + \gamma_0}{s(s^2 + \gamma_5 s + \gamma_4)}\right](\theta - \theta^*(t)) = 0 \tag{3.57}$$

Clearly, the fast adaptation is to be carried out on the feedforward term $u^*(t)$ and in the feedback controller gain. In this case, the controller parameters can be computed independently of the unknown system parameters Π and Λ simply by placing the poles of the following characteristic polynomial, closely related to the linearized part of the closed-loop system:

$$p(s) = s^6 + \gamma_5 s^5 + \gamma_4 s^4 + \gamma_3 s^3 + \gamma_2 s^2 + \gamma_1 s + \gamma_0 \tag{3.58}$$

3.2.5.2   The Algebraic Online Parameter Identifier

The online computation of the unknown parameters Π and Λ is accomplished as follows. Multiply by t^3 the third system equation in (3.54). Then, integrate on the interval $[0, t]$ the obtained expression three times with respect to time. This yields the following:

$$\frac{L_a \Pi}{K_m} \int^{(3)} (-t)^3 \theta^3 = \int^{(3)} (-t)^3 u$$

$$- \left(\frac{L_a B_m}{K_m} + \frac{R_a \Pi}{K_m} \right) \int^{(3)} (-t)^3 \ddot{\theta}$$

$$- \left(\frac{L_a \Lambda \cos(\theta)}{K_m} + \frac{R_a B_m + K_m^2}{K_m} \right) \int^{(3)} (-t)^3 \dot{\theta}$$

$$- \frac{R_a \Lambda}{K_m} \int^{(3)} (-t)^3 \sin(\theta) \tag{3.59}$$

Integrating three times, we obtain

$$\Pi p_{11}(t) + \Lambda p_{12}(t) = q_1(t) \tag{3.60}$$

where

$$p_{11} = \frac{L_a}{K_m} \left(6 \int^{(3)} \theta - 18 \int^{(2)} t\theta + 9 \int t^2 \theta - t^3 \theta \right)$$

$$+ \frac{R_a}{K_m} \left(-6 \int^{(3)} t\theta + 6 \int^{(2)} t^2 \theta - \int t^3 \theta \right)$$

$$p_{12} = \frac{L_a}{K_m} \left(3 \int^{(3)} t^2 \sin(\theta) - \int^{(2)} t^3 \sin(\theta) \right)$$

$$+ \frac{R_a}{K_m} \left(-\int^{(3)} t^3 \sin(\theta) \right)$$

$$q_1 = -\int^{(3)} t^3 u - \frac{(R_a B_m + K_m^2)}{K_m} \left(3 \int^{(3)} t^2 \theta - \int^{(2)} t^3 \theta \right)$$

$$- \frac{L_a B_m}{K_m} \left(-6 \int^{(3)} t\theta + 6 \int^{(2)} t^2 \theta - \int t^3 \theta \right)$$

Integrating once we can obtain a set of equations for the unknown parameters Π and Λ:

$$\begin{bmatrix} p_{11} & p_{12} \\ p_{21} & p_{22} \end{bmatrix} \begin{bmatrix} \Pi \\ \Lambda \end{bmatrix} = \begin{bmatrix} q_1 \\ q_2 \end{bmatrix} \tag{3.61}$$

where

$$p_{21} = \frac{L_a}{K_m} \left(6 \int^{(4)} \theta - 18 \int^{(3)} t\theta + 9 \int^{(2)} t^2 \theta - \int t^3 \theta \right)$$

$$+ \frac{R_a}{K_m} \left(-6 \int^{(4)} t\theta + 6 \int^{(3)} t^2 \theta - \int^{(2)} t^3 \theta \right)$$

$$p_{22} = \frac{L_a}{K_m} \left(3 \int^{(4)} t^2 \sin(\theta) - \int^{(3)} t^3 \sin(\theta) \right)$$

$$+ \frac{R_a}{K_m} \left(- \int^{(4)} t^3 \sin(\theta) \right)$$

$$q_2 = - \int^{(4)} t^3 u - \frac{(R_a B_m + K_m^2)}{K_m} \left(3 \int^{(4)} t^2 \theta - \int^{(3)} t^3 \theta \right)$$

$$- \frac{L_a B_m}{K_m} \left(-6 \int^{(4)} t\theta + 6 \int^{(3)} t^2 \theta - \int^{(2)} t^3 \theta \right) \tag{3.62}$$

The unknown parameters Π and Λ are obtained as follows:

$$\begin{bmatrix} \Pi \\ \Lambda \end{bmatrix} = \begin{cases} \text{arbitrary} & \text{for } t \in [0, \epsilon) \\ \begin{bmatrix} \dfrac{\left(\frac{1}{s^2} \right)[p_{22}q_1 - p_{12}q_2]}{\left(\frac{1}{s^2} \right)[p_{11}p_{22} - p_{12}p_{21}]} \\[2ex] \dfrac{\left(\frac{1}{s^2} \right)[p_{11}q_2 - p_{21}q_1]}{\left(\frac{1}{s^2} \right)[p_{11}p_{22} - p_{12}p_{21}]} \end{bmatrix} & \text{for } t \geq \epsilon \end{cases} \tag{3.63}$$

3.2.5.3 Simulations

We set initial arbitrary values for the unknown parameters, $\Pi = 0.05$ and $\Lambda = 1$. The parameters of the uncertain pendulum and the DC motor are $L_a = 8.9e - 3$ H, $R_a = 5.6\,\Omega$, $K_m = 0.0603$ N m/A, $B_m = 15.61e - 6$ N m s/rad, $J_m = 15.93e - 6$ N m^2, $m = 0.3$ Kg, $R = 0.5$ m. It was desired to smoothly bring the pendulum angular position from a value of 0 rad toward a final value of 2 rad in 3 s. A 20th-order Bézier polynomial, smoothly interpolating between the initial and final values, was prescribed as the desired angular-velocity reference trajectory $\theta^*(t)$. The output feedback controller gains were obtained using the following design values: $\varsigma = 1.5$ and $\omega_n = 400$.

Figure 3.16 shows the trajectory-tracking performance of the uncertain inverted pendulum with a DC motor, under noise-free conditions. The identification process of the unknown parameters Π and Λ is carried out in a short amount of time, approximately 0.5 s, with a waiting time $\epsilon = 0.4$S.

Figure 3.17 shows the trajectory-tracking performance of the uncertain inverted pendulum with a DC motor under noise conditions on the interval $[-0.01, 0.01]$. The identification process of the unknown parameters Π and Λ is again carried out in a short amount of time, approximately 0.5 s, and they are adapted to the controller.

3.2.6 Identification and Control of a Convey Crane

The convey crane, shown in Figure 3.18, is assumed to carry a symmetrical load, supported by a fixed-length string, which is kept under tension at all times. The car moves on a

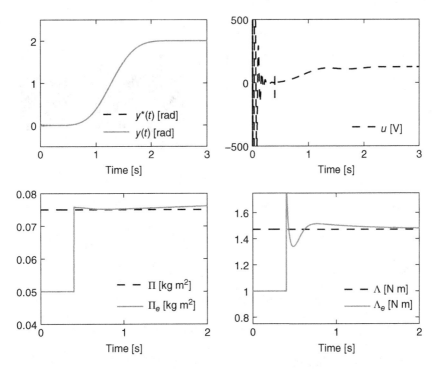

Figure 3.16 Trajectory tracking and parameter identification of the uncertain pendulum with a DC motor without noise

horizontal rail under the action of an external force F. The forces due to the air resistance are neglected. The behavior of the convey crane is thus characterized, on the moving plane, by the Euler–Lagrange equations

$$(m + M)\ddot{x} + mL\ddot{\theta} \cos\theta - mL\dot{\theta}^2 \sin\theta = F \tag{3.64}$$

$$\ddot{x}\cos\theta + L\ddot{\theta} + g\sin\theta = 0 \tag{3.65}$$

where M and m, respectively, stand for the mass of the trolley and the load, L is the fixed length of the string holding the load, x represents the horizontal displacement of the trolley, and θ represents the angle of the string with respect to the vertical axis y. The parameter g is the acceleration due to gravity and F is the horizontal input force applied to the cart. The parameters L and g are assumed known.

The control problem consists of moving the load from an initial position to a final position in such a way as to avoid dangerous residual oscillations. In order to simplify the controller design, the system dynamics can be linearized around the equilibrium point $\bar{\theta} = 0$, $\bar{x} = X$, $\bar{F} = 0$ as follows:

$$\begin{align} (M + m)\ddot{x} + mL\ddot{\theta} &= u \\ \ddot{x} + L\ddot{\theta} + g\theta &= 0 \end{align} \tag{3.66}$$

where $u = F$. The linearized system is controllable, and hence flat, with (incremental) flat output given by the projection of the mass position on the horizontal axis. This is given by

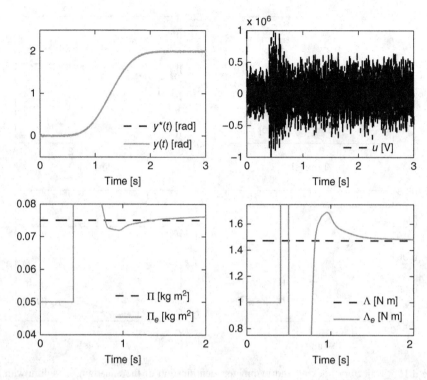

Figure 3.17 Trajectory tracking and parameter identification of the uncertain pendulum with a DC motor with noise

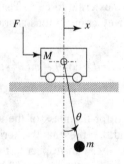

Figure 3.18 A simplified diagram of a planar convey crane

$\vartheta = x + L\theta$. All system variables can be written in terms of the flat output ϑ and a finite number of its time derivatives. Indeed,

$$x = \vartheta + \frac{L}{g}\ddot{\vartheta} \qquad\qquad \theta = -\frac{1}{g}\ddot{\vartheta}$$

$$\dot{x} = \dot{\vartheta} + \frac{L}{g}\vartheta^{(3)} \qquad\qquad \dot{\theta} = -\frac{1}{g}\vartheta^{(3)}$$

$$u = (M + m)\ddot{\vartheta} + \frac{ML}{g}\vartheta^{(4)}$$

We formulate the control problem as follows: *Devise a feedback control law u that forces the flat output signal $\vartheta(t)$ to follow a given rest-to-rest reference trajectory $\vartheta^*(t)$ in spite of the lack of knowledge about the plant parameters M and m. The flat output reference trajectory is consistent with a terminal oscillation-free maneuver of the load.*

3.2.6.1　Design of a GPI Controller

Letting the initial conditions be zero, the transfer function of the linearized system is easily obtained from the differential parametrization of the incremental control input

$$G(s) = \frac{\gamma_0}{s^2(s^2 + \gamma_1^2)} \tag{3.67}$$

with $\gamma_0 = \frac{g}{ML}$, $\gamma_1^2 = \frac{g(M+m)}{ML}$.

The tracking-error dynamics, corresponding to the linearized system, is readily found to be given by $e_\vartheta(s) = G(s)e_u(s)$ with $e_\vartheta = \vartheta - \vartheta^*(t)$ and $e_u = u - u^*(t)$, where $\vartheta^*(t)$ and $u^*(t)$ are the desired nominal trajectories for the incremental flat output and of the incremental control, respectively. In order to ameliorate the effect of constant disturbances and neglected initial conditions, the tracking-trajectory control law is designed using a classical compensation form of the GPI controller including an additional integral control action. We propose the following incremental flat output reference trajectory-tracking controller:

$$e_{u(s)} = -\left(\frac{1}{\gamma_0}\right)\left(\frac{k_4 s^4 + k_3 s^3 + k_2 s^2 + k_1 s + k_0}{s\left(s^3 + k_7 s^2 + k_6 s + k_5\right)}\right) e_{\vartheta(s)} \tag{3.68}$$

This feedback network determines the following closed-loop characteristic polynomial: $s^8 + k_7 s^7 + (k_6 + \gamma_1^2)s^6 + (k_5 + \gamma_1^2 k_7)s^5 + (k_4 + \gamma_1^2 k_6)s^4 + (k_3 + \gamma_1^2 k_5)s^3 + k_2 s^2 + k_1 s + k_0$, whose coefficients can be chosen at will so as to obtain a Hurwitz closed-loop characteristic polynomial for the linearized system. We determine the controller coefficients by equating the characteristic polynomial to the following desired polynomial: $(s^2 + 2\zeta\omega_n s + \omega_n^2)^4$. Taking into account that ζ and ω_n are both positive quantities, the expressions for the controller gains are

$$
\begin{aligned}
k_7 &= 8\zeta\omega_n \\
k_6 &= (24\zeta^2\omega_n^2 + 4\omega_n^2) - \gamma_1^2 \\
k_5 &= (32\zeta^3\omega_n^3 + 24\zeta\omega_n^3) - \gamma_1^2 k_7 \\
k_4 &= (16\zeta^4\omega_n^4 + 48\zeta^2\omega_n^4 + 6\omega_n^4) - \gamma_1^2 k_6 \\
k_3 &= (32\zeta^3\omega_n^5 + 24\zeta\omega_n^5) - \gamma_1^2 k_5 \\
k_2 &= 24\zeta^2\omega_n^6 + 4\omega_n^6 \\
k_1 &= 8\zeta\omega_n^7 \\
k_0 &= \omega_n^8
\end{aligned}
\tag{3.69}
$$

To implement the proposed controller (3.68), we use the following scheme:

$$u = u^* - \left(\frac{1}{\gamma_0}\right)(k_4 s^4 + k_3 s^3 + k_2 s^2 + k_1 s + k_0)\xi \tag{3.70}$$

$$\xi = \frac{(\vartheta - \vartheta^*)}{s(s^3 + k_7 s^2 + k_6 s + k_5)} \tag{3.71}$$

The transfer function (3.71) leads, in the time domain, to a controller expression which can be written as follows:

$$\xi^{(4)} + k_7 \xi^{(3)} + k_6 \ddot{\xi} + k_5 \dot{\xi} = (\vartheta - \vartheta^*) \tag{3.72}$$

From (3.72) and letting $\xi = \xi_1, \dot{\xi} = \xi_2, \ddot{\xi} = \xi_3$, and $\xi^{(3)} = \xi_4$, the following linear state system realization of the controller is easily synthesized:

$$\begin{aligned}
\dot{\xi}_1 &= \xi_2 \\
\dot{\xi}_2 &= \xi_3 \\
\dot{\xi}_3 &= \xi_4 \\
\dot{\xi}_4 &= -k_7 \xi_4 - k_6 \xi_3 - k_5 \xi_2 + (\vartheta - \vartheta^*)
\end{aligned} \tag{3.73}$$

Finally, rewriting (3.70) in the time domain, the implementation of the GPI controller is achieved using (3.73) and

$$\begin{aligned}
u = u^* - \left(\frac{1}{\gamma_0}\right) & \left[(k_3 - k_4 k_7)\xi_4 + (k_2 - k_4 k_6)\xi_3 \right. \\
& \left. + (k_1 - k_4 k_5)\xi_2 + k_0 \xi_1 + k_4(\vartheta - \vartheta^*) \right]
\end{aligned}$$

where u^* is given by

$$u^*(t) = \frac{\gamma_1^2}{\gamma_0} \dot{\vartheta}^*(t) + \frac{1}{\gamma_0}[\vartheta^*(t)]^{(4)}$$

and $[\vartheta^*(t)]^{(i)}$ is the offline computed ith derivative of the nominal reference trajectory for the incremental flat output. For a desired rest-to-rest maneuver, a smooth desired trajectory for the incremental flat output is proposed using an interpolating Bézier polynomial of the form

$$\vartheta^*(t) = \begin{cases}
\hbar_0 & \text{for } t < t_0 \\
\hbar_0 + (\hbar_f - \hbar_0)\phi(t) & \forall t \in [t_0, t_f] \\
\hbar_f & \text{for } t > t_f
\end{cases}$$

$$\phi(t, t_0, t_f) = \left(\frac{t - t_0}{t_f - t_0}\right)^r \sum_{i=0}^{r} (-1)^i \beta_i \left(\frac{t - t_0}{t_f - t_0}\right)^i$$

where the initial and final values of the flat output are denoted by $\hbar_0$ and $\hbar_f$, respectively. The polynomial function of time $\phi(t)$ exhibits a sufficient number of zero derivatives at times t_0 and t_f, while also satisfying $\phi(t_0, t_0, t_f) = 0$ and $\phi(t_f, t_0, t_f) = 1$. Since the fourth derivative of ϑ^* is necessary for the tracking control task, we propose $r = 8$ in order to obtain a very smooth trajectory. Therefore, in this case, the interpolation function ϕ and its derivatives are

$$\phi^*(t) = \chi^8(\beta_0 - \beta_1 \chi + \beta_2 \chi^2 - \cdots + \beta_8 \chi^8)$$

$$\dot{\phi}(t)^* = \chi^7 \eta(8\beta_0 - 9\beta_1 \chi + 10\beta_2 \chi^2 - \cdots + 16\beta_8 \chi^8)$$

$$\ddot{\phi}^*(t) = \chi^6 \eta^2(56\beta_0 - 72\beta_1 \chi + \cdots + 240\beta_8 \chi^8)$$

$$\vdots \quad \vdots$$

$$[\phi^*(t)]^{(4)} = \chi^4 \eta^4(1680\beta_0 - 3024\beta \chi + \cdots + 43680\beta_8 \chi^8)$$

with $\chi = \left(\frac{t-t_0}{t_f-t_0}\right)$, $\eta = \left(\frac{1}{t_f-t_0}\right)$, $\beta_0 = 12\,870$, $\beta_1 = 91\,520$, $\beta_2 = 288\,288$, $\beta_3 = 524\,160$, $\beta_4 = 600\,600$, $\beta_5 = 443\,520$, $\beta_6 = 205\,920$, $\beta_7 = 54\,912$, and $\beta_8 = 6435$.

3.2.6.2 Obtaining an Algebraic Identifier

A very fast parameter estimation is necessary for the implementation of the GPI controller (3.68). The required knowledge of the exact values of M and m can be satisfied using an online algebraic parameter estimator processing the input–output data arising from the nonlinear system. We can obtain an algebraic identifier of these parameters from the linear system, but this identifier is only valid around a neighborhood of the equilibrium $(\bar{\theta}, \bar{x}, \bar{F})$ and its behavior could be affected drastically by the nonlinearity of the system. Then, we are interested in obtaining an algebraic estimator from the nonlinear model, specifically from the equation (3.64). So, defining new parameters $\beta_1 = m + M$ and $\beta_2 = mL$, the equation (3.64) is rewritten as

$$\beta_1\ddot{x} + \beta_2\left(\ddot{\theta}\cos\theta - \dot{\theta}^2\sin\theta\right) = F \tag{3.74}$$

Note that we can simplify $\frac{d^2}{dt^2}(\sin\theta) = \cos\theta\ddot{\theta} - \sin\theta\dot{\theta}^2$, therefore

$$\beta_1\ddot{x} + \beta_2\frac{d^2}{dt^2}(\sin\theta) = F \tag{3.75}$$

Applying the algebraic methodology, we have

$$\beta_1\int^{(2)} t^2\ddot{x} + \beta_2\int^{(2)} t^2\frac{d^2}{dt^2}(\sin\theta) = \int^{(2)} t^2 F \tag{3.76}$$

Integrating (3.76) by parts and expanding terms, we have a regressor form to obtain β_1 and β_2:

$$\beta_1\left[xt^2 - 4\int tx + 2\int^{(2)}x\right] + \beta_2\left[t^2\sin\theta - 4\int t\sin\theta + 2\int^{(2)}\sin\theta\right] = \int^{(2)}t^2 F \tag{3.77}$$

Finally, the estimates of γ_0 and γ_1^2 are given by

$$\gamma_0 = \frac{g}{L\beta_1 - \beta_2} \tag{3.78}$$

$$\gamma_1^2 = \frac{g\beta_1}{L\beta_1 - \beta_2} \tag{3.79}$$

3.2.6.3 Simulation Results

The parameters used in these simulations were: $M = 1.155$ kg, $m = 0.48$ kg, $L = 0.3$ m, $g = 9.81$ m/s^2, $\zeta = 0.95$, and $\omega_n = 7$. The initial conditions were set to be $x_0 = -0.01$ m and $\theta_0 = -0.005$ rad. The simulation results are presented in Figures 3.19 and 3.20. It was desired to track a smooth trajectory in the range $[0, 0.8]$, in a period of $[3, 5]$ s, which produced an indirect smooth displacement of the trolley from 0 to 0.8 m with a controlled oscillation of the load.

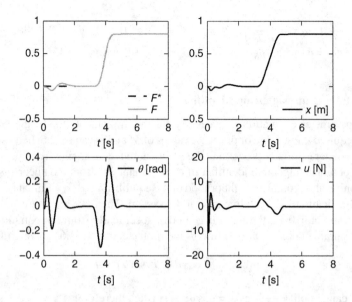

Figure 3.19   Closed-loop behavior of the convey crane

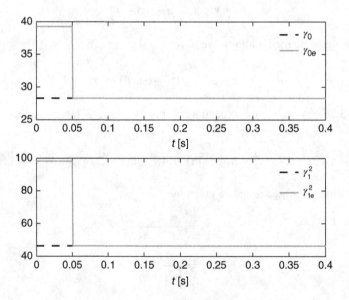

Figure 3.20   Parameter identification

At the beginning of the simulation some preset initial parameters were used in the control, but after a small period of time these values were updated by the algebraic estimator, allowing a successful performance of the system. The control signal is soft and reasonable. Figure 3.20 shows a fast and exact parameter estimation.

3.2.7 Identification of a Magnetic Levitation System

Consider a magnetic levitation system consisting of an iron ball in a vertical magnetic field created by a single electromagnet, as shown in Figure 3.21. The objective of the system is to control the vertical position of the ball by adjusting the current in the electromagnet, through the input voltage V_u.

The ball position is read by an optic transducer which can discriminate the vertical movements of the ball from the horizontal ones.

In medium to high-power applications, the voltage V_u is generated using a rectifier that includes a capacitance. The dynamics of this actuator can be described by a RC circuit given in the work of Astolfi and Ortega (2003). Nonetheless, for the sake of simplicity, it is considered that the rectifier supplies the coil with a current which is proportional to the command voltage of the actuator.

Thus, the simplified model of the sketch can be written as follows:

$$M\ddot{z} = Mg - F_m$$

$$F_m = k_m \frac{i^2}{z^2}$$

$$i = k_a V_u \tag{3.80}$$

where M is the mass of the ball, g is the gravity constant, F_m is the magnetic force generated by the coil, i is the current, z is the vertical position of the ball, k_m is the magnetic constant, k_a is the input conductance, and V_u is the input voltage command.

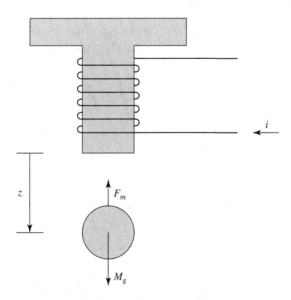

Figure 3.21 Sketch of the magnetic levitation process

Considering the equations (3.80), and dividing by the mass M, we can write the model of the magnetic levitation system as

$$\ddot{z} + \alpha \frac{V_u^2}{z^2} = g \tag{3.81}$$

where $\alpha = \frac{k_m k_a^2}{M}$ is a constant parameter. We denote by α_e the estimate of the parameter α.

3.2.7.1 A Certainty-Equivalence Controller

Consider the following PD controller as the input command $U = V_u^2$:

$$U = \frac{z^2}{\alpha_e^2}[(K_p(z - z^*(t)) + K_d(\dot{z} - \dot{z}^*(t)) - \ddot{z}^*(t)) + g] \tag{3.82}$$

Notice that under perfect cancelation of the actual parameters α and g with the estimated parameter α_e, the closed-loop tracking-error signal $e = z - z^*(t)$ is governed by the linear differential equation

$$e^{(2)} + K_p e + K_d e = 0 \tag{3.83}$$

The design parameters $\{K_p, K_d\}$ are chosen so that the characteristic polynomial $p(s) = s^2 + K_d s + K_p$ becomes a Hurwitz polynomial. This is achieved by choosing $K_p = \omega_n^2$ and $K_d = 2\zeta\omega_n$, with ω_n and ζ both positive.

3.2.7.2 The Online Parameter Identifier

To show the construction of the identifier, consider the equation (3.81). First, multiply the above equation by $(t)^2$, which yields $t^2\ddot{z} + \alpha t^2 \frac{V_u^2}{z^2} = t^2 g$. Then, integrate this expression twice, as follows:

$$\int^{(2)} t^2\ddot{z} + \alpha \int^{(2)} t^2 \frac{V_u^2}{z^2} = \int^{(2)} t^2 g \tag{3.84}$$

After integration by parts we obtain, from the above equation, an estimate α_e of the parameter α:

$$\alpha_e = \frac{\int^{(2)} t^2 g - \left(t^2 z - 4\int tz + 2\int^{(2)} z\right)}{\int^{(2)} t^2 \frac{U}{z^2}} \tag{3.85}$$

3.2.7.3 Simulations

We set an initial arbitrary value for the estimated parameter, $\alpha_e = 0.005$. The simulation parameters are: $M = 0.020$ kg, $k_m = 8.248e - 5$ N(m/A)^2, $k_a = 0.397\ \Omega^{-1}$, $g = 9.81$ m/s^2, $\zeta = 0.9$, and $w_n = 50$. It was desired to follow a rest-to-rest position from 0.005 m to 0.025 m in 3 s. A smooth reference trajectory was implemented by means of a 20th-order Bézier polynomial. The initial condition of the ball was set to $z_0 = 0.007$ m.

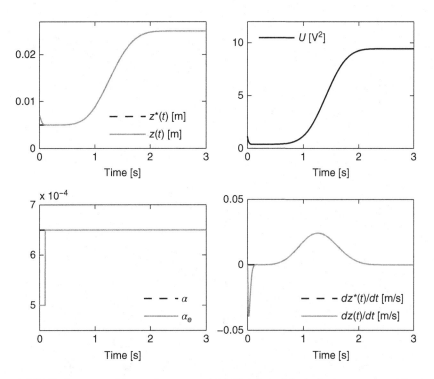

Figure 3.22 Trajectory tracking and parameter identification of the levitation process without noise

Figure 3.22 shows the trajectory-tracking performance of the magnetic levitation system under noise-free conditions. The identification process of the unknown parameter α is carried out in a short amount of time, approximately 0.1 s.

Figure 3.23 shows the trajectory-tracking performance of the magnetic levitation system under noisy conditions. The noise was set to be a sequence of random variables uniformly distributed on the interval $[-0.0003, 0.0003]$. The identification process of the unknown parameter α is again carried out in a short amount of time, approximately 0.3 s. After this instant of time, the estimated value of α is replaced in the feedback controller.

3.3 An Alternative Construction of the System of Linear Equations

After the algebraic technique is applied to a linearly identifiable system, we obtain an expression free from any time derivatives and independent of initial conditions. This expression has the following form:

$$p_1(t)\theta_1 + p_2(t)\theta_2 + \cdots + p_m(t)\theta_m = q(t) \tag{3.86}$$

This can be written in regressor form, as

$$P(t)\Theta = q(t) \tag{3.87}$$

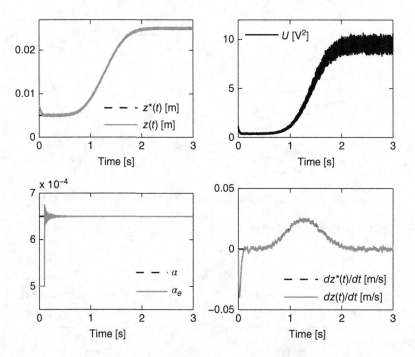

Figure 3.23   Trajectory tracking and parameter identification of the levitation process with noise

where $P(t) = [p_1(t)\, p_2(t) \, \cdots \, p_m(t)]$ is a vector of functions of time, $\Theta = [\theta_1\, \theta_2 \, \cdots \, \theta_m]^T$ is the vector of unknown constant parameters, and $q(t)$ is a scalar function of time.

In order to obtain the vector of unknown parameters Θ, it is necessary to build a system of linearly independent equations. There are several ways to do this from equation (3.87). One form, as shown before, consists of generating $m - 1$ equations by repeated integration of (3.87). In each integration, a new linearly independent equation is obtained. However, often this set of equations is not well conditioned from a numerical viewpoint. Therefore, a new approach to generate a system of linearly independent equations is required.

It is common to formulate the identification of the unknown parameters Θ as an optimization problem of an error function $e(\widehat{\Theta}, t)$ according to some integral error criterion.

Typically, the optimization criterion is proposed as the following cost function:

$$J(\widehat{\Theta}, t) = \frac{1}{2} \int_0^t e^2(\widehat{\Theta}, \sigma) d\sigma \tag{3.88}$$

where

$$e(\widehat{\Theta}, t) = P(t)\widehat{\Theta} - q(t) \tag{3.89}$$

Here, $\widehat{\Theta}$ is the vector of estimated parameters. Thus, the optimization problem is stated as

$$\min_{\widehat{\Theta}} \frac{1}{2} \int_0^t (P(\sigma)\widehat{\Theta} - q(\sigma))^2 d\sigma \tag{3.90}$$

In order to minimize the cost function (3.88), the gradient method is applied. Then, differentiating $J(\widehat{\Theta}, t)$ with respect to the vector of estimated parameters $\widehat{\Theta}$ yields

$$\nabla_{\widehat{\Theta}} J(\widehat{\Theta}, t) = \frac{\partial}{\partial \widehat{\Theta}} \frac{1}{2} \int_0^t e^2(\widehat{\Theta}, \sigma) d\sigma = \frac{1}{2} \int_0^t \frac{\partial e^2}{\partial \widehat{\Theta}} d\sigma \tag{3.91}$$

$$= \int_0^t \frac{\partial e}{\partial \widehat{\Theta}} e(\widehat{\Theta}, \sigma) d\sigma \tag{3.92}$$

Note that $\partial(P(t)\widehat{\Theta} - q(t))/\partial\widehat{\Theta} = P^T(t)$, thus

$$\nabla_{\widehat{\Theta}} J(\widehat{\Theta}, t) = \int_0^t P^T(\sigma)[P(\sigma)\widehat{\Theta} - q(\sigma)] d\sigma \tag{3.93}$$

Given that $J(\widehat{\Theta}, t)$ is a convex function, it has a global minimum that satisfies $\nabla_{\widehat{\Theta}} J(\widehat{\Theta}, t) = 0$ for all t. We then have,

$$\left[\int_0^t P^T(\sigma)P(\sigma) d\sigma \right] \widehat{\Theta} = \int_0^t P^T(\sigma)q(\sigma) d\sigma \tag{3.94}$$

Notice that $P(t)P^T(t) \in R^{m \times m}$ and $P^T(t)q(t) \in R^{m \times 1}$. The expression (3.94) is a set of m linearly independent equations for $t > 0$. Therefore, the vector of estimated parameters $\widehat{\Theta}$ can be obtained from the formula

$$\widehat{\Theta} = \left[\int_0^t P^T(\sigma)P(\sigma) d\sigma \right]^{-1} \int_0^t P^T(\sigma)q(\sigma) d\sigma \tag{3.95}$$

Since $\widehat{\Theta}$ is independent of time, the above formula is valid for any arbitrary small time integration interval $[0, \epsilon]$, with $\epsilon > 0$. The estimation of Θ can be achieved quite fast.

Remark 3.3.1 *The notion of nonlinear invariant filtering was introduced in the work of Belkoura et al. (2008), where some nonlinear operators were applied to reduce the singularities of the identifier. In our approach we define a cost function which permits an optimal filtering beside the singularity avoidance.*

3.3.1 Genesio–Tesi Chaotic System

The dynamics of the Genesio–Tesi chaotic system (Genesio and Tesi, 1992) is given by the following set of equations:

$$\dot{x}_1 = x_2$$
$$\dot{x}_2 = x_3$$
$$\dot{x}_3 = -cx_1 - bx_2 - ax_3 + x_1^2$$
$$y = x_1 \tag{3.96}$$

where a, b, c are unknown constant parameters. In terms of output, the system dynamics is described as follows:

$$y^{(3)} = -a\ddot{y} - b\dot{y} - cy + y^2 \tag{3.97}$$

3.3.1.1 Parameter Estimation

We can express (3.97) in the following form:

$$\begin{bmatrix} -\ddot{y} & -\dot{y} & -y \end{bmatrix} \begin{bmatrix} a & b & c \end{bmatrix}^T = y^{(3)} - y^2 \tag{3.98}$$

which implies a linear equation in three unknown parameters. Multiplying out (3.98) by t^3 and integrating with respect to t three times, we obtain

$$\begin{bmatrix} p_1(t) & p_2(t) & p_3(t) \end{bmatrix} \begin{bmatrix} a & b & c \end{bmatrix}^T = q_1(t) \tag{3.99}$$

$$p_1(t) \triangleq -6 \int^{(3)} (ty) + 6 \int^{(2)} (t^2 y) - \int (t^3 y)$$

$$p_2(t) \triangleq 3 \int (t^2 y) - \int^{(2)} (t^3 y)$$

$$p_3(t) \triangleq - \int^{(3)} (t^3 y)$$

$$q_1(t) \triangleq - \int^{(3)} (6y + t^3 y^2) + 18 \int^{(2)} (ty) - 9 \int (t^2 y) + t^3 x$$

which places the system in a regressor form: $q(t) = p(t)\theta$, where $p(t) = [p_1(t) \ p_2(t) \ p_3(t)]$, $\theta = [a \ b \ c]^T$, and $q(t) = q_1(t)$. Multiplying (3.99) on the left by $p^T(t)$ and integrating yields

$$\begin{bmatrix} \int p_1^2(t) & \int p_1(t)p_2(t) & \int p_1(t)p_3(t) \\ \int p_1(t)p_2(t) & \int p_2^2(t) & \int p_2(t)p_3(t) \\ \int p_1(t)p_3(t) & \int p_2(t)p_3(t) & \int p_3^2(t) \end{bmatrix} \begin{bmatrix} a \\ b \\ c \end{bmatrix} = \begin{bmatrix} \int p_1(t)q_1(t) \\ \int p_3(t)q_1(t) \\ \int p_3(t)q_1(t) \end{bmatrix}$$

The estimation of the parameter vector θ is obtained as follows:

$$\begin{bmatrix} \hat{a} \\ \hat{b} \\ \hat{c} \end{bmatrix} = \begin{bmatrix} \int p_1^2(t) & \int p_1(t)p_2(t) & \int p_1(t)p_3(t) \\ \int p_1(t)p_2(t) & \int p_2^2(t) & \int p_2(t)p_3(t) \\ \int p_1(t)p_3(t) & \int p_2(t)p_3(t) & \int p_3^2(t) \end{bmatrix}^{-1} \begin{bmatrix} \int p_1(t)q_1(t) \\ \int p_3(t)q_1(t) \\ \int p_3(t)q_1(t) \end{bmatrix} \tag{3.100}$$

3.3.1.2 Numerical Results

To test the effectiveness of the proposed method, we simulated the parameter estimation process for the Genesio–Tesi system. In this instance, the fourth-order Runge–Kutta method is used to solve the system with a step size of 0.1 ms. The initial conditions for the system were set as $x_1(0) = -1$, $x_2(0) = -2$, and $x_3(0) = 1$. The system parameters were chosen to be: $a = 1.2$, $b = 2.92$, $c = 6$, which ensures the chaos property of the system response. Figure 3.24 shows the identification process for each parameter and the system output response.

3.3.2 The Ueda Oscillator

In this example we look at the Ueda oscillator (Linz, 2008), considered as a one-polynomial Duffing chaotic-driven dissipative oscillator. This system is described by the following

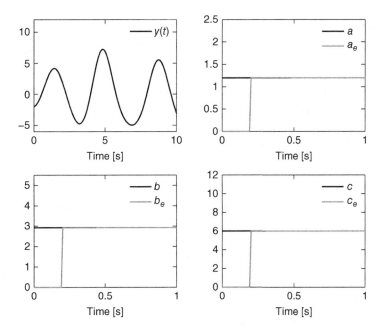

Figure 3.24 Parameter estimation

fourth-order nonlinear differential equation:

$$x^{(4)} = -\mu x^{(3)} - (\omega_c + 3x^2)\ddot{x} - (\omega_c \mu + 6x\dot{x})\dot{x} - \omega_c x^3 \tag{3.101}$$

where $\mu, \omega_c \in \mathbb{R}^+$ are unknown parameters.

The last system can be represented in the following linear regressor form:

$$x^{(4)} + 6x\dot{x}^2 + 3x^2\ddot{x} = \begin{bmatrix} -x^{(3)} & -\ddot{x} - x^3 & -\dot{x} \end{bmatrix} \begin{bmatrix} \mu \\ \omega_c \\ \omega_c \mu \end{bmatrix} \tag{3.102}$$

3.3.2.1 Parameter Estimation

To estimate the system parameters ω_c, μ, and $\omega_c \mu$ it is necessary to multiply (3.102) by t^4 and integrate, with respect to t, four times in order to eliminate the time-derivative terms. We have

$$\left(\int^{(4)} t^4 \left(x^{(4)} + 6x\dot{x}^2 + 3x^2\ddot{x} \right) \right) =$$

$$\left[-\left(\int^{(4)} t^4 x^{(3)} \right) \quad -\left(\int^{(4)} t^4 \ddot{x} - x^3 \right) \quad -\left(\int^{(4)} t^4 \dot{x} \right) \right] \begin{bmatrix} \mu \\ \omega_c \\ \omega_c \mu \end{bmatrix} \tag{3.103}$$

In this case, some of the integrals are directly evaluated by means of an integration by parts procedure. We obtain

$$\int^{(4)} t^4 x^{(4)} = 24 \left(\int^{(4)} x \right) - 96 \left(\int^{(3)} tx \right) + 72 \left(\int^{(2)} t^2 x \right) - 16 \left(\int t^3 x \right) + t^4 x$$

$$-\int^{(4)} t^4 x^{(3)} = 24 \left(\int^{(4)} tx \right) - 36 \left(\int^{(3)} t^2 x \right) + 12 \left(\int^{(2)} t^3 x \right) - \left(\int t^4 x \right)$$

$$-\int^{(4)} t^4 \ddot{x} = -12 \left(\int^{(4)} t^2 x \right) + 8 \left(\int^{(3)} t^3 x \right) - \left(\int^{(2)} t^4 x \right)$$

$$-\int^{(4)} t^4 \dot{x} = 4 \left(\int^{(4)} t^3 x \right) - \left(\int^{(3)} t^4 x \right)$$

However, the integration process for the expression $6x\dot{x}^2 + 3x^2\ddot{x}$ needs additional steps. By simple inspection, $6x\dot{x}^2 + 3x^2\ddot{x}$ can be rewritten as

$$6x\dot{x}^2 + 3x^2\ddot{x} = 3\frac{d}{dt}(x^2 \dot{x}) \tag{3.104}$$

Then, the integration by parts process yields

$$\int^{(4)} \left(3t^4 \frac{d}{dt}(x^2 \dot{x}) \right) = \frac{1}{3} \left(\int^{(2)} t^4 x^3 \right) - \frac{8}{3} \left(\int^{(3)} t^3 x^3 \right) + 4 \left(\int^{(4)} t^2 x^3 \right) \tag{3.105}$$

The regressor, after the integration process, has the form

$$[p_1(t) \; p_2(t) \; p_3(t)] \begin{bmatrix} \mu \\ \omega_c \\ \omega_c \mu \end{bmatrix} = q_1(t) \tag{3.106}$$

where

$$p_1(t) = 24 \left(\int^{(4)} tx \right) - 36 \left(\int^{(3)} t^2 x \right) + 12 \left(\int^{(2)} t^3 x \right) - \left(\int t^4 x \right)$$

$$p_2(t) = 4 \left(\int^{(4)} t^3 x \right) - \left(\int^{(3)} t^4 x \right)$$

$$p_3(t) = -12 \left(\int^{(4)} t^2 x \right) + 8 \left(\int^{(3)} t^3 x \right) - \left(\int^{(2)} t^4 x \right) - \left(\int^{(4)} t^4 x^3 \right)$$

$$q_1(t) = 24 \left(\int^{(4)} x \right) - 96 \left(\int^{(3)} tx \right) + 72 \left(\int^{(2)} t^2 x \right) - 16 \left(\int t^3 x \right) + t^4 x$$

$$+ \frac{1}{3} \left(\int^{(2)} t^4 x^3 \right) - \frac{8}{3} \left(\int^{(3)} t^3 x^3 \right) + 4 \left(\int^{(4)} t^2 x^3 \right)$$

Now, the regressor is free from time-derivative terms. Multiplying (3.106) on the left by $\begin{bmatrix} p_1(t) & p_2(t) & p_3(t) \end{bmatrix}^T$ and integrating yields

$$\begin{bmatrix} \int p_1^2(t) & \int p_1(t)p_2(t) & \int p_1(t)p_3(t) \\ \int p_1(t)p_2(t) & \int p_2^2(t) & \int p_2(t)p_3(t) \\ \int p_1(t)p_3(t) & \int p_2(t)p_3(t) & \int p_3^2(t) \end{bmatrix} \begin{bmatrix} \mu \\ \omega_c \\ \omega_c\mu \end{bmatrix} = \begin{bmatrix} \int p_1(t)q_1(t) \\ \int p_3(t)q_1(t) \\ \int p_3(t)q_1(t) \end{bmatrix}$$

The estimation of the parameter vector is obtained as follows:

$$\begin{bmatrix} \widehat{\mu} \\ \widehat{\omega_c} \\ \widehat{\omega_c\mu} \end{bmatrix} = \begin{bmatrix} \int p_1^2(t) & \int p_1(t)p_2(t) & \int p_1(t)p_3(t) \\ \int p_1(t)p_2(t) & \int p_2^2(t) & \int p_2(t)p_3(t) \\ \int p_1(t)p_3(t) & \int p_2(t)p_3(t) & \int p_3^2(t) \end{bmatrix}^{-1} \begin{bmatrix} \int p_1(t)q_1(t) \\ \int p_3(t)q_1(t) \\ \int p_3(t)q_1(t) \end{bmatrix} \qquad (3.107)$$

3.3.2.2 Numerical Simulations

In this subsection, the identification scheme is tested by means of some numerical simulations for the Ueda oscillator with the following parameters: $\omega_c = 30$, $\mu = 5$. A time step of 0.1 ms was used in a fourth-order Runge–Kutta method. Figure 3.25 shows the identification results for each parameter and the output signal.

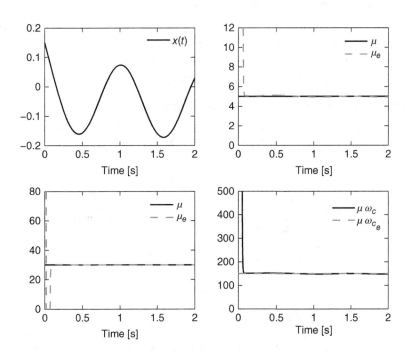

Figure 3.25   Parameter estimation

3.3.3 Identification and Control of an Uncertain Brushless DC Motor

Brushless motors provide better performance than conventional DC motors in highly demanding industrial applications by achieving a longer lifetime and less noisy signals. Their features make them suitable for automotive, aerospace, industrial, and commercial applications due to their high power density, high torque-to-inertia ratio, quiet operation, and lower maintenance cost (Solomon and Famouri, 2009). To take further advantage of the aforementioned characteristics, the controller design must be provided with accurate model parameters invoking novel approaches to online parameter identification. Additionally, an online parameter-identification approach allows us to quickly autotune the controller, reducing significant overshoots, or oscillations, in regulation or trajectory-tracking tasks.

The parameter identification of brushless motor models has been tackled from a wide variety of approaches. In the literature one may find, among many other approaches, methods ranging from experimental, offline, model validations (Schaible and Szabados, 1999) to adaptive methods (Ge and Cheng, 2005), neural networks approaches (El-Sharkawi *et al.*, 1994), and nonlinear identification methods based on wavelets (Rasouli and Karrari, 2004). Most of these methodologies require measurements of state variables, others demand high computational costs.

In this section we propose a generalized proportional integral controller in combination with a parameter-identification scheme for the solution of a reference trajectory-tracking problem in a brushless motor system undergoing unknown but bounded load-disturbance inputs.

3.3.3.1 Preliminaries and Problem Formulation

Here we present the mathematical model of a trapezoidal back electromotive force (emf) synchronous motor. This machine is also known as a brushless DC (BLDC) motor. A comprehensive treatment of its construction along with a detailed description of the stator magnetic field, stator flux linkage, emf in the stator windings, and related torques, is presented in the outstanding book by Chiasson (2005).

Under balanced stator currents, the mathematical model of the brushless motor is given by

$$\frac{di_{S1}}{dt} = \frac{1}{L_S + M}(e_p\omega_R e(\theta_R) - R_S i_{S1} + u_{S1}) \tag{3.108}$$

$$\frac{di_{S2}}{dt} = \frac{1}{L_S + M}(e_p\omega_R e(\theta_R - 2\pi/3) - R_S i_{S2} + u_{S2}) \tag{3.109}$$

$$\frac{di_{S3}}{dt} = \frac{1}{L_S + M}(e_p\omega_R e(\theta_R - 4\pi/3) - R_S i_{S3} + u_{S3}) \tag{3.110}$$

$$\frac{d\omega_R}{dt} = -\frac{e_p}{J}[e(\theta_R)i_{S1} + e(\theta_R - 2\pi/3)i_{S2}$$
$$+ e(\theta_R - 4\pi/3)i_{S3}] - \tau_L/J \tag{3.111}$$

$$\frac{d\theta_R}{dt} = \omega_R \tag{3.112}$$

where e_p is the coefficient of mutual inductance between the stator and the rotor, R_S is the resistance in each stator phase, L and M are the inductance and self-inductance of each winding,

J is the rotor load inertia, u_{S1}, u_{S2}, u_{S3} are the voltages applied to each stator phase, i_{S1}, i_{S2}, i_{S3} represent the current in phases 1, 2, 3, respectively, θ_R is the angular displacement of the shaft of the motor, and ω_R denotes its angular velocity. The total induced torque in the machine is given by

$$\tau = -e_p(e(\theta_R)i_{S1} + e(\theta_R - 2\pi/3)i_{S2} + e(\theta_R - 4\pi/3)i_{S3}) \tag{3.113}$$

The dynamics related to the angular velocity is thus given by

$$\frac{d\omega_R}{dt} = \frac{\tau}{J} - \frac{\tau_L}{J} \tag{3.114}$$

Given the construction features of the stator windings and of the poles of the permanent magnet in the rotor, the magnetic field, produced by stator currents, and the emf in the stator windings, produced by the rotor's magnetic field, are dependent on $e(\theta)$. This is described by the following *trapezoidal* profile:

$$e(\theta_R) = \begin{cases} \frac{6\theta_R}{\pi}, & -\pi/6 \leq \theta_R \leq \pi/6 \\ 1, & \pi/6 \leq \theta_R \leq 5\pi/6 \\ -\frac{6(\theta_R - \pi)}{\pi}, & 5\pi/6 \leq \theta_R \leq 7\pi/6 \\ -1, & 7\pi/6 \leq \theta_R \leq 11\pi/6 \end{cases}$$

In our case, none of the motor parameters R_S, $L_S + M$, e_p, or J is assumed to be known. The variables u_{S1}, u_{S2}, u_{S3} act as the control inputs to the system, and ω_R is considered to be the output. In addition, the stator currents i_{S1}, i_{S2}, i_{S3} and the angular displacement θ_R are regarded as measurable states.

3.3.3.2 Problem Formulation

Given a feasible angular velocity reference trajectory, denoted by $\omega_R^*(t)$, for the completely uncertain BLDC motor model (3.108)–(3.112), devise a feedback controller which processes the angular velocity, the angular displacement, the stator currents (possibly subject to additive, zero-mean noise), the desired reference trajectory itself, and the nominal control inputs, to force $\omega_R(t)$ to asymptotically track the given reference trajectory $\omega_R^*(t)$ in spite of the constant, unknown perturbation load input torque τ_L.

3.3.3.3 Online Parameter Identification

Precise knowledge of all the parameters allows us to implement a high-performance control strategy. It is not difficult to realize that the set of parameters R_S, $L_S + M$, e_p, J is all that is needed to actually control the system and solve the given angular velocity reference trajectory-tracking task. Since equations (3.108)–(3.110) are balanced, the sum of the three current signals is equal to zero. This fact forces us to take just one of this set of equations in the identification scheme.

To obtain fast algebraic estimates for the lumped parameters R_S, $L_S + M$, e_p, J, we proceed as follows. Multiply equation (3.108) by $(L_S + M)t$ and then integrate the resulting expression, over the interval $[0, t]$, which leads to

$$(L_S + M) \left(\int t \frac{di_{S1}}{dt} \right) =$$

$$e_p \left(\int t\omega_R e(\theta_R) \right) - R_S \left(\int t i_{S1} \right) + \left(\int t u_{S1} \right) \qquad (3.115)$$

Equation (3.115) may be rewritten as

$$(L_S + M) \left(t i_{S1} - \int i_{S1} \right) + R_S \left(\int t i_{S1} \right) +$$

$$e_p \left(- \int t\omega_R e(\theta_R) \right) = \left(\int t u_{S1} \right) \qquad (3.116)$$

which is an expression free from the initial condition $i_{S1}(0)$.

The resulting equation is linear in the parameters. Rewrite (3.116) as

$$(L_S + M)p_1(t) + R_S p_2(t) + e_p p_3(t) = q(t) \qquad (3.117)$$

where $p_1(t) = t i_{S1} - \int i_{S1}, p_2(t) = \int t i_{S1}, p_3(t) = - \int t\omega_R e(\theta_R)$, and $q(t) = (\int t u_{S1})$.

Let

$$\gamma = \left[(L_S + M) \ R_S \ e_p \right]^T$$

$$p = \left[p_1(t) \ p_2(t) \ p_3(t) \right]$$

then we have the following linear expression:

$$p(t)\gamma = q(t) \qquad (3.118)$$

Rewrite equation (3.111) as follows:

$$\frac{d\omega_R}{dt} = -\frac{e_p}{J}\tau_u - \frac{1}{J}\tau_L \qquad (3.119)$$

where $\tau_u = [e(\theta_R)i_{S1} + e(\theta_R - 2\pi/3)i_{S2}e(\theta_R - 4\pi/3)i_{S3}]$. This signal can be obtained from current measurements and knowledge of the angle of the rotor.

A single time differentiation in (3.119) annihilates the constant load torque. We have

$$\frac{d^2\omega_R}{dt^2} = -\frac{e_p}{J}\frac{d\tau_u}{dt} \qquad (3.120)$$

Multiplying (3.120) by t^2 and then integrating twice with respect to time leads to

$$\left(\int^{(2)} t^2 \frac{d^2\omega_R}{dt^2} \right) = -\frac{e_p}{J} \left(\int^{(2)} t^2 \frac{d\tau_u}{dt} \right) \qquad (3.121)$$

Using the integration by parts formula we have

$$-2\left(\int^{(2)}\omega_R\right) + 4\left(\int t\omega_R\right) - t^2\omega_R =$$

$$\frac{e_p}{J}\left[2\left(\int^{(2)}t\tau_u\right) + \left(\int t^2\tau_u\right)\right] \qquad (3.122)$$

From (3.122) it is possible to solve directly for e_p/J. However, for a better conditioning of the online identifier, we perform a least-squares invariant filtering scheme. Rewrite the last equation as

$$N(t) = \frac{e_p}{J}D(t) \qquad (3.123)$$

with $N(t) = -2(\int^{(2)}\omega_R) + 4(\int t\omega_R) - t^2\omega_R$ and $D(t) = \left[2\left(\int^{(2)}t\tau_u\right) + \left(\int t^2\tau_u\right)\right]$. Integrate the expression after multiplication throughout by the denominator signal $D(t)$. The proposed algebraic identifier, with least-squares features, is given by

$$\widehat{\left(\frac{e_p}{J}\right)} = \frac{\int_0^t N(\sigma)d(\sigma)d\sigma}{\int_0^t D^2(\sigma)d\sigma} \qquad (3.124)$$

which is indeterminate at $t = 0$. This procedure yields a remarkably better conditioning of the identifier than with the usual direct procedure. The above invariant filtering acts as an effective, desirable handling of the measurement noise. Additional integrations can be performed to further decrease the classical SNR.

3.3.3.4 Controller Design

The proposed control scheme consists of a two-stage feedback controller design procedure (Figure 3.26). The first stage regulates the angular position of the motor shaft to track the reference signal, $\omega_R^*(t)$, by considering the current I_p as an auxiliary control input. For this stage, the control strategy is implemented by means of a GPI controller. As a result of the first, outer-loop, design stage a set of desirable current trajectories is obtained. The obtained currents are then taken as output references for the second multi-variable design stage, or inner-loop stage. The inner-loop design stage designs the feedback GPI controller to force the actual currents to track the current references obtained in the first stage. In the second, inner-loop, stage the voltages u_{S1}, u_{S2}, u_{S3} act as the control input variables.

3.3.3.5 Outer-Loop Controller Design Stage

The key idea for a simple approach to establish the control strategy consists of selecting the stator reference currents in order to obtain a smooth torque at the motor shaft. In particular, for a constant reference angular velocity, we want a constant torque.

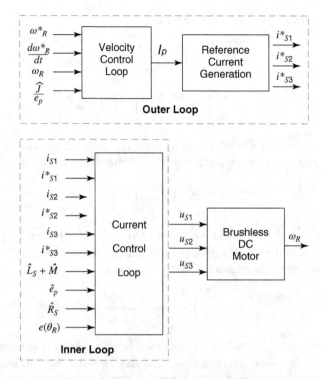

Figure 3.26   Inner and outer control loops

The current references i_{S1}^*, i_{S2}^*, i_{S3}^* are chosen as (according to Chiasson, 2005)

$$i_{S1}^*(\theta_R) = I_p i_S(\theta_R)$$
$$i_{S2}^*(\theta_R) = I_p i_S(\theta_R - 2\pi/3) \tag{3.125}$$
$$i_{S3}^*(\theta_R) = I_p i_S(\theta_R - 4\pi/3)$$

where I_p is an auxiliary control input (to be defined later in this subsection) and i_S is given by

$$i_S(\theta_R) = \begin{cases} 0 & \text{for} & -\pi/6 \leq \theta_R \leq \pi\,(6) \\ -1 & \text{for} & \pi/6 \leq \theta_R \leq 5\pi/6 \\ 0 & \text{for} & 5\pi/6 \leq \theta_R \leq 7\pi/6 \\ 1 & \text{for} & 7\pi/6 \leq \theta_R \leq 11\pi/6 \end{cases} \tag{3.126}$$

To determine the torque produced using these current references, consider the electrical power absorbed by the back emfs which is given, after simple manipulations, by

$$e_{S1}i_{S1} + e_{S2}i_{S2} + e_{S3}i_{S3} = -2e_p \omega_R I_p \tag{3.127}$$

That is, under a proper tracking of the stator currents (inner current control loops working properly), the torque is simply given by

$$\tau = 2e_p I_p \tag{3.128}$$

and the velocity dynamics reduces to

$$\frac{d\omega_R}{dt} = \frac{\tau}{J} - \frac{\tau_L}{J} = \frac{2e_p}{J}I_p - \frac{\tau_L}{J} \tag{3.129}$$

which is in the same form as the current commanded DC motor with constant torque.

Thus, as performed in DC brushed servo systems, I_p is designed as the analog of a "DC" current to achieve the desired servo system performance. The last fact allows us to propose the corresponding outer-loop control law for I_p while using the estimated parameters:

$$I_p = \frac{1}{2}\left(\frac{\widehat{J}}{e_p}\right)\left[\frac{d\omega_R^*}{dt} - k_1 e_\omega - k_0 \int e_\omega(\tau)d\tau\right] \tag{3.130}$$

$$e_\omega = \omega - \omega_R^*(t), k_0, k_1 \in \mathbb{R}$$

or, with an abuse of notation, in operational calculus terms:

$$I_p = \frac{1}{2}\left(\frac{\widehat{J}}{e_p}\right)\left[\frac{d\omega_R^*}{dt} - \frac{k_1 s + k_0}{s}\left(\omega - \omega_R^*(t)\right)\right] \tag{3.131}$$

The proposed controller is robust with respect to constant-perturbation torque inputs and it depends on an estimate of the lumped parameter e_p/J.

The closed-loop tracking-error characteristic polynomial is given by

$$P_\omega(s) = s^2 + k_1 s + k_0 \tag{3.132}$$

where the controller parameters are chosen so that the closed-loop tracking-error dynamics characteristic polynomial is a Hurwitz polynomial.

3.3.3.6 Inner-Loop Controller Design Stage

Consider the equation governing the dynamics of the current, i_{S1}, given by (3.108). The proposed controller relies on accurate and fast estimations of the lumped parameters R_S, $L_S + M$, and e_p. We propose using the estimated parameters in conjunction with $e(\theta_R)$, to cancel out the nonlinearities as well as the parameter-dependent terms, thus reducing the control problem to the rather simple one dealing with a single pure integrator system, where a linear control strategy is quite effective. We have

$$u_{S1} = (\hat{L}_S + \hat{M})\left[\frac{di_S^*}{dt} - \frac{\hat{e}_p}{\hat{L}_S + \hat{M}}\omega_R e(\theta_R)\right.$$

$$\left.+ \frac{\hat{R}_S}{\hat{L}_S + \hat{M}}i_{S1} - \frac{l_1 s + l_0}{s}\left(i_{S1} - i_{S1}^*(t)\right)\right] \tag{3.133}$$

The time differentiation of i_S^* is given by

$$\frac{di_S^*}{dt} = i_S(\theta_R)\frac{dI_p}{dt} + \frac{di_S(\theta_R)}{dt}I_p \tag{3.134}$$

Owing to the nature of the motor, the term $\frac{di_S(\theta_R)}{dt}I_p$ is purely a train of impulses which may generate some singularities in the time derivative of the reference. To overcome this fact, it is more convenient to eliminate this term from the feedforward input. This action improves the control action, achieving better tracking results without eliminating essential information in the control input. Thus, we propose the alternative feedforward input

$$\left(\frac{di_{S1}}{dt}\right)^* = i_S(\theta_R)\frac{dI_p}{dt} \tag{3.135}$$

Finally, the control input for the first stator current is expressed as

$$
\begin{aligned}
u_{S1} =& (\hat{L}_S + \hat{M})\left[\left(\frac{di_{S1}}{dt}\right)^* - \frac{\hat{e}_p}{\hat{L}_S + \hat{M}}\omega_R e(\theta_R)\right. \\
& \left. + \frac{\hat{R}_S}{\hat{L}_S + \hat{M}}i_{S1} - \frac{l_1 s + l_0}{s}\left(i_{S1} - i_{S1}^*(t)\right)\right]
\end{aligned} \tag{3.136}
$$

For this case, the closed-loop tracking-error characteristic polynomial is given by

$$P_{i1}(s) = s^2 + l_1 s + l_0 \tag{3.137}$$

If $l_1, l_0 > 0$, the Hurwitz condition is satisfied and the tracking-error dynamics is exponentially stable. The same approach is proposed for current control in phases 2 and 3.

3.3.3.7   Numerical Example

To show the effectiveness of the proposed approach, we performed some numerical simulations considering numerical parameters of a commercially available BLDC motor. In this case, the peak voltage of the motor was 200 V and the peak current was 10 A. The stator resistance was set to be $R_S = 0.7\ \Omega$; the inductance and self-inductance were, respectively, $L = 2.72$ mH and $M = 0.15$ mH. The coefficient of mutual inductance was $e_p = 0.5128$ N m/A. We considered a rotor load inertia value of $J = 0.0002$ kg m^2. Finally, the load torque was assumed to be $\tau_L = 0.01$ N m. The measurable states are affected by additive zero-mean noise signals with SNR of 20 dB. The proposed scheme was carried out in two steps: identification and control. In the first step, an open-loop input sufficiently rich in harmonic components was applied to ensure the conditioning of the identifier algebraic system of equations. The generation of the input signals u_{S1}, u_{S2}, u_{S3} is based on the principle of the trapezoidal trend exhibited by this class of motors in a constant-angular-velocity regime. We decided to apply trapezoidal profiles according to the nominal motor parameters; for this instance, an artificial constant velocity reference $\omega_{ref} = 600$ rad/s was taken. The voltage inputs for the identification process (denoted dy $\bar{u}$) were given by

$$
\begin{aligned}
\bar{u}_{S1} &= 70e(\omega_{ref}t) \\
\bar{u}_{S2} &= 70e(\omega_{ref}t - 2\pi/3) \\
\bar{u}_{S3} &= 70e(\omega_{ref}t - 4\pi/3)
\end{aligned} \tag{3.138}
$$

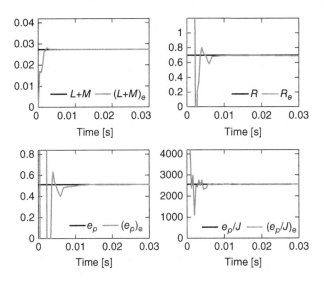

Figure 3.27   Parameter identification

We applied these open-loop control signals during a time interval of 0.05 ms, obtaining fast and good parameter estimates, as depicted in Figure 3.27. In the second part of the scheme, the estimated parameters are used in the devised control inputs given in (3.131), (3.136). The gain parameters of the outer-loop controls were chosen so that the closed-loop tracking-error characteristic polynomials (3.132) exhibited the form $s^2 + 2\zeta_\omega \omega_{n\omega} + \omega_{n\omega}^2$, with $\zeta_\omega = 1$, $\omega_{n\omega} = 100$. Analogously, the inner-loop control gains were set to match the closed-loop error-characteristic polynomial (3.137) with a second-order stable error dynamics dominated by $P_{id}(s) = s^2 + 2\zeta_i \omega_{ni} s + \omega_{ni}^2$, with $\zeta_i = 4$, $\omega_{ni} = 300$. To illustrate the capacity of the control scheme under variations of the trajectory to be tracked, the reference velocity trajectory was defined by the following function:

$$
\omega_R^* = \begin{cases} 300 & 0 \le t < 0.25 \\ 100 & 0.25 \le t < 0.5 \\ 100 + 260\sin(20(t - 0.5)) & 0.5 \le t \end{cases}
$$

As shown in Figures 3.28, 3.29, after the transient response, we can appreciate an excellent velocity tracking where the load torque perturbation input effect is negligible. The obtained voltage inputs are depicted in Figure 3.30.

3.3.4   Parameter Identification and Self-tuned Control for the Inertia Wheel Pendulum

Since its introduction, the inertia wheel pendulum has been an active object of research because it is considered one of the most interesting academic systems (being at the same time a non-minimum-phase under-actuated system with state restrictions due to singularities). Several methodologies have been proposed to solve the problem of swinging up and balancing the

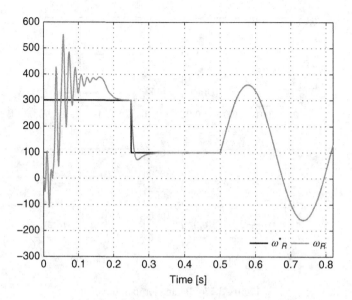

Figure 3.28 Velocity control results

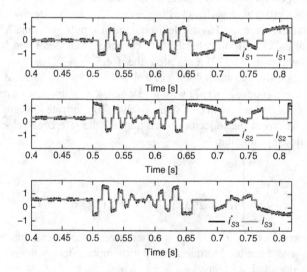

Figure 3.29 Current-tracking control

pendulum, from the original switched strategy (Spong *et al.*, 2001), to the global coordinate change of coordinates approach (Olfati-Saber, 2001), passivity-based control (Ortega *et al.*, 2002), generalized proportional integral control (Hernández and Sira-Ramírez, 2003), energy-based control (Abrahantes *et al.*, 2007; Åstrom and Furuta, 2000), and saturation functions (Ye *et al.*, 2008), among others. An improvement of the control scheme given in Spong *et al.*, (2001) is given in Freidovich *et al.* (2009). Other alternative applications of the

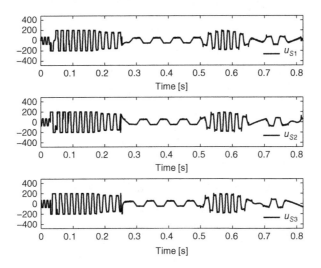

Figure 3.30 Voltage control inputs

pendulum, instead of stabilizing on a vertical position, are the robust control over a trajectory given by a limit cycle as reported in Andary *et al.* (2009).

However, the problem of simultaneous parameter identification and control has been explored less. A complete guide to physical parameter identification based on physical measurements is provided in Block *et al.* (2007); in contrast, in Abrahantes *et al.* (2007), there is an offline parameter identification procedure based on the discrete-time least-squares algorithm. Both cases obtain good results in an offline sense. The mathematical model of the system allows linear parametrization with respect to a set of lumped parameters, achieving the weak algebraic identifiability condition; this condition permits us to use the algebraic parameter-identification approach, which will be developed in this section.

3.3.4.1 Model Description

Figure 3.31 depicts the inertia wheel pendulum. The mathematical model of the system is given by

$$(I_1 + m_1 l_{c1}^2 + m_2 l_1^2)\ddot{\theta}_1 + (m_1 l_{c1} + m_2 l_1)g \sin \theta_1 = -\tau \tag{3.139}$$

$$I_2(\ddot{\theta}_1 + \ddot{\theta}_2) = \tau \tag{3.140}$$

where θ_1, θ_2 denote the pendulum and wheel angular position, τ is the wheel torque input, m_1, I_1 represent the pendulum mass and inertia, and m_2, I_2 are the wheel mass and inertia, respectively. l_1 and l_{c1} are the pendulum length and the position of the center of mass, while the gravity constant is given by $g = 9.81$ m/s^2.

Since the torque input is supplied by a DC motor, it is proportional to the current input, that is,

$$\tau = ki \tag{3.141}$$

with k being the back emf constant for the current motor denoted by i.

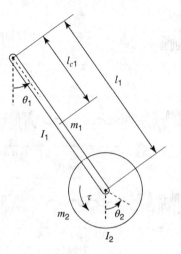

Figure 3.31   The inertia wheel pendulum

The DC motor dynamics is given by

$$L\frac{di}{dt} + Ri = v - k\dot\theta_2 \tag{3.142}$$

where the motor inductance is L, R is the armature resistance, and v the voltage input.

Since the contribution of the term $L\frac{di}{dt}$ is minimal, we will assume that $L\frac{di}{dt} \approx 0$. Thus, equation (3.142) turns into

$$Ri = v - k\dot\theta_2 \tag{3.143}$$

Using (3.143), (3.141), the torque input is given by

$$\tau = \frac{k(v - k\dot\theta_2)}{R} \tag{3.144}$$

By using (3.139), (3.140) and (3.144), the model of the inertia wheel pendulum in terms of the pendulum angle, the rotor angle, the motor voltage, and their time derivatives is given by

$$-\frac{R(I_1 + m_1 l_{c1}^2 + m_2 l_1^2)}{k}\ddot\theta_1 - \frac{R(m_1 l_{c1} + m_2 l_1)g}{k}\sin\theta_1 + k\dot\theta_2 = v \tag{3.145}$$

$$\frac{RI_2}{k}(\ddot\theta_1 + \ddot\theta_2) + k\dot\theta_2 = v \tag{3.146}$$

which, according to the previous parameter definitions, may be rewritten as

$$-\alpha_1\ddot\theta_1 - \alpha_2\sin\theta_1 + \alpha_3\dot\theta_2 = v \tag{3.147}$$

$$\beta_1(\ddot\theta_1 + \ddot\theta_2) + \beta_2\dot\theta_2 = v \tag{3.148}$$

with $\alpha_1 = \frac{R(I_1 + m_1 l_{c1}^2 + m_2 l_1^2)}{k}$, $\alpha_2 = \frac{R(m_1 l_{c1} + m_2 l_1)g}{k}$, $\alpha_3 = \beta_2 = k$, and $\beta_1 = \frac{RI_2}{k}$.

3.3.4.2   Online Parameter Identification

Since equations (3.147), (3.148) can be expressed in regressor form with respect to the lumped parameters, the system is weakly linearly identifiable, that is, it is possible to obtain the lumped parameters $\alpha_1, \alpha_2, \alpha_3, \beta_1, \beta_2$. However, separated parameters like R, m_1, m_2 can not be obtained. The last condition allows the application of the algebraic method. Here, we generate two linear systems of equations, one for the unknown parameters $\alpha_1, \alpha_2, \alpha_3$ and another for β_1 and β_2 respectively. The systems are independent of the plant initial conditions, and should also rely only on knowledge of the signals θ_1, θ_2, and v (i.e., the input and outputs). We perform, to get rid of initial conditions and constant perturbations, the following steps:

Consider equation (3.147). Since the highest-order time derivative of the expression is two, let us multiply by t^2 and then integrate twice with respect to t to get rid of any effects of the initial conditions. We have

$$-\alpha_1 \int_0^t \int_0^\rho \sigma^2 \frac{d^2\theta_1(\sigma)}{d\sigma^2} - \alpha_2 \int_0^t \int_0^\rho \sigma^2 \sin\theta_1(\sigma)d\sigma d\rho$$

$$+\alpha_3 \int_0^t \int_0^\rho \sigma^2 \frac{d\theta_2(\sigma)}{d\sigma}d\sigma d\rho = \int_0^t \int_0^\rho \sigma^2 v(\sigma)d\sigma d\rho \qquad (3.149)$$

After integrating by parts, we have

$$\alpha_1 \left(-t^2\theta_1 + 4\int_0^t \rho\theta_1(\rho)d\rho - 2\int_0^t \int_0^\rho \theta_1(\sigma)d\sigma d\rho\right) \alpha_2 \left(-\int_0^t \int_0^\rho \sigma^2 \sin\theta_1(\sigma)d\sigma d\rho\right)$$

$$+\alpha_3 \left(\int_0^t \rho^2\theta_2(\rho)d\rho - 2\int_0^t \int_0^\rho \sigma\theta_2(\sigma)d\sigma d\rho\right) = \int_0^t \int_0^\rho \sigma^2 v(\sigma)d\sigma d\rho$$

$$(3.150)$$

Analogously, multiplying equation (3.148) by t^2 and then integrating twice with respect to t, we have

$$\beta_1 \left(t^2(\theta_1 + \theta_2) - 4\int_0^t \rho(\theta_1(\rho) + \theta_2(\rho))d\rho + 2\int_0^t \int_0^\rho (\theta_1(\sigma) + \theta_2(\sigma))d\sigma d\rho\right)$$

$$+\beta_2 \left(\int_0^t \rho^2\theta_2(\rho)d\rho - 2\int_0^t \int_0^\rho \sigma\theta_2(\sigma)d\sigma d\rho\right) = \int_0^t \int_0^\rho \sigma^2 v(\sigma)d\sigma d\rho \qquad (3.151)$$

To obtain the set parameters $\alpha_1, \alpha_2, \alpha_3$, two equations in addition to (3.150) are necessary to construct an algebraic linear equation system (one equation in addition to (3.151) is necessary to obtain a linear system for estimating β_1, β_2 respectively). We will apply the least-squares invariant filtering methodology proposed in Section 3.3.

Equation (3.150) has the form

$$p_{\alpha 1}(t)\alpha_1 + p_{\alpha 2}(t)\alpha_2 + p_{\alpha 3}(t)\alpha_3 = q_\alpha(t) \qquad (3.152)$$

where

$$p_{\alpha 1}(t) = -t^2\theta_1 + 4\int_0^t \rho\theta_1(\rho)d\rho - 2\int_0^t \int_0^\rho \theta_1(\sigma)d\sigma d\rho$$

$$p_{\alpha 2}(t) = -\int_0^t \int_0^\rho \sigma^2 \sin\theta_1(\sigma)d\sigma d\rho$$

$$p_{\alpha 3}(t) = \int_0^t \rho^2 \theta_2(\rho) d\rho - 2 \int_0^t \int_0^\rho \sigma \theta_2(\sigma) d\sigma d\rho$$

$$q_\alpha(t) = \int_0^t \int_0^\rho \sigma^2 v(\sigma) d\sigma d\rho$$

and equation (3.151) can be regarded as

$$p_{\beta 1}(t)\beta_1 + p_{\beta 2}(t)\beta_2 = q_\beta(t) \tag{3.153}$$

with

$$p_{\beta 1}(t) = t^2(\theta_1 + \theta_2) - 4 \int_0^t \rho(\theta_1(\rho) + \theta_2(\rho)) d\rho + 2 \int_0^t \int_0^\rho (\theta_1(\sigma) + \theta_2(\sigma)) d\sigma d\rho$$

$$p_{\beta 2}(t) = \int_0^t \rho^2 \theta_2(\rho) d\rho - 2 \int_0^t \int_0^\rho \sigma \theta_2(\sigma) d\sigma d\rho$$

$$q_\beta(t) = \int_0^t \int_0^\rho \sigma^2 v(\sigma) d\sigma d\rho$$

The parameter estimation of α_1, α_2, α_3 is given as follows:

$$\begin{bmatrix} \hat{\alpha}_1 \\ \hat{\alpha}_2 \\ \hat{\alpha}_3 \end{bmatrix} = \left[\int_0^t \begin{bmatrix} p_{\alpha 1}(\sigma) \\ p_{\alpha 2}(\sigma) \\ p_{\alpha 3}(\sigma) \end{bmatrix} \begin{bmatrix} p_{\alpha 1}(\sigma) & p_{\alpha 2}(\sigma) & p_{\alpha 3}(\sigma) \end{bmatrix} d\sigma \right]^{-1} \left[\int_0^t \begin{bmatrix} p_{\alpha 1}(\sigma) \\ p_{\alpha 2}(\sigma) \\ p_{\alpha 3}(\sigma) \end{bmatrix} q_\alpha(\sigma) d\sigma \right] \tag{3.154}$$

and the expression for the algebraic estimators of β_1, β_2 is

$$\begin{bmatrix} \hat{\beta}_1 \\ \hat{\beta}_2 \end{bmatrix} = \left[\int_0^t \begin{bmatrix} p_{\beta 1}(\sigma) \\ p_{\beta 2}(\sigma) \end{bmatrix} \begin{bmatrix} p_{\beta 1}(\sigma) & p_{\beta 2}(\sigma) \end{bmatrix} d\sigma \right]^{-1} \left[\int_0^t \begin{bmatrix} p_{\beta 1}(\sigma) \\ p_{\beta 2}(\sigma) \end{bmatrix} q_\beta(\sigma) d\sigma \right] \tag{3.155}$$

respectively.

3.3.4.3 The Sentinel Parameter as a Criterion to Assess Identifier Convergence

Here, we advocate proposing a methodology to define whether the identifier has reached the actual estimated value by means of an extra parameter called the "sentinel," introduced in Section 2.3, which consists of modifying the system in such a way that we include an extra artificial parameter that is known (sentinel); then, the algebraic method is applied and the convergence of the parameter estimates will be given when the estimated value of the sentinel has converged. The procedure is as follows.

To illustrate the identification procedure, we implement the estimated parameters in a self-tuned control strategy to swing up and stabilize the pendulum, as shown in the following section. Consider equation (3.147); taking a time derivative, we have

$$-\alpha_1 \theta_1^{(3)} - \alpha_2 \frac{d}{dt} \sin \theta_1 + \alpha_3 \ddot{\theta}_2 = \dot{v} \tag{3.156}$$

Multiplying (3.147) by a known constant $\alpha_4 = c_1$ (where α_4 is denoted as a "sentinel parameter"), then

$$\alpha_4[-\alpha_1 \ddot{\theta}_1 - \alpha_2 \sin \theta_1 + \alpha_3 \ddot{\theta}_2] = \alpha_4 v$$

Since the value of α is known, we can substitute its numerical value on the left-hand side of the equation to obtain

$$-\alpha_1 c_1 \ddot{\theta}_1 - \alpha_2 c \sin \theta_1 + \alpha_3 c \dot{\theta}_2 = \alpha_4 v$$

Adding the last equation to (3.156), we have

$$-\alpha_1 [\theta_1^{(3)} + c_1 \ddot{\theta}_1] - \alpha_2 \left[\frac{d}{dt} \sin \theta_1 + c_1 \sin \theta_1 \right] + \alpha_3 [\ddot{\theta}_2 + c_1 \dot{\theta}_2] - \alpha_4 v = \dot{v}$$

Applying the algebraic method, we obtain the following regressor equation:

$$\alpha_1 p_{\alpha_1}(t) + \alpha_2 p_{\alpha_2}(t) + \alpha_3 p_{\alpha_3}(t) + \alpha_4 p_{\alpha_4}(t) = q_\alpha(t) \tag{3.157}$$

with

$$p_{\alpha 1}(t) = t^3 \theta_1 - 9 \int t^2 \theta_1 + 18 \int^{(2)} t \theta_1 - 6 \int^{(3)} \theta_1 + c_1 \int t^3 \theta_1 - 6c_1 \int^{(2)} t^2 \theta_1$$

$$+ 6c_1 \int^{(3)} t \theta_1$$

$$p_{\alpha 2}(t) = - \int^{(2)} t^3 \sin \theta_1 + \int^{(3)} t^2 \sin \theta_1 - c_1 \int^{(3)} t^3 \sin \theta_1$$

$$p_{\alpha 3}(t) = - \int t^3 \theta_2 + 6 \int^{(2)} t^2 \theta_2 - 6 \int^{(3)} t \theta_2 - c_1 \int^{(2)} t^3 \theta_2 + 3c_1 \int^{(3)} t^2 \theta_2$$

$$p_{\alpha 4}(t) = \int^{(3)} t^3 v$$

$$q_\alpha(t) = - \int^{(2)} t^3 v + 3 \int^{(3)} t^2 v$$

Performing the same procedure for equation (3.148), using as sentinel $\beta_3 = c_2$, we have

$$\beta_1 p_{\beta_1}(t) + \beta_2 p_{\beta_2}(t) + \beta_3 p_{\beta_3}(t) = q_\beta(t) \tag{3.158}$$

with

$$p_{\beta 1}(t) = -t^3(\theta_1 + \theta_2) + 9 \int t^2(\theta_1 + \theta_2) - 18 \int^{(2)} t(\theta_1 + \theta_2) + 6 \int^{(3)} (\theta_1 + \theta_2)$$

$$+ c_2 \int t^3(\theta_1 + \theta_2) - 6c_2 \int^{(2)} t^2(\theta_1 + \theta_2) + 6c_2 \int^{(3)} t(\theta_1 + \theta_2)$$

$$p_{\beta 2}(t) = - \int t^3 \theta_2 + 6 \int^{(2)} t^2 \theta_2 - 6 \int^{(3)} t \theta_2 - c_2 \int^{(2)} t^3 \theta_2 + 3c_2 \int^{(3)} t^2 \theta_2$$

$$p_{\beta 3}(t) = \int^{(3)} t^3 v$$

$$q_\beta(t) = - \int^{(2)} t^3 v + 3 \int^{(3)} t^2 v$$

From (3.157), (3.158), it is possible to build other linear systems of equations to obtain the estimated parameters, including the sentinel parameter. In this case, the estimation of the sentinel gives information on the convergence of the rest of the parameters, which allows us to make a disconnection of the identifier. The automatic disconnection of the parameter-estimation subsystem can be implemented based on the monitoring of the introduced sentinel parameter with a disconnection criterion. The proposed criterion (denoted as *IAE*) relates to the fact that the sentinel has reached its actual value, but, to ensure that the sentinel has both reached the value and remains it, the criterion is based on the integral error for a sliding window. That is:

$$IAE_\alpha = \int_{t-\kappa}^{t} |\alpha_4(\sigma) - c_1| d\sigma \tag{3.159}$$

for the first sentinel and

$$IAE_\beta = \int_{t-\kappa}^{t} |\beta_3(\sigma) - c_2| d\sigma \tag{3.160}$$

for the second sentinel, respectively. In order to ensure the quality of estimation, a policy may be applied over the criterion. A single disconnection, governed by a small constant tolerance, is performed assuming that the parameters obtained are sufficiently accurate for the correct operation of the designed control system. The process terminates when the sentinel stays within a minimal specified tolerance.

3.3.4.4   A Certainty-Equivalence Controller in Classical Form

Let us assume momentarily, taking the paradigm of the classical adaptive feedback control (Åstrom and Wittenmark, 1995), that all the system parameters are perfectly known and no perturbation inputs are present. Under these circumstances, consider the switching control strategy proposed by Spong *et al.* (2001), consisting of two steps: a swinging-up control and a balance control, both switched by an angle condition. Since the mathematical model is given in terms of the voltage input instead of the torque, we have to adapt the control input by means of an auxiliary control. A brief description of the control strategy is given as follows.

Consider the equations (3.147), (3.148). Let us propose the following auxiliary control input u:

$$u = v - \alpha_3 \dot{\theta}_r \tag{3.161}$$

Substituting in (3.147), (3.148), we have

$$-\alpha_1 \ddot{\theta}_1 - \alpha_2 \sin \theta_1 = u \tag{3.162}$$

$$\beta_1(\ddot{\theta}_1 + \ddot{\theta}_2) = u \tag{3.163}$$

3.3.4.5   Swinging-up Control

This controller injects some energy into the system, which will be used to swing the pendulum until it reaches an angle close to the vertical position ($\theta_1 \approx \pi$). For the swinging-up control, the system is taken as a parallel interconnection of the pendulum subsystem (3.162) and the wheel subsystem (3.163), respectively. Both systems are passive. For the wheel subsystem, the output is taken as $y_1 = \dot{\theta}_1 + \dot{\theta}_2$ with storage function $S_1 = \frac{\beta_1}{2}(\dot{\theta}_1 + \dot{\theta}_2)^2$. We have

$$\dot{S}_1 = (\dot{\theta}_1 + \dot{\theta}_2)\beta_1(\ddot{\theta}_1 + \ddot{\theta}_2) = y_1 u$$

For the pendulum, consider the following energy function: $E = \frac{1}{2}\alpha_1\dot{\theta}_1^2 + \alpha_2(1 - \cos\theta_1)$. Its corresponding time derivative is given by $\dot{E} = \alpha_1\ddot{\theta}_1\dot{\theta}_1 + \alpha_2(\sin\theta_1)\dot{\theta}_1 = -u\dot{\theta}_1$. Taking E_{ref} as a reference energy constant, the storage function is proposed as $S_2 = \frac{1}{2}(E - E_{ref})^2$. Then

$$\dot{S}_2 = -\dot{\theta}_1(E - E_{ref})u = y_2 u$$

to obtain the output function $y_2 = -\dot{\theta}_1(E - E_{ref})$.

The parallel system is passive, with the output function $y = k_v y_1 + k_e y_2$ and the storage function $S = k_v S_1 + k_e S_2$, where k_v, k_e are adjustable gains to modify the transient response of the system. The control input u is chosen as

$$u = -k_u y = -k_u \left(k_v(\dot{\theta}_1 + \dot{\theta}_2) - k_e \left(E - E_{ref} \right) \dot{\theta}_1 \right) \qquad (3.164)$$

which leads to an asymptotic stable system using the time derivative of the storage function and LaSalle's theorem (Block *et al.*, 2007).

3.3.4.6 Balancing Control

This control acts when the pendulum is almost in a vertical position, using linear techniques. Consider the linearization of (3.162), (3.163) around the equilibrium point $\theta_1 = \pi$:

$$\ddot{\theta}_1 - \frac{\alpha_2}{\alpha_1}\theta_1 = -\frac{1}{\alpha_1}u \qquad (3.165)$$

$$\ddot{\theta}_1 + \ddot{\theta}_2 = \frac{1}{\beta_1}u \qquad (3.166)$$

This linear system is controllable. To avoid problems due to high velocities in the inertia wheel, the following control law is given to stabilize the pendulum but keep the wheel velocity regulated in a reference value $\dot{\theta}_2^*$:

$$u = -k_{pp}\theta_1 - k_{dp}\dot{\theta}_1 + k_{dr}(\dot{\theta}_2^* - \dot{\theta}_2) \qquad (3.167)$$

With some algebraic manipulations, the closed-loop response for θ_1 is governed by the following characteristic equation:

$$P_e(s) = s^3 + \left(\frac{-k_{dp} + k_{dr}}{\alpha_1} + \frac{k_{dr}}{\beta_1} \right)s^2 + \left(-\frac{\alpha_2 + k_{pp}}{\alpha_1} \right)s - \frac{\alpha_2 k_{dr}}{\alpha_1 \beta_1} = 0 \qquad (3.168)$$

The pole placement has to be such that the characteristic equation ensures asymptotic stability of the underlying linear system. This can be achieved by forcing the desired tracking-error dynamics to coincide with a prescribed Hurwitz polynomial $P_d(s)$ of the form

$$P_d(s) = (s^2 + 2\zeta\omega_n s + \omega_n^2)(s + p)$$
$$= s^3 + (2\zeta\omega_n + p)s^2 + (\omega^2 + 2\zeta\omega_n p)s + \omega_n^2 p \qquad (3.169)$$

In contrast, the closed-loop dynamics of θ_2 is given by

$$\ddot{\theta}_2 + b_r k_{dr} \dot{\theta}_2 = -\ddot{\theta}_1 - b_r k_{dp} \dot{\theta}_1 - b_r k_{pp} \theta_1 + b_r k_{dr} \dot{\theta}_2^* \qquad (3.170)$$

Considering that θ_1 tends to zero as time elapses, the dynamics of θ_2 in a stable regime is given by

$$\ddot{\theta}_2 + b_r k_{dr} \dot{\theta}_2 = b_r k_{dr} \dot{\theta}_2^* \qquad (3.171)$$

Thus, while $b_r k_{dr} > 0$, $\dot{\theta}_2$ will tend asymptotically to $\dot{\theta}_2^*$ as time elapses.

3.3.4.7   Experimental Results

Some experiments were carried out to assess the performance of the identification and self-tuned control scheme. The experimental platform was on the basis of the Mechatronics Control Kit model M-1 from Mechatronic Systems Inc. This device is provided with two incremental encoders of 1000 counts/rev for measurement and a 24V DC motor as the actuator. The original platform uses a DSP board with parallel port interface; we substituted the interface to devise the controller in a MATLAB®-xPC Target environment using, as interface, a National Instruments PCI-6602 data acquisition card, with a sampling period $T = 0.1$ ms. The control strategy was performed considering "wrong" parameters (85% of their nominal value) during the time interval $[0, \epsilon]$, where we set $\epsilon = 1$ s. The angular velocity estimations were given by the average of the last three discrete-time differentiation terms (according to the Euler backward differentiation rule). The angle condition for controller switching was given as follows:

$$\begin{cases} \text{Swinging-up control} & |(\theta_1 \mod 2\pi) - \pi| < 0.3 \\ \text{Balancing control} & |(\theta_1 \mod 2\pi) - \pi| \geq 0.3 \end{cases}$$

The certainty-equivalence switched control parameters are given as follows. For the swinging-up control, the energy reference was $E_{ref} = 1.05$ J, the control constants were $k_e = 4000$, $k_v = 4$, $k_u = 0.4$. For the balancing control, the velocity reference for the wheel was $\dot{\theta}_2^* = 300$ rad/s and the desired closed-loop response parameters were given by $\zeta = 3$, $\omega_n = 6.5$, $p = 0.5$. The parameter-identification results are depicted in Figures 3.32 and 3.33, where the estimates are reached within less than 1 s. On the contrary, Figure 3.34 shows good behavior in the swinging up and stabilization of the pendulum. Notice that in the stable position, the wheel velocity keeps a constant desired reference value.

3.3.5   Algebraic Parameter Identification for Induction Motors

A fast, online algebraic identification scheme is proposed for the determination of some relevant induction motor parameters. The estimated values, given by the algebraic identification methodology, are tested here in a self-tuning output feedback control, subject to constant perturbation load torques while solving a reference trajectory-tracking task. An output feedback controller of the classical field-oriented, proportional integral (PI) type is proposed for the perturbed-output trajectory-tracking problem. Experimental results validate the effectiveness of the proposed approach.

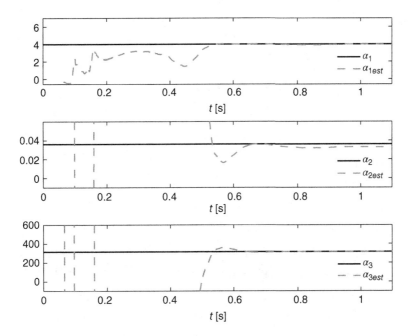

Figure 3.32   Parameter estimation results: first dynamical equation

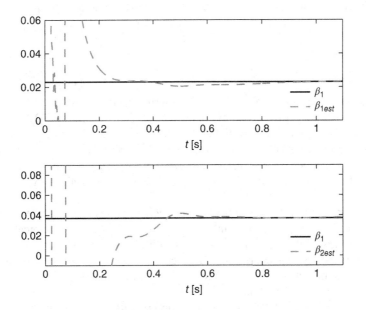

Figure 3.33   Parameter estimation results: second dynamical equation

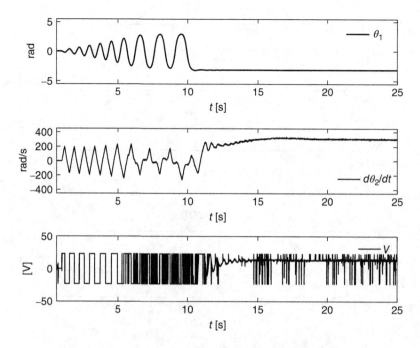

Figure 3.34   Pendulum angle and wheel velocity regulation

The problem of controlling the angular velocity, or the generated torque, in an induction motor is closely related to accurate knowledge of the parameters in the mathematical model. Good knowledge of the machine parameters will allow for high-performance control laws.

There exist a wide variety of parameter-identification approaches for induction motors; ranging from those which focus on the rotor time constant, to those which try to obtain the complete set of parameters (Wang *et al.*, 2005). Some other schemes are advocated for the estimation of parameters which can be subject to variation during operation. A comprehensive review of parameter estimation over induction motors is given in Toliyat *et al.* (2003).

The problem of simultaneous identification and control for uncertain systems, from a deterministic perspective, has been, generally speaking, tackled via adaptive control and robust identification techniques (see Chen and Gu, 2000). Another scheme, in the spirit of classical adaptive feedback control, is the certainty-equivalence control method (Åstrom and Wittenmark, 1995).

In this section, the problem of parameter identification in induction motors is solved online using two different linear system equations which involve both mechanical and electrical parameters. To assess the performance of the identifiers, we evaluate classical identification error indexes to measure the degree of confidence in the estimated parameters under the validity of the adopted model. We show that the methodology estimates the actual parameter values within a rather short time interval.

We properly combine the algebraic identification method for fast online computation of unknown system parameters with the field-oriented PI output feedback controller scheme. This combination of techniques is proposed for the solution of a reference trajectory-tracking task in an unknown, constant-load perturbed induction motor system. The PI controller is regarded as a certainty-equivalence controller in a self-tuning scheme.

3.3.5.1   Problem Formulation

Consider the following dynamic model of a two-phase induction motor[1] controlled by the phase voltages u_{Sa} and u_{Sb} with state variables given by: θ, describing the rotor angular position; ω, being the rotor angular velocity; ψ_{Ra} and ψ_{Rb}, representing the unmeasured rotor fluxes; i_{Sa} and i_{Sb}, being the stator currents.

$$\frac{d\theta}{dt} = \omega$$

$$\frac{d\omega}{dt} = \mu(i_{Sb}\psi_{Ra} - i_{Sa}\psi_{Rb}) - f\omega - \frac{\tau_L}{J}$$

$$\frac{d\psi_{Ra}}{dt} = -\eta\psi_{Ra} - n_p\omega\psi_{Rb} + \eta M i_{Sa}$$

$$\frac{d\psi_{Rb}}{dt} = -\eta\psi_{Rb} + n_p\omega\psi_{Ra} + \eta M i_{Sb} \tag{3.172}$$

$$\frac{di_{Sa}}{dt} = \eta\beta\psi_{Ra} + \beta n_p\omega\psi_{Rb} - \gamma i_{Sa} + \frac{u_{Sa}}{\sigma L_S}$$

$$\frac{di_{Sb}}{dt} = \eta\beta\psi_{Rb} - \beta n_p\omega\psi_{Ra} - \gamma i_{Sb} + \frac{u_{Sb}}{\sigma L_S}$$

Here, $\eta := \frac{R_R}{L_R}$, $\beta := \frac{M}{\sigma L_R L_S}$, $\mu := \frac{n_p M}{J L_R}$, $\gamma := \frac{M^2 R_R}{\sigma L_R^2 L_S} + \frac{R_S}{\sigma L_S}$, and $\sigma := 1 - \frac{M^2}{L_R L_S}$. R_R and R_S are, respectively, the rotor and stator resistances while L_R and L_S represent, respectively, the rotor and stator inductances. M is the mutual inductance constant, J is the moment of inertia, f is the viscous friction coefficient, and n_p is the number of pole pairs. The signal τ_L is the unknown load torque perturbation input.

We have the following assumptions:

A1 The stator currents i_{Sa}, i_{Sb}, the stator voltages u_{Sa}, u_{Sb}, as well as the rotor angular position, θ, are available for measurement. The integer number n_p is known.

A2 The initial values of the stator currents i_{Sa}, i_{Sb} are zero.

A3 The rotor flux linkages ψ_{Ra}, ψ_{Rb} are not available for measurement and, therefore, their initial values are unknown.

A4 The torque input is unknown but constant.

Assume momentarily, in the context of classical adaptive feedback control (Åstrom and Wittenmark, 1995), that all system parameters are perfectly known and no disturbance inputs (different from the torque load) are present. Under these circumstances, we will consider a certainty-equivalence control. Recall that most of the proposed control schemes are of field-oriented type (Leonhard, 2001). A quite common control strategy consists of devising a two-design stage; an outer-loop controller followed by an inner-loop controller. The external loop controls the angular velocity from the stator currents. The internal loop forces the stator currents to track the references given by the external loop.

Using the direct quadrature (DQ) reference framework, having orientation with respect to the rotor flux angle $\rho = \arctan(\psi_{Rb}/\psi_{Ra})$, i_d and i_q correspond to the transforms of the currents i_{Sa},

---

[1] Or, the two-phase equivalent model of a three-phase motor.

i_{Sb}; analogously, the voltage variables u_d, u_q and the rotor flux ψ_d are obtained. The velocity control (external loop) is governed by a reference value of i_q, denoted by i_q^*, where the angular velocity reference trajectory is denoted by $\omega^*(t)$ and satisfies $\omega^*(t) = d\theta^*(t)/dt$. In particular:

$$i_q^* = \frac{1}{\mu\psi_d^*}[k_{q0}(\theta^*(t) - \theta(t)) + k_{q1}(\omega^*(t) - \omega(t)) + (f/J)\omega + \dot{\omega}^*] \tag{3.173}$$

The flux reference ψ_d^* is given in terms of the nominal motor current i_{d0} and M, that is, $\psi_d^* = Mi_{d0}$. The flux regulation is carried out by means of i_d. We specify the required current as the following controller with pre-compensation:

$$\eta M i_d^* = k_{d0}\int_0^t (\psi_d^*(\tau) - \psi_d(\tau))d\tau + k_{d1}(\psi_d^*(t) - \psi_d(t))$$

The current control is a high-gain PI controller of the form

$$\frac{u_d}{\sigma L_S} = K_{dI}\int_0^t (i_d^*(\tau) - i_d(\tau)) + K_{dP}(i_d^*(t) - i_d(t)) \tag{3.174}$$

$$\frac{u_q}{\sigma L_S} = K_{qI}\int_0^t (i_q^*(\tau) - i_q(\tau)) + K_{qP}(i_q^*(t) - i_q(t)) \tag{3.175}$$

Applying high gains to K_{dI}, K_{dP}, K_{qI}, and K_{dP} forces us to track i_d and i_q against their corresponding references i_d^* and i_d^*, respectively. And with the proper choice of gains k_{q0} and k_{q1}, ω tracks $\omega^*(t)$ even with a constant load torque τ_L acting on the motor. Finally, the proper choice of k_{d0} and k_{d1} guarantees flux magnitude regulation.

Since the fluxes are not measurable, it is necessary to implement a flux observer. In particular, the real-time flux simulator

$$\frac{d\hat{\rho}}{dt} = n_p\omega + \eta M(-i_{Sa}\sin(\hat{\rho}) + i_{Sb}\cos(\hat{\rho}))/\hat{\psi}_d \tag{3.176}$$

$$\frac{d\hat{\psi}_d}{dt} = -\eta\hat{\psi}_d + \eta M(-i_{Sa}\cos(\hat{\rho}) + i_{Sb}\sin(\hat{\rho})) \tag{3.177}$$

guarantees accurate estimations in a simple manner. The stability of this observation scheme has been explored widely in the literature, see Leonhard (2001).

To perform this observer-based control scheme, we need precise knowledge of the lumped parameters n_p, f/J, σL_S, M, η, μ.

A separate, online, fast and precise parameter-identification process will be proposed to obtain, quite robustly with respect to possible zero-mean noise-measurement processes, a parameter-estimate set, which will replace the unknown parameters in a controller identical to the certainty-equivalence controller (3.173)–(3.177), except that it is written in terms of estimated parameter values rather than the actual parameters.

Given the uncertain induction motor model (3.172), and the control law (3.173)–(3.177), devise an online parameter-identification procedure to obtain a self-tuned control for accurate velocity tracking and flux magnitude regulation considering the assumptions A1–A4.

3.3.5.2 Parameter Identification

To obtain the set of parameters, we need to construct an algebraic system of equations concerning the *lumped parameters* and some functions of the *measurable variables*. Since the rotor flux linkages ψ_{Ra} and ψ_{Rb} are not assumed to be measured, it is necessary to eliminate them and their derivatives.

Performing simple algebraic manipulations over the motor equations (3.172) leads to the following expressions for $d\psi_{Ra}/dt$ and $d\psi_{Rb}/dt$:

$$\frac{d}{dt}\psi_{Ra} = \frac{L_R}{M}(u_{Sa} - R_S i_{Sa}) - \frac{\sigma L_S L_R}{M}\frac{di_{Sa}}{dt} \qquad (3.178)$$

$$\frac{d}{dt}\psi_{Rb} = \frac{L_R}{M}(u_{Sb} - R_S i_{Sb}) - \frac{\sigma L_S L_R}{M}\frac{di_{Sb}}{dt} \qquad (3.179)$$

Integrating (3.178), (3.179) on the interval $[0, t]$, we obtain a linear integral reconstructor for the rotor fluxes[2] independent of the rotor resistance R_R:

$$\psi_{Ra} = \frac{L_R}{M}\int_0^t (u_{Sa}(\tau) - R_S i_{Sa}(\tau))d\tau - \frac{\sigma L_S L_R}{M}i_{Sa} + \psi_{Ra}(0) \qquad (3.180)$$

$$\psi_{Rb} = \frac{L_R}{M}\int_0^t (u_{Sb}(\tau) - R_S i_{Sb}(\tau))d\tau - \frac{\sigma L_S L_R}{M}i_{Sb} + \psi_{Rb}(0) \qquad (3.181)$$

Expressions (3.178)–(3.181) will be used to eliminate any dependence on the rotor flux variables in the system equations.

We propose two linear systems for the parameter-determination task. For the first, consider the angular velocity dynamics from (3.172) in which we have substituted the expressions for the fluxes (3.180) and (3.181). We obtain

$$\frac{J}{n_p}\frac{d^2\theta}{dt^2} + \frac{f}{n_p}\frac{d\theta}{dt} + R_S\zeta_1 - \frac{M}{L_R}\psi_{Ra}(0)i_{Sb} + \frac{M}{L_R}\psi_{Rb}(0)i_{Sa} = \zeta_2 - \tau_L/n_p \qquad (3.182)$$

where $\zeta_1 = i_{Sb}\int_0^t i_{Sa}(\lambda)d\lambda - i_{Sa}\int_0^t i_{Sb}(\lambda)d\lambda$, $\zeta_2 = i_{Sb}\int_0^t u_{Sa}(\lambda)d\lambda - i_{Sa}\int_0^t u_{Sb}(\lambda)d\lambda$. Notice that $\zeta_1(t)$ and $\zeta_2(t)$ are free from time-derivative expressions.

A single time differentiation in (3.182) annihilates the constant load torque:

$$\frac{J}{n_p}\frac{d^3\theta}{dt^3} + \frac{f}{n_p}\frac{d^2\theta}{dt^2} + R_S\frac{d\zeta_1}{dt} - \frac{M}{L_R}\psi_{Ra}(0)\frac{di_{Sb}}{dt} + \frac{M}{L_R}\psi_{Rb}(0)\frac{di_{Sa}}{dt} = \frac{d\zeta_2}{dt} \qquad (3.183)$$

It is clear that (3.183) may generate a linear system for the unknown parameters. However, this would depend on non-measurable variables, such as the iterated time derivatives of the angular position. If we integrated (3.183) directly, the resulting expression would depend on the initial condition terms, which, in general, are not available either. To overcome these difficulties in obtaining a system in terms of available data, we propose using the algebraic identification method of Fliess and Sira-Ramírez (2003), which consists of the following steps.

---

[2] Although the expression can be used like a flux observer, the only proposition for the present work is to provide a relation that depends only on the measured variables without derivative terms.

Given that the order of the equation is three, let us first multiply (3.183) by the factor $(-t)^3$ and then proceed to integrate, three times, the resulting expression over the interval $[0, t]$ using the integration by parts formula. This leads to

$$\frac{J}{n_p}p_{1,1}(t) + \frac{f}{n_p}p_{1,2}(t) + R_S p_{1,3}(t) - \frac{M}{L_R}\psi_{Ra}(0)p_{1,4}(t) + \frac{M}{L_R}\psi_{Rb}(0)p_{1,5}(t) = q_1(t) \tag{3.184}$$

where

$$p_{1,1}(t) = \left(\int^{(3)} (-t^3)\frac{d^3\theta}{dt^3}\right) = 6\left(\int^{(3)}\theta\right) - 18\left(\int^{(2)} t\theta\right) + 9\left(\int t^2\theta\right) - t^3\theta$$

$$p_{1,2}(t) = \left(\int^{(3)} (-t^3)\frac{d^2\theta}{dt^2}\right) = -6\left(\int^{(3)} t\theta\right) + 6\left(\int^{(2)} t^2\theta\right) - \left(\int t^3\theta\right)$$

$$p_{1,3}(t) = \left(\int^{(3)} (-t^3)\frac{d\zeta_1}{dt}\right) = 3\left(\int^{(3)} t^2\zeta_1\right) - \left(\int^{(2)} t^3\zeta_1\right)$$

$$p_{1,4}(t) = -\left(\int^{(2)} (-t^3)\frac{di_{Sb}}{dt}\right) = -3\left(\int^{(3)} t^2 i_{Sb}\right) + \left(\int^{(2)} t^3 i_{Sb}\right)$$

$$p_{1,5}(t) = \left(\int^{(2)} (-t^3)\frac{di_{Sa}}{dt}\right) = 3\left(\int^{(3)} t^2 i_{Sa}\right) - \left(\int^{(2)} t^3 i_{Sa}\right)$$

$$q_1(t) = \left(\int^{(3)} (-t^3)\frac{d\zeta_2}{dt}\right) = 3\left(\int^{(3)} t^2\zeta_2\right) - \left(\int^{(2)} t^3\zeta_2\right)$$

The resulting equation (3.184) is linear in the unknown parameters. We rewrite it as

$$\mathbf{p}_1(t)\Theta_1 = q_1(t) \tag{3.185}$$

where

$$\mathbf{p}_1(t) = \begin{bmatrix} p_{1,1}(t) & p_{1,2}(t) & p_{1,3}(t) & p_{1,4}(t) & p_{1,5}(t) \end{bmatrix} \text{ and}$$

$$\Theta_1 = \begin{bmatrix} \dfrac{J}{n_p} & \dfrac{f}{n_p} & R_S & \dfrac{M}{L_R}\psi_{Ra}(0) & \dfrac{M}{L_R}\psi_{Rb}(0) \end{bmatrix}^T$$

We have five unknown parameters and just one equation. We need four other linearly independent equations to construct an identifier. To generate the remaining equations one possibility, advocated in Fliess and Sira-Ramírez (2003), lies in sequentially integrating (3.184) with respect to time to generate new linearly independent equations. An alternative procedure, which will be used in this chapter consists of obtaining a formula for Θ_1 through some further algebraic manipulations as follows.

Pre-multiply both terms of equation (3.185) by the column vector $\mathbf{p}_1^T(t)$. Integrate both terms of the equation during an arbitrary interval of time $[0, t]$, extracting the constant vector of parameters Θ_1 out of the integration to obtain

$$\left[\int_0^t \mathbf{p}_1^T(\lambda)\mathbf{p}_1(\lambda)d\lambda\right]\Theta_1 = \int_0^t \mathbf{p}_1^T(\lambda)q_1(\lambda)d\lambda$$

Clearly, the matrix $M_{\mathbf{pp1}} = \left[\int_0^t \mathbf{p}_1^T(\lambda)\mathbf{p}_1(\lambda)d\lambda \right]$ is a symmetric and positive-definite matrix for any $t \geq \epsilon > 0$. Symmetry is obvious; positive definiteness is proved by considering the following triple product: $z_T M_{\mathbf{pp1}} z$ for any arbitrary, nonzero, vector z of the same dimension as $\mathbf{p}1(t)$. We have $z^T M_{\mathbf{pp1}} z = \int_0^t (z^T \mathbf{p}_1^T(\lambda)\mathbf{p}_1(\lambda)z)d\lambda = \int_0^t (\mathbf{p}_1(\lambda)z)^2 d\lambda \geq 0$. The matrix $M_{\mathbf{pp1}}$ is hence invertible for any strictly positive time t. Note that the solution for Θ_1, obtained by elementary algebraic manipulations,

$$\hat{\Theta}_1 = \left[\int_0^t \mathbf{p}_1^T(\lambda)\mathbf{p}_1(\lambda)d\lambda \right]^{-1} \left[\int_0^t \mathbf{p}_1^T(\lambda)q_1(\lambda)d\lambda \right] \tag{3.186}$$

is the same classical solution given by the least-squares method, which may be viewed as an optimization problem consisting of minimizing an integral, square error criterion. That is,

$$J(\Theta_1, t) = \frac{1}{2} \int_0^t \varepsilon^2(\Theta_1, \lambda)d\lambda \tag{3.187}$$

$$\varepsilon(\Theta_1, t) = \mathbf{p}_1(t)\Theta_1 - q_1(t)$$

$$\hat{\Theta}_1 = \arg \left\{ \min_{\Theta_1} J(\Theta_1, t) \right\}$$

Let us apply the integral flux reconstructor in the third equation of (3.172) (it can be used also in the fourth equation, without loss of generality), which corresponds to the dynamics of ψ_{Ra}. We obtain an expression free from rotor flux terms, as follows:

$$u_{Sa} + n_p\omega \int_0^t u_{Sb} = \sigma L_S \left[\frac{di_{Sa}}{dt} + n_p\omega i_{Sb} \right] + R_S \left[n_p\omega \int_0^t i_{Sb} \right] + -\frac{M}{L_R}\psi_{Rb}(0)n_p\omega$$

$$\left(R_S + \sigma L_S \frac{R_R}{L_R} + \frac{M^2 R_R}{L_R^2} \right) i_{Sa} - \frac{R_R}{L_R} \left[\int_0^t u_{Sa} \right] + \frac{R_R}{L_R}R_S \left[\int_0^t i_{Sa} \right] - \frac{R_R}{L_R}\frac{M}{L_R}\psi_{Ra}(0)$$

$$\tag{3.188}$$

The system is linear in the parameters but is over-parameterized, meaning that the lumped parameters are not independent of each other. In contrast some problems may arise in (3.188) when we have a constant velocity; in this case, the last two terms are constant and then they become linearly dependent. It is possible to obtain a specific regressor system version for constant velocities, limiting the regressor system to this case (see Oteafy et al., 2009).

Since (3.188) has first-order time-derivative terms, according to the algebraic parameter-estimation procedure, it is just necessary to multiply (3.188) by $-t$ and then integrate by parts with respect to the time to avoid any derivative terms in the equation. We have

$$\mathbf{p}_3\Theta_2 = q_3 \tag{3.189}$$

where

$$\Theta_2 = \left[\sigma L_S \ R_S \ \left(R_S + \sigma L_S \frac{R_R}{L_R} + \frac{M^2 R_R}{L_R^2} \right) \ \frac{R_R}{L_R} \ \frac{R_R}{L_R}R_S \ \frac{R_R}{L_R}\frac{M}{L_R}\psi_{Ra}(0) \ \frac{M}{L_R}\psi_{Rb}(0) \right]^T$$

$$\mathbf{p}_3 = \begin{bmatrix} p_{3,1} & p_{3,2} & p_{3,3} & p_{3,4} & p_{3,5} & p_{3,6} & p_{3,7} \end{bmatrix}$$

with

$$p_{3,1} = -ti_{Sa} + (\int i_{Sa}) - (\int n_p t\hat{\omega}i_{Sb}), \quad p_{3,2} = -n_p(\int_0^t i_{Sb})(t\theta - (\int \theta)) + n_p\{(\int t\theta i_{Sb}) - [\int i_{Sb}$$

$$(\int \theta)]\}, \quad p_{3,3} = (\int -ti_{Sa}), \quad p_{3,4} = (\int t(\int_0^t u_{Sa})), \quad p_{3,5} = -(\int t(\int_0^t i_{Sa})), \quad p_{3,6} = \frac{t^2}{2}, \quad p_{3,7} =$$

$$n_p(t\theta - (\int \theta)), \quad q_3 = -(\int tu_{Sa}) - n_p(\int_0^t u_{Sb})(t\theta - (\int \theta)) + n_p\{(\int t\theta u_{Sb}) - [\int u_{Sb}(\int \theta)]\}.$$

For the term $\int n_p t\hat{\omega}i_{Sb}$, it is necessary to use an estimate of the velocity, $\hat{\omega}$, provided by an observer; in particular, a simple discrete-time derivation, in the Euler sense, of the angular position works fine for that purpose.

Finally, solving for Θ_2, in accordance with (3.186) we have

$$\hat{\Theta}_2 = \left[\int \mathbf{p}_3^T \mathbf{p}_3\right]^{-1} \left[\int \mathbf{p}_3^T q_3\right]$$

Remark 3.3.2 *Since the parameters M, L_R, L_S, σ, and R_R are not directly identifiable, they are normally approximated by assuming that $L_S = L_R$ so that we can propose a simple procedure to obtain them from those parameters that are directly identifiable from Θ_2: (1) $L_S = \sigma L_S + M^2/L_R$, (2) $L_R = L_S$, (3) $M = \sqrt{L_R(M^2/L_R)}$, (4) $\sigma = (\sigma L_S)/L_S$, (5) $R_R = (R_R/L_R)L_R$.*

3.3.6 A Criterion to Determine the Estimator Convergence: The Error Index

In Chapter 2, the sentinel parameter was proposed as an alternative to decide experimentally the instant at which the estimated parameters have converged to acceptable values (see also García-Rodríguez et al., 2009). An alternative approach, based on the condition number involved in the identifier equation system, is given in Trapero-Arenas et al. (2008). Here, we will use the integral square error criterion, which provides an additional effective tool to determine the instant at which the estimated parameters can be used for control purposes. This criterion consists of observing the fit of the estimated model to the plant behavior as the identifier obtains the set of estimated parameters. A detailed analysis of optimization criterion sensibility to the parameter variation is provided in Chiasson (2005).

To introduce the error index formulation, let us define $M_{\mathbf{pp}} \triangleq \int_0^t \mathbf{p}^T(\lambda)\mathbf{p}(\lambda)d\lambda$, $M_{\mathbf{pq}} \triangleq \int_0^t \mathbf{p}^T(\lambda)q(\lambda)d\lambda$, $M_{qq} \triangleq \int_0^t q(\gamma)^2 d\gamma$, $M_{q\mathbf{p}} \triangleq \int_0^t q(\gamma)\mathbf{p}(\gamma)d\gamma$, where $M_{\mathbf{pp}} \in \mathbb{R}^{m \times m}$, $M_{\mathbf{pq}}, M_{q\mathbf{p}} \in \mathbb{R}^m$, and $M_{qq} \in \mathbb{R}$, denoting by m the number of unknown parameters. The cost function (3.187) can be rewritten as follows:

$$\mathcal{J}(\Theta, t) = \frac{1}{2}(M_{qq} - M_{q\mathbf{p}}\Theta - \Theta^T M_{q\mathbf{p}}^T + \Theta^T M_{\mathbf{pp}}\Theta)$$

$$= \frac{1}{2}\left(M_{qq} - M_{q\mathbf{p}}M_{\mathbf{pp}}^{-1}M_{q\mathbf{p}}^T\right.$$

$$\left. + \left(\Theta - M_{\mathbf{pp}}^{-1}M_{\mathbf{pq}}\right)^T M_{\mathbf{pp}}\left(\Theta - M_{\mathbf{pp}}^{-1}M_{\mathbf{pq}}\right)\right)$$

where $M_{\mathbf{pp}}$ is a symmetric positive semi-definite matrix. Assuming that the system is sufficiently well numerically conditioned, such that $M_{\mathbf{pp}}$ becomes invertible (and also positive definite), it is satisfied that $(\Theta - M_{\mathbf{pp}}^{-1}M_{\mathbf{pq}})^T M_{\mathbf{pp}}(\Theta - M_{\mathbf{pp}}^{-1}M_{\mathbf{pq}}) \geq 0$ for all $\Theta \in \mathbb{R}^m$. This term is zero if and only if $\Theta - M_{pp}^{-1} M_{pq} = [0 \ \cdots \ 0]$. Let us define the *residual error* as

$$\mathcal{J}(\Theta, t)|_{\Theta = \hat{\Theta} \triangleq M_{pp}^{-1} M_{pq}} = \frac{1}{2} \left(M_{qq} - M_{qp} M_{pp}^{-1} M_{qp}^T \right)$$

Given $M_{qp} M_{pp}^{-1} M_{qp}^T \geq 0$, the cost function is bounded by

$$\mathcal{J}(\hat{\Theta}, t) = \frac{1}{2} \left(M_{qq} - M_{qp} M_{pp}^{-1} M_{qp}^T \right) \leq \frac{1}{2} M_{qq}$$

Notice that $M_{qq} > 0$ for $t \geq 0$. Thus, the *error index* relative to $\mathcal{J}(0, t)$ is given by

$$\text{error } index \triangleq \sqrt{\frac{M_{qq} - M_{qp} M_{pp}^{-1} M_{qp}^T}{M_{qq}}} \leq 1 \tag{3.190}$$

For reliable estimates, this index should be much smaller than 1 for any set of estimates. If the index value is close to 1, we should infer that the assumed model of the system is wrong.

3.3.6.1 Optimization Criterion Sensitivity with Respect to the Variation of Parameter Magnitudes

Another advantage of using (3.187) to obtain the parameter estimates is given by the *sensitivity indices*, which indicate the sensitivity of the residual error with respect to the magnitude variation of each parameter, letting us know indirectly which parameters are being estimated with more accuracy. To introduce these indices, the residual error is written in terms of a variation of the parameter vector, denoted by $\Delta\Theta$, as

$$\mathcal{J}(\hat{\Theta} + \Delta\Theta) = \mathcal{J}(\hat{\Theta}, t) + \Delta\Theta^T M_{pp} \Delta\Theta \tag{3.191}$$

We need to determine the maximum variation of a particular parameter, such that this variation produces a specific change over the residual error. In this case, the problem is, given (3.191), to find the maximum value of $\Delta\theta_i$ such that $\mathcal{J}(\hat{\Theta} + \Delta\Theta) = 2\mathcal{J}(\hat{\Theta}, t)$, or

$$\Delta\Theta^T M_{pp} \Delta\Theta = J(\hat{\Theta}, t) \tag{3.192}$$

It is possible to define a *parametric error index* related to the parameter θ_i as the maximum value $\Delta\theta_i$ such that (3.192) is satisfied. This condition is formulated as a maximization problem

$$\max \Delta\theta_i \quad \text{subject to} \quad \Delta\Theta^T M_{pp} \Delta\Theta = \mathcal{J}(\hat{\Theta}, t) \tag{3.193}$$

whose solution is (Chiasson, 2005)

$$\Delta\theta_i = \sqrt{J(\hat{\Theta}, t)(M_{pp}^{-1})_{ii}} \tag{3.194}$$

with $(M_{ii}^{-1})_{11}$ denoting the element (i, i) of the matrix M_{pp}^{-1}. Here, $\Delta\theta_i$ is the maximum variation of θ_i such that the residual error duplicates. Therefore, a small index indicates that the identifier achieved high-quality estimates.

3.3.6.2 Experimental Results

The motor was a 3-phase induction motor with the following features: 220/440 V 1.1/2.2 A 60 Hz, 0.559 kW, 3415 rpm, and efficiency of 75.1%. Two data acquisition boards, PCI-6025 and PCI-6602 from National Instruments, were used to measure currents, voltages, and the angular position of the motor, as well as to generate six PMW signals (at 16 kHz) to control the 3-phase inverter bridge. The optical encoder has an angular resolution of 0.036°. The MATLAB®-xPC Target was the real-time environment. Data acquisition was performed at a sampling rate of 6.7 kHz in an AMD Athlon XP processor at 1.7 GHz as a target CPU. The execution mode was *single tasking*. The signals were pre-conditioned with a low-pass filter with cut-off frequency of 1 kHz. It was desired to smoothly bring the motor angular velocity from a value of 0 rad/s toward a final value of 45 rad/s in 0.5 s. A 16th-order Bézier polynomial, smoothly interpolating between the initial and final values, was prescribed as the desired angular velocity reference trajectory $\omega^*(t)$.

The initial parameter values of the induction motor model are obtained from steady-state analysis, using an equivalent circuit of the induction motor and the manufacturers' nominal characteristics. For this experiment, we use $R_{S0} = 5.1\,\Omega$, $R_{R0} = 1.785\,\Omega$, $\sigma_0 = 0.0582$, $L_{S0} = L_{R0} = 0.184$ H, $M_0 = 0.178$ H, $J_0 = 0.001$ kg m^2, and $f_0 = 0$ N m s.

The output feedback controller gains were obtained using the following design values: $k_{q0} = 2500$, $k_{q1} = 100$, $k_{d0} = 64$, $k_{d1} = 16$, $K_{dI} = K_{qI} = 2025$, $K_{dP} = K_{qP} = 90$.

Figure 3.35 depicts the fast parameter-estimation process taking place in an interval of duration 2 s. The corresponding parametric error indices are shown in Figure 3.36. We adopt as evaluation period [0, 2] s. The estimators were switched off at 2 s and the computed values

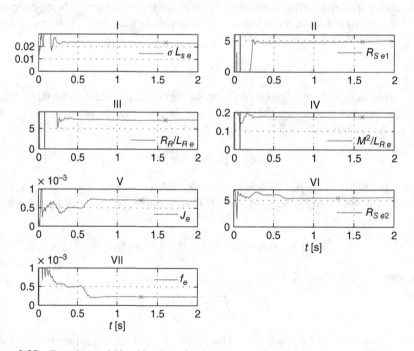

Figure 3.35 Experimental identification of induction motor parameters under feedback control

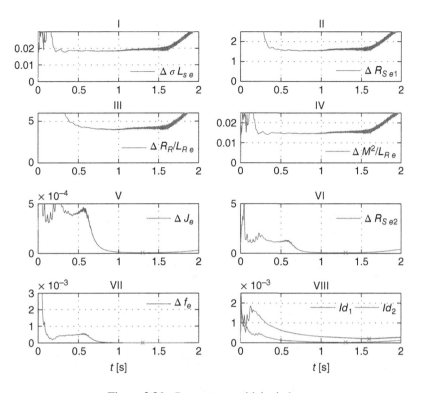

Figure 3.36 Parameter sensitivity indexes

with minimum corresponding parametric error index were saved as parameter estimates for the remaining computation time. The minimum error index for Θ_1, Id_1 occurred at 1.3 s and the minimum error index for Θ_2, Id_2 occurred at 1.62 s therefore these are the instants when the estimated parameter values are taken (see Figure 3.36, part VIII). Locating the minimum value of the error index, it is possible to determine the instant when the induction motor model, with the actual estimated parameters, predicts more accurately the behavior of the plant. The comparison between the magnitudes of the parameter sensitivity indexes shows that the inertia and friction parameters are identified with more accuracy than the stator resistance in the mechanical parameter estimation. In contrast, these indices also show that the estimated values of σL_S and M^2/L_R are more accurate than those obtained for R_S or R_R/L_R in the electric case. The values obtained for these parameters, were $R_S = 5.12\Omega$, $R_R = 2.23\Omega$, $\sigma = 0.0582$, $L_S = L_R = 0.2919$ H, $M = 0.2768$ H, $J = 0.00045$ kg m^2, and $f = 0.0001935$ N m s.

A smooth transition was planned between the initial parameter values and the estimated parameter values obtained by the algebraic identification procedure during the interval [2, 3] s. As shown in Figure 3.37, part I, the velocity tracking is not significantly affected by this change of parameters; this is due to the robustness of the strategy control. However, the stator currents and stator voltages present a notorious attenuation (see Figure 3.37, parts II and III) which means better power management for the same tracking task.

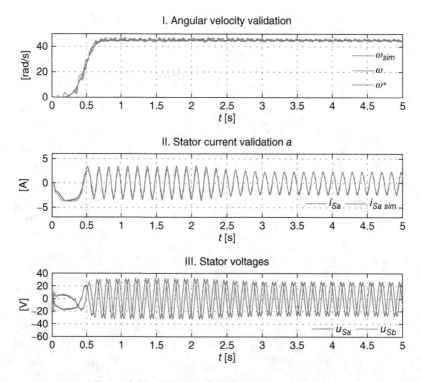

Figure 3.37   Model validation and stator voltages

Figure 3.37, part I, depicts the performance of the self-tuning controller in achieving the desired angular velocity profile. The system was subject to a load torque perturbation input of 0.3 N m by means of a coupled DC motor.

In agreement with the parametric error indices, a simple form to evaluate the identified parameters is to simulate the induction motor model (3.172) using the estimated parameters with the inputs given by the measured stator voltages. We then compare the simulation outputs with the actual measurements. In this experiment, we used i_{Sa} and ω as the validation variables. The simulated and measured values of i_{Sa} are depicted in Figure 3.37, part II, which shows a good match between them. In the same way, the simulated velocity goes according to the measured velocity; see Figure 3.37, part I. In both cases, the steady-state correspondence is better than the transient correspondence.

3.3.6.3   Discussion

The effectiveness of the algebraic parameter identification method was analyzed, and tested in an actual test bed, in the self-tuning output feedback control of an unknown induction motor. We have shown that the angular velocity reference trajectory-tracking task for an induction motor can be handled efficiently via a combination of output feedback control and the algebraic identification method.

The results indicate that the method can be used in control applications, with a variety of strategies leading to more accurate results in the control of motors with uncertain parameters. The methodology can be adapted for time-varying load torque, assuming that the load torque input is approximately and locally modeled by a family of polynomial signals as in a Taylor polynomial approximation.

The algebraic identification methodology is restricted to appropriate values of velocity trajectories. It is possible to program the system identification to be enabled in the presence of important velocity variations, which induce a temporal conditioning of the associated parameter equation system, achieving good estimations of the related parameters. Also, the system can be designed for constant-velocity conditions.

Since the parameters η and R_S are time varying, with slight variations, the approach can also be adapted to estimate time-varying parameters. The parameters can be assumed to be piecewise constant.

The computational complexity of the algebraic methods is overcome by the capacity of the embedded systems, or the parallel processing, present in some platforms, offering more implementation possibilities. It is important to point out that the highest computational cost occurs at the beginning of the process, which implies a better task scheduling when starting the identification process.

3.4 Remarks

- In this chapter we have demonstrated that the algebraic approach for online parameter identification can be extended to the case of nonlinear systems with relative ease, provided the systems considered are either linearly identifiable or weakly linearly identifiable.
- One additional limitation of the method, which will soon be lifted, is the need to have access to *all* state variables of the system. In particular, the flat outputs and a finite number of their time derivatives are required for the proposed online parameter-identification scheme.

References

Abrahantes, M., Mulder, J., and Butter, K. (2007) Modeling, identification and control of an under actuated inertial wheel pendulum. *Proceedings of the 39th Southeastern Symposium on System Theory*, pp. 1–5, Macon. GA.

Andary, S., Chemori, A., and Krut, S. (2009) Estimation-based disturbance rejection in control for limit cycle generation on inertia wheel inverted pendulum testbed. *Proceedings of the 2009 IEEE/RSJ International Conference on Intelligent Robots and Systems*, pp. 1302–1307, St. Louis, MO.

Astolfi, A. and Ortega, R. (2003) Immersion and invariance: A new tool for stabilization and adaptive control of nonlinear systems. *IEEE Transactions on Automatic Control* **48**(4), 590–606.

Åstrom, K., and Furuta, K. (2000) Swinging up a pendulum by energy control. *Automatica* **36**, 287–295.

Åstrom, K., and Wittenmark, B. (1995) *Adaptive Control*, 2nd edn. Addison-Wesley, New York.

Belkoura, L., Richard, J., and Fliess, M. (2008) A convolution approach for delay systems identification. *Proceedings of the 17th IFAC World Congress*, vol. 17, pp. 6325–6329, South Korea.

Block, D., Åstrom, K., and Spong, M. (2007) *The Reaction Wheel Pendulum*, vol. 1 in Synthesis Lectures on Control and Mechatronics. Morgan & Claypool, San Rafael, CA.

Chen, J., and Gu, G. (2000) *Control-Oriented System Identification: An H-infinity Approach*. Wiley Interscience, New York.

Chiasson, J. (2005) *Modeling and High-Performance Control of Electric Machines*. John Wiley & Sons, Inc., New York.

El-Sharkawi, M., El-Samany, A., and El-Sayed, M. (1994) High performance drive of dc brushless motors using neural network. *IEEE Transactions on Energy Conversion* **9**(2), 317–322.

Fliess, M., and Sira-Ramírez, H. (2002) On the non-calibrated visual based control of planar manipulators: An on line algebraic identification approach. *IEEE-SMC-2002*, Tunisia.

Fliess, M., and Sira-Ramírez, H. (2003) An algebraic framework for linear identification. *ESAIM, Control, Optimization and Calculus of Variations* **9**(1), 151–168.

Fliess, M., Join, C., and Sira-Ramirez, H. (2008) Non-linear estimation is easy. *International Journal of Modelling, Identification and Control* **4**(1), 12–27.

Freidovich, L., Hera, PL., Mettin, U., Robertsson, A., Shiriaev, A. and Johansson, R. (2009) Shaping stable periodic motions of inertia wheel pendulum: Theory and experiment. *Asian Journal of Control* **11**(5), 548–556.

García-Rodríguez, C., Cortés-Romero, J., and Sira-Ramírez, H. (2009) Algebraic identification and discontinuous control for trajectory tracking in a perturbed 1-dof suspension system. *IEEE Transactions on Industrial Electronics* **56**(9), 3665–3674.

Ge, Z. and Cheng, J. (2005) Chaos synchronization and parameter identification of three time scales brushless dc motor system. *Chaos, Solitons and Fractals* **24**, 597–616.

Genesio, R., and Tesi, A. (1992) Harmonic balance methods for the analysis of chaotic dynamics in nonlinear systems. *Automatica* **28**(3), 531–548.

Hagenmeyer, V., and Delaleau, E. (2003) Exact feedforward linearization based on differential flatness. *International Journal of Control* **76**, 537–556.

Hernández, V., and Sira-Ramírez, H. (2002) Sliding mode generalized PI tracking control of a dc–motor–pendulum system. *Proceedings of the 7th IEEE Variable Structure Control Workshop*, Sarajevo, Bosnia-Herzegovina.

Hernández, V., and Sira-Ramírez, H. (2003) Generalized PI control for swinging up and balancing the inertia wheel pendulum. *Proceedings of the American Control Conference*, pp. 2809–2814, Denver, CO.

Khalil, HK. (2002) *Nonlinear Systems*, 3rd edn. Prentice-Hall, Upper Saddle River, NJ.

Leonhard, W. (2001) *Control of Electrical Drives*, 3rd edn. Springer-Verlag, Berlin.

Linz, SJ. (2008) On hyperjerky systems. *Chaos, Solitons and Fractals* **37**(3), 741–747.

Ma, BL., Tso, SK., and Xu, WL. (2002) Adaptive/robust time-varying stabilization of second-order non-holonomic chained form with input uncertainties. *International Journal of Robust and Nonlinear Control* **12**(15), 1299–1316.

Middleton, R., and Goodwin, GC. (1990) *Digital Control and Estimation: A Unified Approach*. Prentice-Hall, Englewood Cliffs, NJ.

Morales, R., Feliu, V., and Sira-Ramírez, H. (2010) Nonlinear control for magnetic levitation systems based on fast on-line algebraic identification of the input gain. *IEEE Transactions on Control Systems Technology* **19**(4), 757–771.

Olfati-Saber, R. (2001) Global stabilization of a flat underactuated system: The inertia wheel pendulum. *Proceedings of the 40th IEEE Conference on Decision and Control*, pp. 3764–3765, Orlando, FL.

Ortega, R., Spong, M., Gómez-Estern, F., and Blankenstein, G. (2002) Stabilization of a class of underactuated mechanical systems via interconnection and damping assignment. *IEEE Transactions on Automatic Control* **47**(8), 1218–1233.

Oteafy, A., Chiasson, J., and Bodson, M. (2009) Online identification of the rotor time constant of an induction machine. *Proceedings of the American Control Conference*, pp. 4373–4378, St. Louis, MO.

Rasouli, M., and Karrari, M. (2004) Nonlinear identification of a brushless excitation system via field tests. *IEEE Transactions on Energy Conversion* **19**(4), 733–740.

Schaible, U., and Szabados, B. (1999) Dynamic motor parameter identification for high speed flux weakening operation of brushless permanent magnet synchronous machines. *IEEE Transactions on Energy Conversion* **14**(3), 486–492.

Solomon, O., and Famouri, P. (2009) Control and efficiency optimization strategy for permanent magnet brushless ac motors. *Proceedings of the IEEE International Symposium on Industrial Electronics*, pp. 505–512, Seoul, Korea.

Spong, M., Corke, P., and Lozano, R. (2001) Nonlinear control of the reaction wheel pendulum. *Automatica* **37**(11), 1845–1851.

Toliyat, H., Levi, E., and Raina, M. (2003) A review of RFO induction motor parameter estimation techniques. *IEEE Transactions on Energy Conversion* **18**(2), 271–283.

Trapero-Arenas, J., Mboup, M., Pereira-Gonzalez, E., and Feliu, V. (2008) On-line frequency and damping estimation in a single-link flexible manipulator based on algebraic identification. *16th Mediterranean Conference on Control and Automation, 2008*, pp. 338–343, Ajaccio Corsica, France.

Wang, K., Chiasson, J., Bodson, M., and Tolbert, L. (2005) A nonlinear least-squares approach for estimation of the induction motor parameters. *IEEE Transactions on Automatic Control* **50**(10), 1622–1628.

Ye, H., Liu, G., Yang, C., and Gui, W. (2008) Stabilisation designs for the inertia wheel pendulum using saturation techniques. *International Journal of Systems Science* **39**(12), 1203–1214.

4

Algebraic Parameter Identification in Discrete-Time Systems

4.1 Introduction

The main objective of this chapter is the development and adaptation of the algebraic method of parameter identification for the class of discrete-time linearly parameterizable systems. In general, discrete-time systems arise from two specific instances: the systems are either naturally discrete or they represent a discretization of continuous systems. In some cases, discretization may be carried out exactly (as in the linear systems case). In nonlinear systems, the exact discretization problem is considerably more difficult and not always feasible. Approximate discretization thus generates the most common class of sampled time systems. The speed of the sampling process motivates the use of alternative schemes of system representation, thanks to the so-called Goodwin's unified approach.

The importance of studying the discrete-time version of continuous systems dates back to the use of digital data-processing systems, which are the most common signal-processing tool in control and estimation problems. In general, control processes are implemented via digital systems. Some mechatronic and electromechanical systems (stepping motors, production processes) exhibit a discrete, or combination of discrete and continuous, process. They are usually called "hybrid" in nature and we shall not deal with this interesting class of systems in this book. Discrete-time systems thus motivate the setting of parallel techniques for parameter identification.

4.2 Algebraic Parameter Identification in Discrete-Time Systems

The algebraic parameter-identification methodology uses similar principles to the continuous-time counterpart, based on the possibility of rearranging the system as a linear regression of the parameter vector θ, including a class of structured disturbances (expressible as families of time polynomials). The main difference between this methodology and the continuous-time method is the treatment given in terms of the algebraic theory of

Algebraic Identification and Estimation Methods in Feedback Control Systems, First Edition.
Hebertt Sira-Ramírez, Carlos García-Rodríguez, Alberto Luviano-Juárez, John Alexander Cortés-Romero.
© 2014 John Wiley & Sons, Ltd. Published 2014 by John Wiley & Sons, Ltd.

discrete-time systems (which can be found in the appendix). The problem formulation will be given for the linear case, but it may be extended for some nonlinear cases using similar assumptions to those in the last chapter.

4.2.1   Main Purpose of the Chapter

In this chapter, we present the discrete-time counterpart to the online continuous-time linear parametric identification approach[1] (Fliess and Sira-Ramírez, 2003). As in the continuous-time version, the approach is based on the algebraic theory of modules, which has become classic for discrete-time linear systems (see, e.g., Bourlès and Fliess, 1997; Fliess and Glad, 1993; Fliess and Marquez, 2000, 2001; Fliess *et al.*, 1995), and on difference algebra (see Cohn, 1965; Fliess, 1992).

The approach may be summarized as follows:

- We introduce *linear identifiability* and *weak linear identifiability*, which are commonly encountered in practice.
- Most usual deterministic structured perturbations may be eliminated via algebraic manipulations.
- We achieve identification in a closed loop by calculating the unknown parameters, which are solutions of algebraic equations, in a small time interval.
- It does not resort to probabilistic formulations, and the exact values of the unknown parameters are not asymptotically obtained.[2]

Several case studies and their computer simulations are contained in this chapter, which demonstrate the efficiency of the method and its robustness with respect to a large variety of perturbations.

The chapter is organized as follows. Section 4.2.1 presents an introductory example on the discrete algebraic methodology. Sections 4.2.2 to 4.2.5 present some other realistic case studies: an economic model, heating of a slab dynamics, sampled control of a hard disk drive, a discrete-time chaotic system, a visual servo-tracking problem and a rolling mill, beside an experimental frequency-estimation problem under the exact discretization approach. The discrete counterpart of a proportional integral control (proportional additive) for those systems demonstrates the usefulness of our online identification scheme. Numerical simulations demonstrate the robustness of the approach with respect to additive random noises. This part is written in such a way that any reader who is not familiar with the algebraic language may nevertheless grasp the essential features of the proposed techniques. Appendix B is devoted to a general module-theoretic exposition of discrete-time linear time-invariant systems. It contains a short summary of the main properties of principal ideal rings, which are already familiar in the control literature (see, e.g., Kailath, 1979). Section III defines various types of identifiability, and especially linear identifiability and weak linear identifiability, by utilizing elementary

---

[1] We are not adopting the scheme proposed in Fliess *et al.* (2008), which incorporates an operational calculus framework for the discrete-time parameter identification, since working directly with the difference equations allows a straightforward methodology to obtain the identified parameters within a less complex procedure.

[2] This is a major difference from the vast literature on identification and adaptive control (see, e.g., Åstrom and Wittenmark, 1995; Caines, 1988; Goodwin and Sin, 1984; Isermann, 1987; Landau, 1990; Ljung, 1987; Richalet, 1998; Söderström and Stoica, 1989 and the references cited therein).

facts from field theory (Jacobson, 1974) and difference algebra (Cohn, 1965). In Section IV we show how to eliminate classic deterministic perturbations and, thus, how to avoid one of the major sources of bias in other approaches. This method rests on skew difference polynomial rings (McConnell and Robson, 1987), the properties of which are briefly reviewed. The conclusion lists some possible extensions.

4.2.2 Problem Formulation and Assumptions

Consider a linear time-invariant system with one input, denoted by u_k, and one output, say y_k, represented by an nth-order difference equation

$$y_{k+n} + a_{n-1}y_{k+n-1} + \ldots + a_1y_{k+1} + a_0y_k = b_m u_{k+m} + \ldots + b_0 u_k + \eta_k + w_k \qquad (4.1)$$

having the following hypotheses:

- $m < n$.
- The control input and the output, u_k, y_k, and their past data values, are available for measurement.
- η_k is an additive disturbance input of polynomial type.
- w_k is a possible additive noise input of deterministic nature with zero mean.

The system (4.1) can be rewritten as follows:

$$y_k + a_{n-1}y_{k-1} + \ldots + a_1y_{k-n+1} + a_0y_{k-n} = b_m u_{k+m-n} + \ldots + b_0 u_{k-n} + \eta_{k-n} + w_{k-n} \qquad (4.2)$$

which is expressed in causal form. The linear regression is given as follows:

$$\begin{bmatrix} y_{k-1} & y_{k-2} & \cdots & y_{k-n} & u_{k+m-n} & \cdots & u_{k-n} \end{bmatrix} \begin{bmatrix} a_{n-1} \\ \vdots \\ a_0 \\ b_m \\ \vdots \\ b_0 \end{bmatrix} = y_k + \eta_{k-n} + w_{k-n} \qquad (4.3)$$

Since the system is linear with respect to the vector of parameters $\theta^T = [a_{n-1}, \ldots, a_0, b_m, \ldots, b_0]$, it is said to be linearly identifiable. This fact allows the application of the algebraic parameter-estimation methodology.

Consider system (4.3). Given the measurements of u_k, y_k, devise a parameter-identification scheme of the vector θ such that, despite the presence of both additive noisy input and structural disturbances, the set of parameters can be estimated, in a non-asymptotic manner, by means of algebraic manipulations.

Remark 4.2.1 *Notice that the same scheme can be extended to a class of nonlinear systems which can be expressed in a linear regression form, as shown in Chapter 3. The nonlinear system can then be written as*

$$P(y_{k-1}, \ldots, y_k, u_{k-n+m}, \ldots, u_{k-n})^T \theta = Q(y_k, \ldots, y_{k-n}, u_{k-n+m}, \ldots, u_{k-n}) \qquad (4.4)$$

where $\theta \in \mathbb{R}^h$, $h \in \mathbb{N}$

4.2.3 An Introductory Example

The purposes and results of this chapter will be better understood by starting with the following elementary first-order SISO system:

$$y_{k+1} + cy_k = au_{k+1} + bu_k + \xi \mathbf{1}_k \tag{4.5}$$

where

- a, b, c are unknown constant parameters,
- $\xi \mathbf{1}_k$ is a constant perturbation of unknown intensity ($\mathbf{1}_k$ is the unit-step Heaviside function).

We do not exclude that system (4.5) might be unstable, i.e., $|c| > 1$, and a non-minimum phase, i.e., $|b/a| > 1$. We only assume that the parameters a, b, c satisfy the controllability/observability condition $ac - b \neq 0$.

Note that $t(\mathbf{1}_k - \mathbf{1}_{k-1}) = 0$ for all $k = 0, \pm 1, \pm 2, \ldots$. Thus, taking a backward time shift of both sides of equation (4.5), subtracting them from equation (4.5), and multiplying both sides by k yields an expression which is free from the perturbation. By taking successive backward time shifts of this resulting equation, we obtain the following linear identifier for the unknown coefficients:

$$P(k, k-1, k-2, k-3, k-4) \begin{bmatrix} a_e \\ b_e \\ c_e \end{bmatrix} = q(k, k-1, k-2, k-3) \tag{4.6}$$

where

$$P(k, k-1, k-2, k-3, k-4)$$
$$= \begin{bmatrix} u_k - u_{k-1} & u_{k-1} - u_{k-2} & y_{k-2} - y_{k-1} \\ u_{k-1} - u_{k-2} & u_{k-2} - u_{k-3} & y_{k-3} - y_{k-2} \\ u_{k-2} - u_{k-3} & u_{k-3} - u_{k-4} & y_{k-4} - y_{k-3} \end{bmatrix}$$

$$q(k, k-1, k-2, k-3) = \begin{bmatrix} y_k - y_{k-1} \\ y_{k-1} - y_{k-2} \\ y_{k-2} - y_{k-3} \end{bmatrix}$$

After the fifth clock pulse has elapsed at $k = 4$, we can use a certainty-equivalent controller, which is designed as if the parameters a, b, and c were already perfectly known, with the system parameter values replaced by the values a_e, b_e, and c_e.

Now, via the identifier (4.6), we have been able to accurately determine the parameters a, b, c of system (4.5), on the basis of measured past values of the input signal u_k and the output signal y_k. The only remaining task is that of proposing a controller which guarantees the tracking of the given output reference signal y_k^*. This will be done by utilizing a flatness-based linear predictive control.[3]

Set, for the nominal system,

$$x_k = \frac{1}{ac - b}(au_k - y_k) \tag{4.7}$$

---

[3] See Fliess and Marquez (2001) and the references cited therein.

The variable x verifies

$$y_k = ax_{k+1} + bx_k \tag{4.8}$$

and

$$u_k = x_{k+1} + cx_k \tag{4.9}$$

Equations (4.7)–(4.9) mean that x is a *flat output* (cf. Fliess and Marquez, 2001). The steering of x_k toward a trajectory x_k^* such that, in turn, the quantity $ax_{k+1}^* + bx_k^*$ actually coincides with the desired output y_k^* yields via equation (4.9) a certainty-equivalence tracking controller where the "integral" action counteracts the constant perturbation:

$$u = c_e x_k + x_{k+1}^* - k_1(x_k - x_k^*) - k_0 \rho_k$$

$$\rho_{k+1} = \rho_k + (y_k - y_k^*)$$

$$k_1 = \frac{(-a_e p^2 + (2p+1)b_e)}{a_e + b_e}, \qquad k_0 = \frac{p^2 + 2p + 1}{a_e + b_e} \tag{4.10}$$

It should be clear that with x_k being a variable satisfying controllable dynamics, the steering of x toward a trajectory x_k^* that exactly corresponds with the desired output reference signal y_k^* would provide us with a feedback controller devoid of internal instability problems, especially in the case of a non-minimum-phase system.

Note that if the system is minimum phase, the nominal dynamics, obtained from (4.8),

$$\zeta_{k+1}^* = -\frac{b_e}{a_e}\zeta_k^* + \frac{1}{a_e}y_k^*, \qquad \zeta^*(0) = 0$$

$$x_k^* = \zeta_k^* \tag{4.11}$$

stably defines the required trajectory for x_k when y_k is replaced by y_k^* and provided the estimated parameters coincide with the actual system parameters. The minimum-phase output case then offers no particular difficulty. Assume for a moment that the system output is non-minimum phase, then (4.11) can no longer be used as a reference trajectory generator for the flat output x, due to its instability.

Consider then the following associated *stable* linear system, evolving backward in time and representing a nominal trajectory generator for the flat output x_k, as obtained also from (4.8), after a simple time transformation:

$$\zeta_{-k-1}^* = -\frac{a_e}{b_e}\zeta_{-k}^* + \frac{1}{b_e}y_{-k-1}^*, \qquad \zeta^*(0) = 0$$

$$x_k^* = \zeta_{-k}^* \tag{4.12}$$

Note that the nominal output asymptotic-tracking property is verified, since the expression $a_e x_{k+1}^* + b_e x_k^*$ is, in such a case, exactly coincident with the desired reference signal y_k^*, provided a_e, b_e, and c_e coincide with the actual system parameter values.

In the absence of *a priori* knowledge of the actual values of a and b, the minimum or non-minimum phase character of the output is clearly undefined. However, any one of the two reference trajectory generators (4.11) or (4.12) will equally serve to obtain an initial sequence of inputs and outputs that allows us to use the parameter computation formula (4.6).

Once the actual values of the unknown parameters are computed from a few time steps of the actual input–output behavior, the appropriate dynamic reference trajectory generator may be suitably decided and the proposed certainty-equivalence PI feedback controller (4.10) can be updated with the computed parameter values. So, even if we initially try to track a "right or wrong trajectory" with the "wrong controller," the two main ingredients of the tracking controller can be rightfully reset in just a few time steps. After the parameter determination is resolved, the right trajectory will be tracked with the right controller.

Figure 4.1 shows the computer simulations of the perturbed closed-loop system response for an unstable, non-minimum-phase case ($a = 1, b = 2, c = 1.2$) whose output is required to track the signal $y_k^* = \sin(-k) = -\sin(k)$. The values of the unknown parameters a, b, c during the first three clock pulses were arbitrarily set to be $a = 0.5$, $b = 1.6$, and $c = 0.8$ in the proposed certainty-equivalence controller. The external perturbation starts at $t = 30$. Note that at $k = 30$, the unknown parameters have already been identified and the "integral control" action is fully operational, thus counteracting the perturbation.

4.2.4   Samuelson's Model of the National Economy

The behavior of the economic variables is important to understand the economic growth of a country. For this, several models have been developed. Samuelson (1939) presented a discrete-time simplified model of the national economic dynamics. This model is based on the

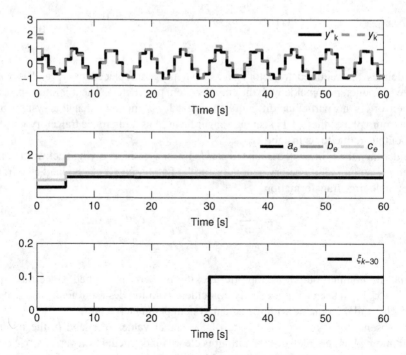

Figure 4.1 Trajectory tracking and parameter identification for unstable non-minimum-phase perturbed linear system

following four economic variables sampled at regular time periods of value k. These periods are usually selected as quarters or years. One considers the following. The national income, denoted Y_k, which represents the total amount earned by all individuals in the economy. The consumption expenditure, C_k, which is the total amount spent by individuals for goods and services. The induced investment, I_k, which is the public and private investment during the period. Finally, the governmental expenditure, G_k, which is the total amount spent by the government in the same period.

Considering the above variables, the national income Y at time k can be written as the sum of three variables:

$$Y_k = G_k + C_k + I_k \tag{4.13}$$

In order to describe the behavior of each variable, some assumptions are taken into account. The first assumption is that the individuals spend a proportional part of the national income of the previous period, that is,

$$C_k = \alpha Y_{k-1} \tag{4.14}$$

where the parameter $0 < \alpha < 1$ is called the *marginal propensity to consume*. For example, if $\alpha = 0.5$, this means that the consumption is equal to half the national income of the previous period. The next assumption is that the induced investment I_k is proportional to the increase in consumption between the previous and the current period. In other words,

$$I_k = \beta[C_k - C_{k-1}] \tag{4.15}$$

where the parameter $\beta > 0$ is called the *growth factor*. Considering equations (4.13)–(4.15), the national income can be rewritten as a second-order linear system as follows:

$$Y_k = \alpha[1 + \beta]Y_{k-1} - \alpha\beta Y_{k-2} + G_k \tag{4.16}$$

The recurrence relation in the income variable is a second-order linear difference equation. In equation (4.16), the government expenditure $G_k = u_k$ can be considered as a control input to the national income Y_k. Nonetheless, when the government expenditure is constant, that is, $G_k = \overline{G}$, equation (4.16) has a unique long-run equilibrium point given by

$$\overline{Y} = \frac{1}{1-\alpha}\overline{G}, \quad \overline{I} = \alpha\overline{Y}, \quad \overline{C} = 0 \tag{4.17}$$

Finally, the second-order linear model of the controlled national income can be written as follows:

$$Y_k = \alpha[1 + \beta]Y_{k-1} - \alpha\beta Y_{k-2} + u_k \tag{4.18}$$

4.2.4.1 Formulation of the Problem

The national income is helpful to assess and compare the progress achieved by a country over a period of time. Assuming the economy is stuck, which is not desirable, in the equilibrium given by equation (4.17), we formulate the following problem:

Assume the constant parameters $\alpha[1 + \beta]$ and $\alpha\beta$ to be unknown. Given the second-order linear system (4.18), devise a feedback control law, u_k, which achieves the tracking of an equilibrium-to-equilibrium national income trajectory, Y_k^*, of the economy during a finite period of time.

4.2.4.2   A Feedback Controller

Assume for a moment that the parameters $\alpha[1 + \beta]$ and $\alpha\beta$ are perfectly known. From equations (4.13)–(4.18), it is easy to see that the flat output is given by the national income $F_k = Y_k$. Therefore, we have the following relations:

$$C_k = \alpha F_{k-1}$$

$$I_k = \alpha\beta(F_{k-1} - F_{k-2})$$

$$Y_k = F_k$$

$$u_k = F_k - \alpha[1 + \beta]F_{k-1} - \alpha\beta F_{k-2}$$

The state-dependent input coordinate transformation

$$u_k = v_k - \alpha[1 + \beta]F_{k-1} - \alpha\beta F_{k-2} \tag{4.19}$$

makes the system (4.18) equivalent to the following system:

$$F_k = v_k \tag{4.20}$$

A feedback controller that achieves the tracking of a desired evolution of the national income F_k^* is

$$v_k = F_k^* - k_2(F_{k-1} - F_{k-1}^*) - k_1(F_{k-2} - F_{k-2}^*) \tag{4.21}$$

and since $u_k^* = F_k^* - \alpha[1 + \beta]F_{k-1}^* + \alpha\beta F_{k-2}^*$ we obtain

$$u_k = u_k^* - (k_2 + \alpha[1 + \beta])(Y_{k-1} - F_{k-1}^*) - (k_1 - \alpha\beta)(Y_{k-2} - F_{k-2}^*) \tag{4.22}$$

The closed-loop behavior of the tracking error $e_k = F_k - F_k^*$ is given by the following linear dynamics:

$$e_{k+2} + k_2 e_{k+1} + k_1 e_k = 0 \tag{4.23}$$

In original variables, the controller is given by

$$u_k = u_k^* - (k_2 + \alpha[1 + \beta])(Y_{k-1} - Y_{k-1}^*) - (k_1 - \alpha\beta)(Y_{k-2} - Y_{k-2}^*) \tag{4.24}$$

where $u_k^* = Y_k^* - \alpha[1 + \beta]Y_{k-1}^* + \alpha\beta Y_{k-2}^*$.

4.2.4.3   An Algebraic Identifier

The developments leading to the identification of the unknown parameters are carried out as follows. First, consider equation (4.18), which can be rewritten as

$$P_k\theta = q_k \tag{4.25}$$

where

$$P_k = \begin{bmatrix} Y_{k-1} & -Y_{k-2} \end{bmatrix}$$

$$q_k = Y_k - u_k$$

$$\theta = \begin{bmatrix} (\alpha[1 + \beta])_e & (\alpha\beta)_e \end{bmatrix}^T$$

In the above expression, we have two unknown parameters $(\alpha[g + \beta])$ and $(\alpha\beta)$. We need to formulate an independent equation to solve for these parameters. We construct the required equation by taking a delay of (4.18). Thus, we obtain the following system:

$$\begin{bmatrix} Y_{k-1} & -Y_{k-2} \\ Y_{k-2} & -Y_{k-3} \end{bmatrix} \begin{bmatrix} (\alpha [1 + \beta])_e \\ (\alpha\beta)_e \end{bmatrix} = \begin{bmatrix} Y_k - u_k \\ Y_{k-1} - u_{k-1} \end{bmatrix} \tag{4.26}$$

An estimate of the parameters $\alpha[g + \beta]$ and $\alpha\beta$ is obtained as follows:

$$\begin{bmatrix} (\alpha [1 + \beta])_e \\ (\alpha\beta)_e \end{bmatrix} = \frac{1}{\Delta} \begin{bmatrix} -Y_{k-3} & Y_{k-2} \\ -Y_{k-2} & Y_{k-1} \end{bmatrix} \begin{bmatrix} Y_k - u_k \\ Y_{k-1} - u_{k-1} \end{bmatrix} \tag{4.27}$$

where $\Delta = [Y_{k-2}]^2 - Y_{k-3}Y_{k-1}$.

4.2.4.4   Simulation Results

The equilibrium achieved, if $u_k = \overline{G} = 10$, is $\overline{Y} = 20$. Then, it is desired to smoothly achieve a new equilibrium for the national income, given by $\overline{Y} = 32$, in 10 periods of time k. For the purpose of simulation, the following parameters were selected: $\alpha = 0.5$, $\beta = 0.7$, $k_1 = 0.34$, and $k_2 = 0.6$, corresponding to the following characteristic polynomial: $p(z) = (z + 0.3)^2 + 0.5^2$. Figure 4.2 shows the equilibrium-to-equilibrium transfer of the national income Y_k during 10 periods of time. The government control expenditure is also shown.

The national induced investment I_k and the national consumption C_k corresponding to the planning horizon are shown in Figure 4.3. Notice that the induced investment remains positive.

Finally, the exact and identified parameters of Samuelson's model of the national economy are shown in Figure 4.4. Note that the identification process converges at the fourth period,

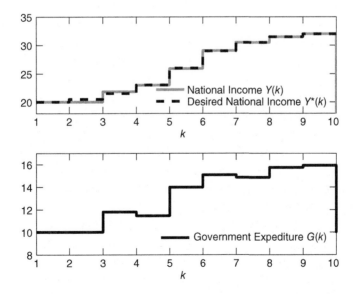

Figure 4.2   Equilibrium-to-equilibrium trajectory of the national income Y and government expenditure $G = u$

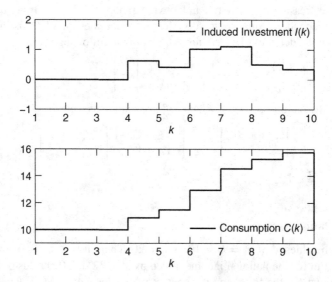

Figure 4.3   National induced investment I and national consumption C

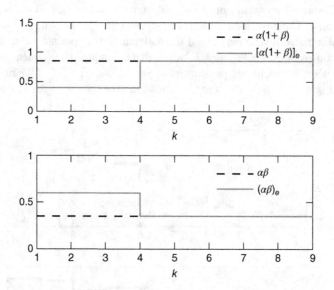

Figure 4.4   Exact and identified parameters of Samuelson's model of the national economy

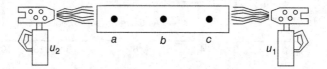

Figure 4.5   Heated slab of material with two boundary controls

which corresponds to the instant at which all the information required for the algebraic identifier becomes available.

4.2.5 Heating of a Slab from Two Boundary Points

According to *Newton's law of cooling*, the dynamic model of the system (Sira-Ramírez and Agrawal, 2004) depicted in Figure 4.5 is described by

$$c_{k+1} = p\, b_k + (1 - 2p)c_k + p\, u_{1,k} \tag{4.28}$$

$$b_{k+1} = p\, a_k + (1 - 2p)b_k + p\, c_k \tag{4.29}$$

$$a_{k+1} = (1 - 2p)a_k + p\, b_k + p\, u_{2,k} \tag{4.30}$$

$$y_{1,k} = c_k$$

$$y_{2,k} = a_k$$

where a_k, b_k, and c_k are the temperatures of points a, b, and c, respectively, at time k. Notice that the b_k temperature monitoring is not available. The control inputs, denoted by $u_{1,k}$ and $u_{2,k}$, represent the external input temperatures. In matrix form, the linear system is written

$$\begin{bmatrix} c_{k+1} \\ b_{k+1} \\ a_{k+1} \end{bmatrix} = \begin{bmatrix} 1 - 2p & p & 0 \\ p & 1 - 2p & p \\ 0 & p & 1 - 2p \end{bmatrix} \begin{bmatrix} c_k \\ b_k \\ a_k \end{bmatrix} + \begin{bmatrix} p & 0 \\ 0 & 0 \\ 0 & p \end{bmatrix} \begin{bmatrix} u_{1,k} \\ u_{2,k} \end{bmatrix}$$

$$\begin{bmatrix} y_{1,k} \\ y_{2,k} \end{bmatrix} = \begin{bmatrix} 1 & 0 & 0 \\ 0 & 0 & 1 \end{bmatrix} \begin{bmatrix} c_k \\ b_k \\ a_k \end{bmatrix} \tag{4.31}$$

4.2.5.1 Formulation of the Problem

Assume the heat transfer coefficient p to be unknown but constant, the temperature b_k not available. Given the linear discrete model (4.31), devise a feedback multi-variable control law $u_{1,k}$, $u_{2,k}$ which achieves tracking of an equilibrium-to-equilibrium temperature trajectory b_k^* in the heated slab of material during a finite period of time.

4.2.5.2 Control Law Design

The flatness of the system is verified thanks to its controllability. The Kalman controllability matrix of (4.31) is given by

$$C_{\mathcal{K}} = [b_1, \quad Ab_1, \quad A^2 b_1, \quad b_2, \quad Ab_2, \quad A^2 b_2]$$

$$= \begin{bmatrix} p & p(1 - 2p) & p(1 - 2p)^2 + p^3 & 0 & 0 & p^3 \\ 0 & p^2 & 2(1 - 2p)p^2 & 0 & p^2 & 2p^2(1 - 2p) \\ 0 & 0 & p^3 & p & p(1 - 2p) & p(1 - 2p)^2 + p^3 \end{bmatrix} \tag{4.32}$$

The Kronecker controllability indices are $\kappa_1 = 2$ and $\kappa_2 = 1$. Indeed, the controllability matrix is just

$$C = [b_1, \quad Ab_1, \quad b_2]$$

$$= \begin{bmatrix} p & p(1-2p) & 0 \\ 0 & p^2 & 0 \\ 0 & 0 & p \end{bmatrix} \tag{4.33}$$

which is non-singular for all positive p.

A set of flat outputs for the linear, time-invariant, controllable MIMO system (4.31) is given by

$$\mathcal{F}_{1,k} = b_k \quad \mathcal{F}_{2,k} = a_k \tag{4.34}$$

Notice that b_k is not available originally as an output of the system; for control purposes, we assume that this temperature is known. Thus, the differential parametrization of the system is

$$c_k = \left(\frac{1}{p} \mathcal{F}_{1,k+1} - (1-2p)\mathcal{F}_{1,k} - p\mathcal{F}_{2,k} \right)$$

$$b_k = \mathcal{F}_{1,k}$$

$$a_k = \mathcal{F}_{2,k}$$

$$u_{2,k} = \left(\frac{1}{p} \mathcal{F}_{2,k+1} - (1-2p)\mathcal{F}_{2,k} - p\mathcal{F}_{1,k} \right)$$

$$u_{1,k} = \frac{1}{p^2} \mathcal{F}_{1,k+2} - 2(1-2p)\mathcal{F}_{1,k+1} + [(1-2p)^2 - p^2]\mathcal{F}_{1,k}$$

$$-p\mathcal{F}_{2,k+1} + (1-2p)p\mathcal{F}_{2,k}) \tag{4.35}$$

Thus, the equilibrium of the system is defined by $\bar{a} = \overline{\mathcal{F}}_2, \bar{b} = \overline{\mathcal{F}}_1, \bar{c} = 2\overline{\mathcal{F}}_1 - \overline{\mathcal{F}}_2, \bar{u}_1 = 3\overline{\mathcal{F}}_1 - 2\overline{\mathcal{F}}_2$, and $\bar{u}_2 = 2\overline{\mathcal{F}}_2 - \overline{\mathcal{F}}_1$. Now, for control purposes, we define

$$\mathcal{F}_{1,k+2} = v_{1,k} \quad \mathcal{F}_{2,k+1} = v_{2,k} \tag{4.36}$$

The following controllers are proposed:

$$u_{2,k} = \frac{1}{p} \left(v_{2,k} - (1-2p)\mathcal{F}_{2,k} - p\mathcal{F}_{1,k} \right) \tag{4.37}$$

$$u_{1,k} = \frac{1}{p^2}(v_{1,k} - 2(1-2p)\mathcal{F}_{1,k+1} + [(1-2p)^2 - p^2]\mathcal{F}_{1,k} \tag{4.38}$$

$$-p\mathcal{F}_{2,k+1} + (1-2p)p\mathcal{F}_{2,k})$$

where

$$v_1(t) = \mathcal{F}_1^*(t+2) - \lambda_1(\mathcal{F}_1(t+1) - \mathcal{F}_1^*(t+1)) \tag{4.39}$$

$$-\lambda_0(\mathcal{F}_1(t) - \mathcal{F}_1^*(t))$$

$$v_2(t) = \mathcal{F}_2^*(t+1) - \kappa_0(\mathcal{F}_2(t) - \mathcal{F}_2^*(t)) \tag{4.40}$$

In terms of the states of the system, we have

$$u_{2,k} = \frac{1}{p}\left(v_{2,k} - (1-2p)a_k - pb_k\right) \tag{4.41}$$

$$u_{1,k} = \frac{1}{p^2}(v_{1,k} - 2p(1-2p)a_k - [(1-2p)^2 + 2p^2]b_k \tag{4.42}$$

$$-2p(1-2p)c_k - p^2 u_{2,k})$$

with

$$v_1(t) = \mathcal{F}^*_{1,k+2} - \lambda_1(pa_k + (1-2p)b_k + pc_k - \mathcal{F}^*_{1,k+1}) \tag{4.43}$$

$$-\lambda_0(b_k - \mathcal{F}^*_1(t))$$

$$v_2(t) = \mathcal{F}^*_{2,k+1} - \kappa_0(a_k - \mathcal{F}^*_{2,k}) \tag{4.44}$$

4.2.6 An Exact Backward Shift Reconstructor

In order to implement the flatness-based controller, an exact backward shift reconstructor of the unmeasured state, b_k, is obtained from the system equations as follows:

$$b_{k+1} = pa_k + (1-2p)b_k + pc_k$$

$$= py_{2,k} + (1-2p)b_k + py_{1,k} \tag{4.45}$$

See Figure 4.6. Taking one backward shift on this last equation and one of the measured variable equations yields

$$b_k = py_{2,k-1} + (1-2p)b_{k-1} + py_{1,k-1}$$

$$y_{1,k} = (1-2p)y_{1,k-1} + pb_{k-1} + pu_{1,k-1}$$

Eliminating b_{k-1} between these two equations yields the exact backward-shifted parametrization of b_k in terms of inputs, outputs, and a finite number of delayed values:

$$\hat{b}_k = py_{2,k-1} + \frac{1-2p}{p}[y_{1,k} - pu_{1,k-1}] + \left[p - \frac{(1-2p)^2}{p}\right]y_{1,k-1} \tag{4.46}$$

4.2.6.1 An Algebraic Estimator for the Heat Transfer Coefficient p

Given that the temperature of the point b is not available to measure, the identifier expression must be written in terms of the temperatures of the points a and c (Figure 4.7). Notice that, if (4.30) is subtracted from (4.28), an expression of this type is obtained:

$$(c_{k+1} - a_{k+1}) = c_k - a_k + p(2(a_k - c_k) + u_{1,k} - u_{2,k})$$

In terms of outputs and a finite number of delayed values:

$$(y_{1,k+1} - y_{2,k+1}) = y_{1,k} - y_{2,k} + p[2(y_{2,k} - y_{1,k}) + u_{1,k} - u_{2,k}]$$

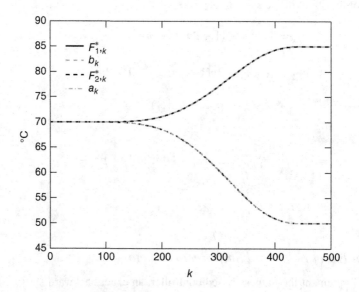

Figure 4.6   Closed-loop behavior of a multi-variable flatness-based control of the heat equation

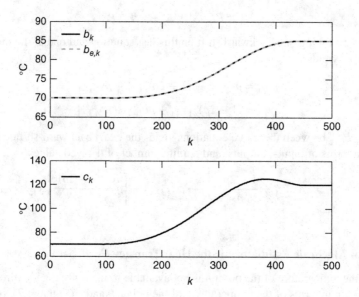

Figure 4.7   Closed-loop estimation of temperature b_k and behavior of temperature c_k

Taking one backward shift on this equation:

$$(y_{1,k} - y_{2,k}) = y_{1,k-1} - y_{2,k-1} + p[2(y_{2,k-1} - y_{1,k-1})$$

$$+ u_{1,k-1} - u_{2,k-1}] \tag{4.47}$$

Thus, the heat transfer coefficient can be estimated via the following formula:

$$p_e = \frac{y_{1,k} - y_{1,k-1} + y_{2,k-1} - y_{2,k}}{2y_{2,k-1} - 2y_{1,k-1} + u_{1,k-1} - u_{2,k-1}} \tag{4.48}$$

4.2.6.2 Simulations Results

We have considered a sample of cubic-shaped aluminum of side $d = 0.1$ m, whose density is $\rho = 2700$ kg/m^3 and specific heat $c = 880$ J/(kg K). When this material is heated to $100\,^{\circ}$C, it exhibits a heat transfer coefficient $p = 0.0053$ (Figure 4.8). The initial conditions for simulation were $a_i = 70.0001\,^{\circ}$C, $b_i = 70.0005\,^{\circ}$C, $c_i = 70\,^{\circ}$C. The control gains $\kappa_0 = 0.003$, $\lambda_1 = 0.01$, and $\lambda_0 = 0.000106$ were chosen from the following desired characteristic polynomials:

$$p_{d,1} = (z + 0.003)$$

$$p_{d,2} = (z^2 + 0.01z + 0.000106)$$

whose roots satisfy $|z_i| < 1$, $i = 1, 2, 3$.

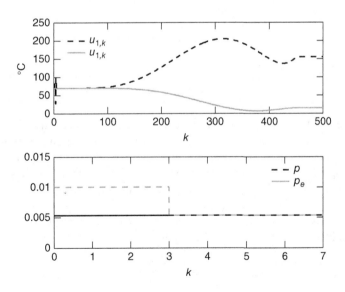

Figure 4.8   Control signals and identification of the heat transfer coefficient p

We set an equilibrium-to-equilibrium transfer for each flat output specified by $y_{1,i} = 70$, $y_{2,i} = 70$ at time $k_i = 50$ and $y_{1,f} = 85$, $y_{2,f} = 50$ at time $k_i = 450$. The desired nominal trajectory is proposed as

$$y_k^* = y_i + (y_f - y_i)\phi(k, k_i, k_f)$$

$$\phi(k, k_i, k_f) = \beta^5(21 - 35\beta + 15\beta^2)$$

$$\beta = \frac{k - k_i}{k_f - k_i}$$

4.3   A Nonlinear Filtering Scheme

Consider the following expression which corresponds to a difference equation with n unknown parameters:

$$\begin{bmatrix} p_{1_k} & p_{2_k} & \cdots & p_{n_k} \end{bmatrix} \begin{bmatrix} \alpha_1 \\ \alpha_2 \\ \vdots \\ \alpha_n \end{bmatrix} = q_k \tag{4.49}$$

where $\alpha_i, i = 1, \ldots, n$ are the parameters to be estimated and $q, p_{i_k}, i = 1, \ldots, n$ are some time functions, respectively. For the sake of simplicity, we use the following vectorial notation:

$$P_k = \begin{bmatrix} p_{1_k} & p_{2_k} & \cdots & p_{n_k} \end{bmatrix} \tag{4.50}$$

$$\theta = \begin{bmatrix} \alpha_1 & \alpha_2 & \cdots & \alpha_n \end{bmatrix}^T \tag{4.51}$$

Equation (4.49) is then expressed as

$$P_k\theta = q_k \tag{4.52}$$

Let us assume that P_k, q_k are affected by an additive zero-mean random signal with unknown statistical parameters. To improve the identification scheme, filtering strategies are a good alternative. In the literature, there is a considerable quantity of digital filtering strategies, from the classical analog-based filters to others based on new developments (De-Freitas and De-Freitas, 2005; Parks and Burrus, 1987). We propose a nonlinear invariant filtering method, and this method will be analyzed in the remainder of this section.

Consider the expression (4.52). Owing to the noisy measurements, there exists an equation error, denoted by ε_k, such that (4.52) takes the form

$$P_k\hat{\theta} = q_k + \varepsilon_k \tag{4.53}$$

where $\hat{\theta}$ is the vector of estimate parameters.

For the minimization of the equation error, we propose the following cost function:

$$J(\hat{\theta}, k) = \frac{1}{2} \sum_{i=0}^{i=k} \varepsilon_i^2 \tag{4.54}$$

Taking the partial derivative with respect to $\hat{\theta}$ in (4.54) to minimize J yields

$$\frac{\partial J}{\partial \hat{\theta}} = \frac{1}{2}\frac{\partial}{\partial \hat{\theta}}\sum_{i=0}^{i=k}\varepsilon^2(iT) = \frac{1}{2}\sum_{i=0}^{i=k}\frac{\partial \varepsilon^2(iT)}{\partial \hat{\theta}} = \sum_{i=0}^{i=k}P_k^T(P_k\hat{\theta} - q_k)) \qquad (4.55)$$

Given that $J(\hat{\alpha}, k)$ is a convex function (Polyak, 1987), it has a global minimum satisfying $\partial J(\hat{\theta}, k)/\partial \hat{\theta} = 0$ for all k. Thus, $\hat{\theta}$ is expressed as

$$\hat{\theta}_k = \left[\sum_{i=0}^{i=k}P_k^T P_k\right]^{-1}\sum_{i=0}^{i=k}P_k^T q_k \qquad (4.56)$$

Remark 4.3.1 *Since there may be a lack of complete measurements for the first sampling times, the above formula is valid once the required measure terms of the identification formula are available. Then, it is necessary to wait this amount of time to start the filtering process.*

Remark 4.3.2 *The proposed nonlinear filtering can be modified, without affecting its invariance property, by means of some nested additional summations, simultaneously applied in both numerator and denominator signals (nested averaging filtering). This can be viewed as an extra low-pass filtering process in order to increase the filtering effect. This extra filtering may be helpful in very noisy environments, achieving better estimations. However, this procedure may produce a short delay in the time estimation. For instance, nonlinear filtering with an additional extra summation becomes*

$$\hat{\theta}_k = \left[\sum_{i=0}^{i=k}\sum_{j=0}^{j=i}P_j^T P_j\right]^{-1}\sum_{i=0}^{i=k}\sum_{j=0}^{j=i}P_j^T q_j \qquad (4.57)$$

4.3.1 Hénon System

This two-dimensional system, introduced by Hénon (1976), is one of the most important examples of chaotic discrete-time mappings. This system has been applied in a wide variety of applications, from secure communications to heart dynamic modeling (Fradkov and Pogromsky, 1998):

$$x_{k+1} = a - x_k^2 + by_k$$

$$y_{k+1} = x_k \qquad (4.58)$$

where a, b are constant unknown parameters and x_k is measurable.

4.3.1.1 Parameter Identification

The system can be rewritten as the following difference equation:

$$x_k = a - x_{k-1}^2 + bx_{k-2} \qquad (4.59)$$

The last equation is linear with respect to its parameters, so to identify a and b it is necessary to obtain another linearly independent equation. Since the system to identify is oscillatory, an alternative to the second linearly independent equation is the time shift of (4.59). We have

$$\begin{bmatrix} 1 & x_{k-2} \\ 1 & x_{k-3} \end{bmatrix} \begin{bmatrix} a \\ b \end{bmatrix} = \begin{bmatrix} x_k + x_{k-1}^2 \\ x_{k-1} + x_{k-2}^2 \end{bmatrix} \tag{4.60}$$

where the estimates of a and b are obtained as

$$\hat{a}_k = \frac{(x_{k-3})(x_k + x_{k-1}^2) - (x_{k-2})(x_{k-1} + x_{k-2}^2)}{x_{k-3} - x_{k-2}}$$

$$\hat{b}_k = \frac{(x_{k-1} + x_{k-2}^2) - (x_k + x_{k-2}^2)}{x_{k-3} - x_{k-1}} \tag{4.61}$$

4.3.1.2 Numerical Results

A numerical simulation of the identification method was performed to verify the numerical response of the method. In this example, we simulated the Hénon system for the parameters $a = 1.4$, $b = 0.3$. The initial conditions for the system were set at $x_0 = 0.5$, $y_0 = 1$. Figure 4.9 depicts the output of the system and the identification process for each parameter in which there is an exact estimation when the algebraic system achieves complete measurement of the algebraic system (4.61).

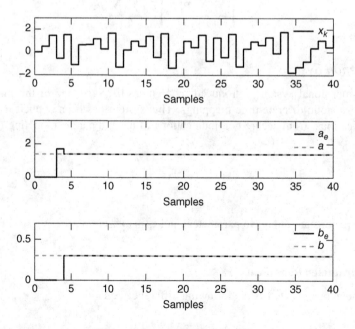

Figure 4.9   Parameter estimation

4.3.1.3   A nonlinear Filtering Approach

Consider equation (4.59). Applying a time shift and subtracting, a is eliminated from the expression resulting in

$$x_k - x_{k-1} + x_{k-2}^2 - x_{k-3}^2 = b(x_{k-3} - x_{k-4}) \tag{4.62}$$

The last equation has the form

$$p_k b = q_k \tag{4.63}$$

where $q_k = x_k - x_{k-1} + x_{k-2}^2 - x_{k-3}^2$, $p_k = x_{k-3} - x_{k-4}$. Applying the scalar instance of the proposed nonlinear filter, the estimation of b is given as follows:

$$\hat{b}_k = \frac{\displaystyle\sum_{i=0}^{i=k} q_i p_i}{\displaystyle\sum_{i=0}^{i=k} p_i^2} \tag{4.64}$$

and finally, the parameter a is estimated as follows:

$$\hat{a}_k = x_k + x_{k-2}^2 + \hat{b}_k x_{k-3} \tag{4.65}$$

4.3.1.4   Numerical Results

A noisy measurement of the Hénon map with zero mean and variance 0.1 was introduced to perform the corresponding algebraic method with nonlinear filtering. The same parameters and initial conditions were employed for this numerical simulation. Figure 4.10 shows the output

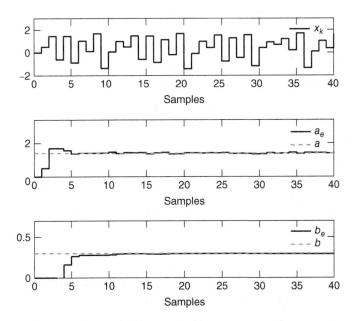

Figure 4.10   Parameter estimation for noisy measurements

and the identification process; in this case, the estimation is not exact due to the noise in the output. However, the filtering scheme allows good convergence in a short amount of time.

4.3.2 A Hard Disk Drive

In this example, we take the problem of identification over a sampled continuous-time system, in which the system has a digital control scheme. Since controlling servo systems requires speed and accuracy, this is a good example to carry out a fast identification scheme. Let us consider the mathematical model of a commercially available hard disk drive with a voice-coil motor actuator (see Venkataramanan *et al.*, 2003), whose continuous transfer function is given by

$$\frac{y(s)}{u(s)} = \frac{a}{s^2} \tag{4.66}$$

where $a = K_t/J_a s^2$, K_t is the torque constant, J_a is the moment of inertia of the actuator mass, u is the voltage input of the actuator, and y represents the position of the R/W head.

The discretized model of (4.66) obtained with a sampling time T is

$$\begin{cases} x_{k+1} = \begin{bmatrix} 1 & T \\ 0 & 1 \end{bmatrix} x_k + \begin{bmatrix} b_1 \\ b_0 \end{bmatrix} u_k \\ y_k = \begin{bmatrix} 1 & 0 \end{bmatrix} x_k \end{cases} \tag{4.67}$$

This state-space model can be represented as the following input–output system:

$$y_{k+2} - 2y_{k+1} + y_k = b_1 u_{k+1} + (Tb_0 - b_1)u_k. \tag{4.68}$$

The problem statement is: From the knowledge of u_k, y_k, find the constant parameter values of b_0, b_1 and then design a control law to track a specific trajectory y_k^*.

4.3.2.1 Parameter Identification

Let us denote the term $Tb_0 - b_1$ as b_3, so that equation (4.68) becomes

$$y_{k+2} - 2y_{k+1} + y_k = b_1 u_{k+1} + b_3 u_k, \tag{4.69}$$

which in causal form is expressed as

$$y_k - 2y_{k-1} + y_{k-2} = b_1 u_{k-1} + b_3 u_{k-2}. \tag{4.70}$$

It is necessary to derive another equation to obtain the unknown parameters. Applying a single time-shift operator to (4.70), we have

$$y_{k-1} - 2y_{k-2} + y_{k-3} = b_1 u_{k-2} + b_3 u_{k-3}. \tag{4.71}$$

From equations (4.70) and (4.71), the parameter estimated values $\hat{b}_1$, $\hat{b}_3$ are calculated as follows:

$$\begin{bmatrix} \hat{b}_1 \\ \hat{b}_3 \end{bmatrix} = \begin{bmatrix} u_{k-1} & u_{k-2} \\ u_{k-2} & u_{k-3} \end{bmatrix}^{-1} \begin{bmatrix} y_k - 2y_{k-1} + y_{k-2} \\ y_{k-1} - 2y_{k-2} + y_{k-3} \end{bmatrix} \tag{4.72}$$

where, after at most three sampling times, the correct parameter estimation values are available.

Once the plant values are identified, it is possible to design a control law to track any admissible trajectory. Next lines will describe the control designing procedure.

4.3.2.2 GPI Control Design

The transfer function of (4.70) in the z domain is given by

$$\frac{y(z)}{u(z)} = \frac{b_1 z^{-1} + b_3 z^{-2}}{1 - 2z^{-1} + z^{-2}} \tag{4.73}$$

In this case, the numerator and denominator polynomials are coprime and the system is controllable; the system is flat and the flat output $x(z)$ is given by

$$x(z) = \frac{1}{1 - 2z^{-1} + z^{-2}} u(z)$$

$$y(z) = (b_1 z^{-1} + b_3 z^{-2}) x(z)$$

The coprimeness of the numerator and denominator implies that there exist two polynomials $p(z), r(z)$ such that

$$x(z) = r(z)y(z) + p(z)y(z)$$

By solving Bezout's identity, $r(z), p(z)$ have the form $r(z) = r_0 + r_1 z^{-1}$, $p(z) = p_0 + p_1 z^{-1}$, where $p_0 = 1$ and r_0, r_1, p_1 satisfy the following equation:

$$\begin{bmatrix} b_1 & 0 & 1 \\ b_3 & b_1 & -2 \\ 0 & b_3 & 1 \end{bmatrix} \begin{bmatrix} r_0 \\ r_1 \\ p_1 \end{bmatrix} = \begin{bmatrix} 2 \\ -1 \\ 0 \end{bmatrix}$$

Suppose it is desired to track a given reference of the flat output signal x_k^*. u_k^* is denoted as an input, which in unperturbed system conditions generates x_k^*. In other words:

$$u_k^* = x_k^* - 2x_{k-1}^* + x_{k-2}^* \tag{4.74}$$

The tracking error is given by

$$e_{x_k} \Delta = x_k - x_k^*$$

A control law is given by

$$u_k = u_k^* - k_1(x_{k-1} - x_{k-1}^*) - k_0(x_{k-2} - x_{k-2}^*) \tag{4.75}$$

and the closed-loop tracking error dynamics is

$$e_{x_k} + (k_1 - 2)e_{x_{k-1}} + (k_0 + 1)e_{x_{k-2}} = 0 \tag{4.76}$$

To obtain an asymptotically stable tracking error, the controller parameters can be tuned such that the characteristic equation $e_x(z)(z^2 + (k_1 - 2)z + k_0 + 1) = 0$ has its roots inside the unit circle. By setting $z^2 + (k_1 - 2)z + k_0 + 1$ to match a desired monic stable polynomial $P_d(z) = (z + p_1)(z + p_0)$, $|p_1| < 1$, $|p_0| < 1$, the correct tracking is achieved.

4.3.2.3 Simulations

A numerical simulation was made to show the identification and controller processes. The parameters and sampling ratio were taken from Venkataramanan *et al.* (2003). In this case, we have the following plant parameters: $b_0 = 6401.3$, $b_1 = 0.3201$, $b_3 = 0.32003$, $T = 0.1$ ms. The trajectory to follow is a rest-to-rest transition sequence between the origin and 1 for the flat output. In a stable state, the following statement is valid. For a stationary flat output $\bar{x}$, the stationary output $\bar{y}$ is given by $\bar{y} = (b_1 + b_3)\bar{x}$. Then, the transition of y_k is given from zero to 0.64013. Since the parameter values can not be computed for $k < 4$, the arbitrary values of $\hat{b}_1 = 0.27$ and $\hat{b}_3 = 0.29$ were taken. The identifier has a lack of persistency after a few times, however, this time it is sufficient to know the parameter estimation values to be used in the control scheme. Figure 4.11 depicts the tracking process of the flat output and the system output y_k, and Figure 4.12 shows the identified values for this process.

4.3.3 The Visual Servo Tracking Problem

4.3.3.1 Modeling

Consider a fixed camera whose line-of-sight axis is surveying a perpendicular plane where a 2-DOF planar manipulator, characterized by the joint angular coordinates $q = (q_1, q_2)$, can move freely about its two-dimensional workspace (Sira-Ramírez and Fliess, 2002). Under the standard simplifying assumption that a sufficiently fast velocity feedback loop regulates the manipulator joint velocity coordinates $\dot{q}$ (see Hsu *et al.*, 2001), the underlying problem is that of generating such joint angular velocity reference signals in correspondence with the end effector trajectory-tracking task, as defined only in the visual camera coordinates.

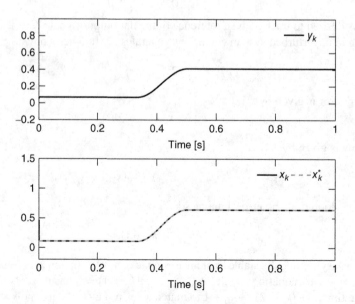

Figure 4.11   Trajectory tracking

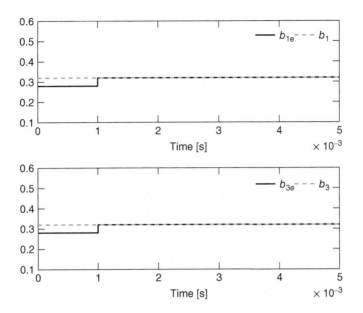

Figure 4.12 Identified parameters

Let $\beta < 0$ be the camera scaling factor of unit length in the image plane, assumed to be unknown and rather large in absolute value (typically -10^3 to -10^5 pixels/m). Let θ denote the unknown angular orientation of the camera with respect to the frame of reference of the manipulator system. We assume that $-\pi < \theta < \pi$. Let y_1, y_2 be the position coordinates of the end effector in the two-dimensional planar image space. The map from the joint robot position vector, q, to the visual frame coordinates, $y = (y_1 \; y_2)$, is given by

$$\begin{pmatrix} y_1 \\ y_2 \end{pmatrix} = \beta \frac{f}{f - Z} \begin{pmatrix} \cos\theta & -\sin\theta \\ \sin\theta & \cos\theta \end{pmatrix} (\kappa(q) - f_0) + c_0 \tag{4.77}$$

where $\kappa(q) \in \mathbf{R}^2$ is the robot *direct kinematics map*. The parameter f is the known camera focal length and Z is the, also known, total distance from the robot workspace to the image plane. The quantities f_0 and c_0 are constant vectors related to center offset and image center values in the workspace coordinates.

Given that the camera data acquisition framework is digitally based, consider then T to be a fixed, small, sampling time. If we assume that y is measured at time intervals of the form kT, $k = 0, 1, 2$, yielding $y_{kT} = y_k = y_{1_k}, \quad y_{2_k}$ then, from (4.77), we obtain the following updating formula relating the values of y at times kT and $(k+1)T$:

$$\begin{pmatrix} y_{1_{k+1}} \\ y_{2_{k+1}} \end{pmatrix} = \begin{pmatrix} y_{1_k} \\ y_{2_k} \end{pmatrix} + \beta \frac{f}{f - Z} \begin{pmatrix} \cos\theta & -\sin\theta \\ \sin\theta & \cos\theta \end{pmatrix} (\kappa(q_{k+1}) - \kappa(q_k)) \tag{4.78}$$

where q_k represents the sampled vector q_{kT}, that is, the joint position at time kT. Using the approximate expansion,

$$\kappa(q_{k+1}) = \kappa(q_{k+1}) + \frac{\partial \kappa(q)}{\partial q}\Big|_{q=q_k}(q_{k+1} - q_k) \tag{4.79}$$

and from the assumption that the control input variable τ will be provided in a piecewise constant, sampled, fashion to the robot velocity controller, that is, $\tau_{kT} = \tau_k$, we obtain the discrete simplified robot dynamics

$$q_{k+1} = q_k + T\tau_k \tag{4.80}$$

Defining the control input vector $u_{kT} = u_k = (u_{1_k}\ u_{2_k})$ as

$$\begin{pmatrix} u_{1_k} \\ u_{2_k} \end{pmatrix} = \frac{\partial k(q)}{\partial q}\Big|_{q=q_k}\tau_k = J(q_k)\tau_k \tag{4.81}$$

where $J(\cdot)$ is known as the analytic robot Jacobian matrix, we obtain, from the substitution of (4.79)–(4.81) into (4.78), the following unperturbed discrete-time visual flow dynamics:

$$\begin{pmatrix} y_{1_{k+1}} \\ y_{2_{k+1}} \end{pmatrix} = \begin{pmatrix} y_{1_k} \\ y_{2_k} \end{pmatrix} + aT\begin{pmatrix} \cos\theta & -\sin\theta \\ \sin\theta & \cos\theta \end{pmatrix}\begin{pmatrix} u_{1_k} \\ u_{2_k} \end{pmatrix} \tag{4.82}$$

where $a = \beta f/(f - Z)$ is a positive, yet uncertain, parameter.

By setting $\eta_k = y_{1_k} + jy_{2_k}$, $v_k = u_{1_k} + ju_{2_k}$, $j = \sqrt{-1}$, equation (4.82) has a simple form with complex numbers:

$$\eta_{k+1} = \eta_k + Tae^{j\theta}v_k \tag{4.83}$$

We further hypothesize the presence of a constant, uncertain, set of bias error inputs described by the complex quantity $\zeta = \zeta_1 + j\zeta_2$, affecting the visual servo dynamics (4.83) from time $k = 0$ on

$$\eta_{k+1} = \eta_k + Tae^{j\theta}v_k + \zeta 1_k \tag{4.84}$$

We refer to model (4.84) as the *perturbed visual flow dynamics*.

4.3.3.2 Problem Formulation

Given the discrete-time complex-valued perturbed visual flow dynamics (4.84) and a desired reference trajectory, η_k^*, find a complex feedback control law for v_k such that $\eta_k \to \eta_k^*$ asymptotically, regardless of the unknown, but constant, values of the real parameters a, θ and the constant value of the complex bias perturbation input ζ.

4.3.3.3 A Certainty Equivalence PI Controller

Set $\xi = ae^{j\theta}$. Thus, $a = |\xi|$, $\theta = \arg(\xi)$. Suppose, just for a moment, that the uncertain parameter ξ is known. Then, a global trajectory-tracking controller of PI type, overcoming the constant perturbation input ζ, is given by

$$v_k = \frac{1}{T\xi}[-\eta_k + \eta_{k+1}^* - k_1(\eta_k - \eta_k^*) - k_0\rho_k]$$

$$\rho_{k+1} = \rho_k + (\eta_k - \eta_k^*) \tag{4.85}$$

The closed-loop complex-valued tracking-error dynamics is given by the evolution of the quantity $e_k = \eta_k - \eta_k^*$. This is clearly given by

$$e_{k+1} + k_1 e_k + k_0 \rho_k = \zeta$$

$$\rho_{k+1} = \rho_k + e_k \tag{4.86}$$

Eliminating the variable ρ, we obtain the perturbation-free closed-loop dynamics

$$e_{k+2} + (k_1 - 1)e_{k+1} + (k_0 - k_1)e_k = 0 \tag{4.87}$$

which can be made to globally and asymptotically approach the origin of the complex line for a suitable set of, possibly complex, design coefficients k_1 and k_0. We may nevertheless take, for simplicity, k_1 and k_0 to be real coefficients.

4.3.3.4 An Algebraic Identifier for the Unknown Calibration Parameters

Rewrite the perturbed visual flow dynamics $\eta_{k+1} = \eta_k + T\xi v_k + \zeta \mathbf{1}_k$ as $\delta^{-1}\eta = \eta + T\xi v + \zeta \mathbf{1}$. Multiplying both sides by $k(1 - \delta)$, we obtain an expression which is free from the perturbation:

$$(1 - \delta)\delta^{-1}\eta_k = (1 - \delta)\eta_k + T\xi(1 - \delta)v_k$$

We readily have the exact value of ξ after multiplying by δ and rewriting the resulting expression in terms of past values of y and u:

$$\xi = \frac{-\eta_k + 2\eta_{k-1} - \eta_{k-2}}{T(v_{k-2} - v_{k-1})} \tag{4.88}$$

The singularity condition $v_{k-1} = v_{k-2}$ is related to a lack of persistency.

4.3.3.5 A Combined Online Identifier–Controller for Visual Servoing

The online parameter identifier is only well defined for $k \geq 2$. Before this instant, the parameter ξ is undetermined and an arbitrary reasonable nonzero estimated value should be used in the certainty-equivalence controller. Once the complex parameter ξ is obtained, by accurately evaluating the previous expression, we use such an online computed value in the proposed PI certainty-equivalence controller (4.89). We next summarize the algebraic identifier–controller with use of the notation $\hat{\xi}$ for the computed value of the unknown parameter ξ:

$$v_k = \frac{1}{T\hat{\xi}}[-\eta_k + \eta_{k+1}^* - k_1(\eta_k - \eta_k^*) - k_0\rho_k]$$

$$\rho_{k+1} = \rho_k + (\eta_k - \eta_k^*)$$

$$\hat{\xi} = \begin{cases} \text{arbitrary nonzero} \\ \quad \text{for } k = 0, 1, 2 \\ \frac{-\eta_k + 2\eta_{k-1} - \eta_{k-2}}{T(-v_{k-1} + v_{k-2})} \\ \quad \text{for } k \geq 3 \end{cases} \tag{4.89}$$

Remark 4.3.3 *Note that the constant value of $\hat{\xi}$ needs to be computed only* once. *Thus, after an accurate computation of the parameter $\hat{\xi}$ has been obtained, we may opt to "switch off" the identifier, say, right after time $k = 3$, and indefinitely use this computed value of the parameter ξ in the controller. However, in many other applications, the uncertain parameter ξ may suddenly change to a new constant value. In such cases, the algebraic identifier may then be used in a repetitive fashion by reinitiating the computations with a new "initial time." In this manner, the certainty-equivalence controller can be properly updated.*

4.3.3.6   Simulation Results

We want to track a 3-leaved rose trajectory given by

$$y^*_{1_k} = \cos(3kT)\cos(kT), \ y^*_{2_k} = \cos(3kT)\sin(kT)$$

which can also be briefly expressed as $\eta^*_k = \cos(3kT)e^{jkT}$. The unknown orientation angle for the camera is $\theta = \pi/2$ rad and the unknown scaling parameter was set to $a = 10^4$. Figure 4.13 shows the performance of the bias-perturbed system controlled variables under the action of the controller–identifier (4.89). Figure 4.13 also shows the unknown bias perturbation inputs to the visual flow system ($\zeta_1 = -0.15$, $\zeta_2 = 0.1$). The true values of the unknown calibration parameters a and θ are determined and used right after the second clock pulse (i.e., from $k = 3$ onwards). The sampling interval T was set to be $T = 0.015$ s. The values of a and θ, used in the controller identifier (4.89) during the identification phase, were arbitrarily set to $a_e = 9000$ and $\theta_e = 1.35$. The controller design coefficients k_1 and k_0 were chosen to be real, so that the closed-loop characteristic polynomial of the tracking error was of the form $z + 0.8z + 0.5825$.

Naturally, we also tested the robustness of the proposed identifier–controller scheme by introducing an unmodeled complex perturbation input to the visual flow dynamics (4.83) as in the following *stochastically perturbed* model:

$$\eta_{k+1} = \eta_k + Tae^{j\theta}v_k + (\zeta + v_k)\mathbf{1}_k \tag{4.90}$$

where $v_k = v_{1_k} + jv_{2_k}$ is a complex stochastic input signal, whose real and imaginary components we simulated by means of a computer-generated zero-mean random process of amplitude 1e−4, approximately. The corresponding performance of the identifier–controller, on the system variables, is shown in the responses depicted in Figure 4.14.

4.3.4   A Shape Control Problem in a Rolling Mill

4.3.4.1   Modeling

Consider the following discrete-time linearized version of a cooling spray system controlling the non-uniform thermal expansion, or thermal camber, across the work roll in a thickness vector "shape" control task for a cold-rolling mill device (see Goodwin *et al.*, 2001). If we let $y_k \in \mathbf{R}^m$ denote the vector of roll thickness measurements and u_k denote the corresponding vector of spray valve positions, we have the following linear multi-variable model of the system:

$$y_{k+1} = by_k + (1 - b)M(\alpha)u_k + \xi\mathbf{1}_k \tag{4.91}$$

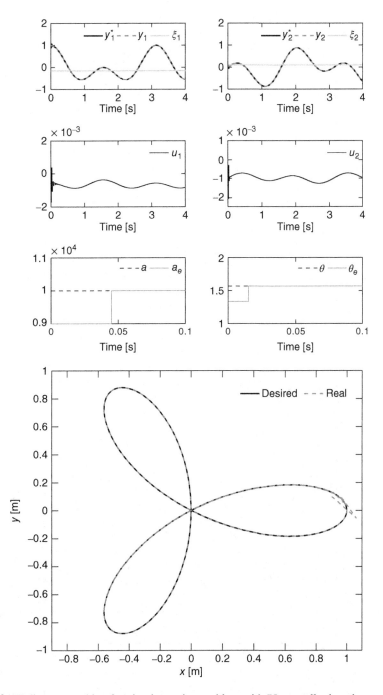

Figure 4.13 Trajectory tracking for visual servoing problem with PI controller based on an algebraic identifier

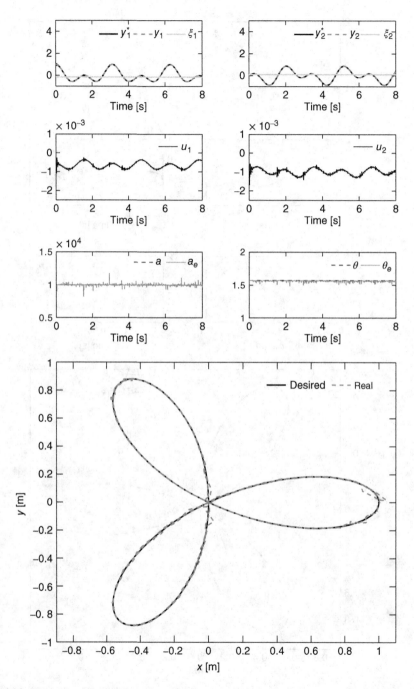

Figure 4.14 Trajectory tracking for stochastically perturbed visual servoing problem with PI controller based on an algebraic identifier

where $b = e^{-T/\tau}$, with $T > 0$ being the sampling period and τ the known, common, scalar time constant of the valves; ξ is a constant, unknown, perturbation input and $M(\alpha)$ is an $m \times m$ matrix of the form

$$
\begin{pmatrix}
1 & \alpha & \alpha^2 & \alpha^3 & \cdots & \\
\alpha & 1 & \alpha & \alpha^2 & \cdots & \\
\alpha^2 & \alpha & 1 & \alpha & \cdots & \\
\vdots & & & \cdots & & \vdots \\
\vdots & & & \alpha & 1 & \alpha \\
& & \cdots & & \alpha & 1
\end{pmatrix}
$$

The *interactivity parameter* α of the spray arrangement, which is assumed to be an unknown constant lying in the interval $(0, 1)$, is clearly weakly linearly identifiable.

4.3.4.2   A Certainty-Equivalence Controller

A certainty-equivalence controller, including "integral action," for the shape control system described above, which drives the thickness vector y_k to zero, is given by

$$
u_k = \frac{1}{1-b} M^{-1}(\alpha)[(\lambda_1 - b)y_k + \lambda_0 \rho_k]
$$

$$
\rho_{k+1} = \rho_k + y_k \tag{4.92}
$$

where λ_1 and λ_0 are scalar parameters such that the following second-order characteristic polynomial $p(z) = z^2 - (1 + \lambda_1)z - (\lambda_0 - \lambda_1)$, describing the set of identical closed-loop decoupled dynamics for the thickness vector components, has all its roots strictly inside the unit circle in the complex plane. Note that λ_1 and λ_0 may also be prescribed as appropriate diagonal matrices with different entries.

We particularized the above-described rolling mill model for the case of nine sprays. Figure 4.15 depicts the closed-loop response of the perturbed system to the proposed controller when the interactivity parameter is known to be $\alpha = 0.5$. The sampling period was set to the value $T = 0.25$, and the time constant of the process τ was set to $\tau = 4\,\mathrm{s}^{-1}$. A constant-step perturbation input ξ, of amplitude -0.05, appears at time $kT = 5$ s. The controller parameters were chosen to be $\lambda_1 = 0.3$ and $\lambda_0 = -0.1225$.

4.3.4.3   An Online Interactivity Parameter Identification Scheme

Rewrite equation (4.91) as $\delta^{-1}y = by + (1-b)M(\alpha)u + \xi \mathbf{1}$. Multiplying out the previous expression by $k(1-\delta)$ (where δ is the time-shift operator), simplifying, and multiplying out by δ we have

$$
(\delta - 1)y_k = b(\delta^2 - \delta)y_k + (1-b)M(\alpha)(\delta^2 - \delta)u_k
$$

Setting $\alpha^j = a_j$, for $j = 1, \ldots, m-1$, the previous vector relation yields a system of $m-1$ linear equations in the $m-1$ unknowns, $a = (a_1, \ldots, a_{m-1})^T$, which is of the form

$$
P(u_{k-1}, u_{k-2})a = q(y_k, y_{k-1}, y_{k-2}, u_{k-1}, u_{k-2}) \tag{4.93}
$$

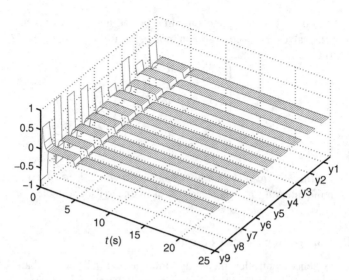

Figure 4.15 Closed-loop response of perturbed shape control model with known interactivity parameter

If the system starts at time $k = 0$, the matrix $P(\cdot)$ becomes full rank at the third clock pulse, $k = 2$. We uniquely determine the value of a at this time instant and, hence, all the components of the matrix $M(\alpha)$ are completely determined for $k \geq 3$. The certainty-equivalent controller (4.92) may then be updated with the computed matrix value, $M(\alpha)$. The right value of the interactivity parameter α is used in the certainty-equivalence controller (4.92), for all times after $k = 3$, as follows:

$$u_k = \frac{1}{1 - b} M^{-1}(\hat{\alpha})((\lambda_1 - b)y_k + \lambda_0 \rho_k) \tag{4.94}$$
$$\rho_{k+1} = \rho_k + y_k$$

$$\hat{\alpha} = \begin{cases} \text{arbitrary nonzero} \\ \quad \text{for} \quad t = 0, 1, 2 \\ (a_j)^{\frac{1}{j}} \text{for any} j \in \{1, \dots, m - 1\} \\ \quad \text{for} \quad t \geq 3 \end{cases}$$

We used the same perturbed model and controller parameters of the cold rolling mill process found in the previous simulation. Figure 4.16 depicts the closed-loop response of the perturbed system to the proposed identifier–controller scheme (4.93)–(4.94). The unknown interactivity parameter α was arbitrarily set to be $\alpha = 0.3$ for the time instants $k = 0, 1, 2$, while the identifier produced the right value of the plant parameter, $\alpha = 0.5$. The transient responses of the two simulation runs are evidently different at the beginning of the process. They already have an identical closed-loop behavior, with the same recovery features, by the time the unknown perturbation step appears.

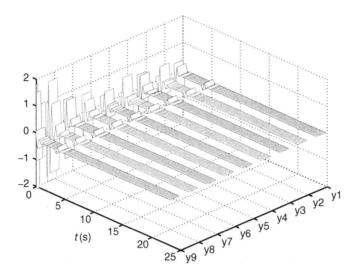

Figure 4.16 Closed-loop response of perturbed shape control model with online identification of interactivity parameter

4.3.5 Algebraic Frequency Identification of a Sinusoidal Signal by Means of Exact Discretization

Consider a biased sinusoidal signal with an exact discretization process

$$x_k = A\sin(\omega kT + \phi) + b \tag{4.95}$$

where $A \in \mathbb{R}^+$ is the amplitude, $\phi \in \mathbb{R}$ is a constant phase component, $T \in \mathbb{R}^+$ is the sampling time, and $\omega \in \mathbb{R}^+$ is the frequency component. We have

$$
\begin{aligned}
x_{k+1} &= A\sin(\omega kT + \phi + \omega T) + b \\
&= A\sin(\omega kT + \phi)\cos(\omega T) + A\cos(\omega kT + \phi)\sin(\omega T)) + b
\end{aligned}
\tag{4.96}
$$

$$
\begin{aligned}
x_{k+2} &= A\sin(\omega(kT + T) + \phi + \omega T) + b \\
&= A\sin(\omega(kT + T) + \phi)\cos(\omega T) + A\cos(\omega(kT + T) + \phi)\sin(\omega T) + b
\end{aligned}
\tag{4.97}
$$

From the last equations we have

$$x_{k+1} - \cos(\omega T)x_k = b(1 - \cos(\omega T)) + A\cos(\omega kT + \phi)\sin(\omega T) \tag{4.98}$$

Taking a time-shift operation and applying some trigonometric identities:

$$
\begin{aligned}
x_{k+2} - \cos(\omega T)x_{k+1} = b(1 - \cos(\omega T)) - A\sin(\omega kT + \phi)\sin^2(\omega T) \\
+ A\cos(\omega kT + \phi)\cos(\omega T)\sin(\omega T)
\end{aligned}
\tag{4.99}
$$

Multiplying (4.98) by $\cos(\omega T)$ and subtracting it from (4.99):

$$x_{k+2} - 2\cos(\omega T)x_{k+1} + \cos^2(\omega T)x_k =$$
$$- A\sin(\omega kT + \phi)\sin^2(\omega T) + b(1 - \cos(\omega T))^2 \quad (4.100)$$

In contrast:

$$-A\sin(\omega kT + \phi)\sin^2(\omega T) = -\sin^2(\omega T)x(kT) + b\sin^2(\omega T)$$

Using this fact in (4.100), we have

$$x_{k+2} - 2\cos(\omega T)x_{k+1} + x_k = 2b(1 - \cos(\omega T)) \quad (4.101)$$

Applying a time-shift operation in (4.101), subtracting it, and multiplying out by the time term k to eliminate the constant term, we have

$$k(x_{k+3} - x_{k+2} + x_{k+1} - x_k - 2\cos(\omega T)[x_{k+2} - x_{k+1}]) = 0 \quad (4.102)$$

or in causal terms

$$k(x_k - x_{k-1} + x_{k-2} - x_{k-3} - 2\cos(\omega T)[x_{k-1} - x_{k-2}]) = 0 \quad (4.103)$$

The frequency can then be estimated as follows:

$$\cos(\omega T)_k = \frac{k}{2k}\left(\frac{x_k - x_{k-1} + x_{k-2} - x_{k-3}}{x_{k-1} - x_{k-2}}\right) = \frac{n_k}{d_k} \quad (4.104)$$

that is,

$$\omega_k = \frac{1}{T}\arccos\left[\frac{k}{2k}\left(\frac{x_k - x_{k-1} + x_{k-2} - x_{k-3}}{x_{k-1} - x_{k-2}}\right)\right] \quad (4.105)$$

Therefore, an exact frequency estimation of the sinusoidal signal can be obtained after three sampling times.

4.3.5.1 Numerical Results

A sinusoidal wave with an amplitude of 20 V, a bias component of 5 V, and a frequency of $2\pi f$ rad/s, $f = 1$ Hz, was taken to carry out the proposed identification procedure. The sampling period was set to be $T = 0.0625$ s (16 samples per period) and there was a measurement additive noise whose SNR was set to 30 dB. The nonlinear invariant filtering was defined by the scheme proposed in Section 4.3, that is

$$\hat{\omega}_k = \frac{\sum_{i=0}^{i=k} n_i d_i}{\sum_{i=0}^{i=k} d_i^2} \quad (4.106)$$

Figure 4.17 shows the signal used and its frequency estimation. Notice that the desired identification value is reached as soon as all the data values are available with good accuracy.

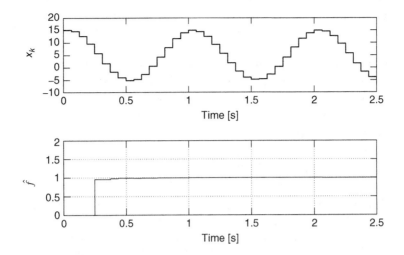

Figure 4.17 Frequency estimation

4.3.5.2 Experimental Results

Finally, to know the behavior of the algebraic approach in an experimental framework, we tested a sinusoidal wave obtained by an oscilloscope and given by

$$x_k = 4 \sin(2\pi f k T + 0.9) + 2 + \xi_k \ [\text{V}]$$

with a sampling value $T = 6.25 \times 10^{-3}$ s and $f = 10$ Hz. Additionally, there was an additive noise component of unknown nature denoted ξ_k. The methodology was the same as in the simulation instance. Figure 4.18 shows the signal and the identified frequency. Despite the fact that there is no information about the statistical properties of the noise, the invariant filtering

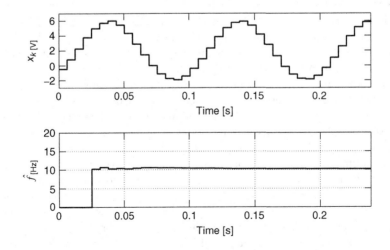

Figure 4.18 Experimental signal and its frequency estimation

has good behavior and, moreover, the estimation process reaches a region close to the real value in a fourth part of the sinusoidal signal period.

Remark 4.3.4 *The problem of frequency estimation in sinusoidal signals in continuous time was solved by Trapero et al. (2007) and an error analysis of the methodology is studied in depth in Liu et al. (2008).*

4.4   Algebraic Identification in Fast-Sampled Linear Systems

In real-world applications, when a controller is implemented using finite-precision hardware, the user is interested in determining constraint aspects such as the minimum precision required and the necessary sample rate to achieve the specified performance and accuracy. Higher sampling frequency is often preferred, since it typically implies a better approximation to the continuous-time results. Most traditional discrete-time algorithms are ill-conditioned when applied to data taken at sampling rates which are high relative to the dynamics of the sampled data. This fact encouraged Goodwin and Middleton (1992) to develop a strategy capable of unifying both continuous and discrete-time formulations. Moreover, this approach overcomes the unstable sampling zero problem as analyzed in Tesfaye and Tomizuka (1995), and the procedure for control gains is enhanced since the stability region increases as the sampling time decreases, avoiding extra reparametrizations as in the Tustin approach. In this type of approach, the authors proposed the use of an operator called the delta operator, defined as follows:

$$\delta = \frac{q-1}{\Delta}$$

where q is the shift operator in the time domain and Δ is the sampling time. This approach has been used with good results regarding robustness and other features, even in a class of nonlinear systems where fast-sampled data are involved, as shown in Chadwick *et al.* (2006) in the context of NARMAX models. The response of the delta operator converges to its continuous counterpart as the sample period tends to zero, which is better suited to applications with sampling frequencies significantly higher than those of the filter poles. In particular, delta-operator realizations are generally accompanied by better round-off noise performance and more robust coefficient and frequency sensitivities (Cheng and Chiu, 2007). In the context of digital filtering design, this operator approach has shown some advantages in power electronic inverter applications to achieve substantial performance benefits compared with equivalent shift-based implementations (Newman and Holmes, 2003).

In this section we look at the problem of controlling an unknown DC motor, taking the certainty-equivalence control approach, in which the control design assumes knowledge of the plant parameters, while an identification algorithm is used to obtain them. For this instance, the proposed controller will be given as the delta-operator-based version of the flatness-based generalized PI control (Fliess and Marquez, 2001).

The identification scheme consists of the adaptation of the algebraic parameter identification technique, using the delta operator (transform) as the main tool. The initial conditions and structured perturbations can be eliminated efficiently by means of algebraic manipulations involving the use of the properties of the delta transform (Goodwin and Middleton, 1992). Equivalently, in the time domain, initial conditions and structured perturbations are handled via the use of the delta operator, with suitable powers of the time variable in the regressor expression.

The use of the delta operator scheme results in a more natural form of expressing a sampled system without the necessity for additional reparametrization in case of high sampling rates, as usually occurs in sampled-time schemes based on the shift operator. The algebraic delta-operator methodology turns out to be more natural as the sampling time approaches zero (Feuer and Goodwin, 1996).

4.4.1   The Delta-Operator Approach: A Theoretical Framework

Some preliminary concepts regarding the delta operator and its properties are given here. The reader is referred to Goodwin and Middleton (1992); Middleton and Goodwin (1990) for detailed information.

Definition 4.4.1 *The domain of possible non-negative "times"* $\Omega^+(\Delta)$ *is defined as follows:*

$$\Omega^+(\Delta) = \left\{ \begin{matrix} \mathbb{R}^+ \cup \{0\} & : \Delta = 0 \\ \{0, \Delta, 2\Delta, 3\Delta, \,\dots\, \} & : \Delta \neq 0 \end{matrix} \right\} \tag{4.107}$$

where Δ *denotes the sampling period in discrete time, or* $\Delta = 0$ *for a continuous-time framework.*

Definition 4.4.2 *A time function* $x(t)$, $t \in \Omega^+(\Delta)$ *is, in general, a mapping from the time to either the real or the complex set. That is,* $x(t) : \Omega^+ \to \mathbb{C}$.

Definition 4.4.3 *The delta operator is defined as follows:*

$$\delta x(t) \triangleq \frac{x(t + \Delta) - x(t)}{\Delta}, \qquad \Delta \neq 0$$

where

$$\lim_{\Delta \to 0^+} \delta(x(t)) = \frac{dx(t)}{dt}$$

Definition 4.4.4 *The unified integration operation* **S** *is given as follows:*

$$\mathbf{S}_{t_1}^{t_2} x(\tau) d\tau = \left\{ \begin{matrix} \int_{t_1}^{t_2} x(\tau) d\tau & : \Delta = 0 \\ \Delta \sum\limits_{l = t_1/\Delta}^{l = t_2/\Delta - 1} x(l\Delta) & : \Delta \neq 0 \end{matrix} \right\}, t_1, t_2 \in \Omega^+$$

The integration operator corresponds to the antiderivative operator.

Definition 4.4.5 (Generalized matrix exponential) *In the case of the unified transform theory, the generalized exponential E is defined as follows:*

$$E(A, t, \Delta) = \left\{ \begin{matrix} e^{At} & : \Delta = 0 \\ (I + A\Delta)^{t/\Delta} & : \Delta \neq 0 \end{matrix} \right\}$$

where $A \in \mathbb{C}^{n \times n}$, I is the identity matrix. The generalized matrix exponential satisfies the conditions to be the fundamental matrix of $\delta x = Ax$ and, thus, the unique solution to

$$\delta x = Ax; \ x(0) = x_0 \tag{4.108}$$

is $x(t) = E(A, t, \Delta)x_0$. The general solution to

$$\delta x = Ax + Bu, \ x(0) = x_0 \tag{4.109}$$

is $x(t) = E(A, t, \Delta)x_0 + \mathbf{S}_0^t E(A, t - \tau - \Delta, \Delta)Bu(\tau)d\tau$.

Definition 4.4.6 *[Stability boundary] The solution of (4.108) is said to be asymptotically stable if and only if, for all x_0, $x(t) \to 0$ as time elapses. The stability arises if and only if $E(A, t, \Delta) \to 0$ as $t \to \infty$, if and only if every eigenvalue of A, denoted λ_i, $i = 1, \dots, n$ satisfies the following condition:*

$$Re\{\lambda_i\} + \frac{\Delta}{2}|\lambda_i|^2 < 0 \tag{4.110}$$

Therefore, the stability boundary is the circle with center $(-1/\Delta, 0)$ and radius $1/\Delta$ (see Figure 4.19). In particular, consider the following nth degree characteristic equation on the complex variable γ (see Definition 4.4.7):

$$\gamma^n + c_{n-1}\gamma^{n-1} + \dots + c_1\gamma + c_0 = 0 \tag{4.111}$$

If all the roots $\lambda_i, i = 1, \dots, n$ of the last equation satisfy the condition (4.110), then the solution of the system associated with (4.111) is asymptotically stable.

Definition 4.4.7 *[**Unified transform theory**] There is a close connection between the forward-shift operator and the Z-transform variable z. Analogously, consider a new transform variable $\gamma \in \mathbb{C}$ associated with the delta operator as $\gamma = (z - 1)/\Delta$. Consider the continuous-time functions $f, g : \mathbb{R} \to \mathbb{R}$. From the Z-transform, the delta transform is derived as follows:*

$$\mathbf{T}(f(k)) = \Delta F(z)|_{z=\Delta\gamma+1} = \Delta \sum_{k=0}^{\infty} f(k)(1 + \Delta\gamma)^{-k} \tag{4.112}$$

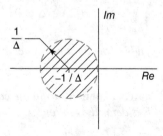

Figure 4.19   Stability region for the delta operator

4.4.2 Delta-Transform Properties

We will just consider the properties to be used throughout this books. A more extensive list of delta-transform properties can be found in Middleton and Goodwin (1990).

(i) Linearity. For any scalars α_1, α_2,

$$\mathbf{T}\{\alpha_1 f(t) + \alpha_2 g(t)\} = \alpha_1 \mathbf{T}\{f(t)\} + \alpha_2 \mathbf{T}\{g(t)\}$$

(ii) Differentiation,

$$\mathbf{T}\{\delta[f(t)]\} = \gamma \mathbf{T}\{f(t)\} - f(0^-)(1 + \Delta\gamma)$$

(iii) Integration,

$$\mathbf{T}\{\mathbf{S}_0^t f(\tau)d\tau\} = \frac{1}{\gamma}\mathbf{T}\{f(t)\}$$

(iv) Frequency differentiation,

$$\mathbf{T}\{t^n f(t)\} = (-1)^n (\Delta\gamma + 1)^n \frac{d^n}{d\gamma^n}[\mathbf{T}\{f(t)\}]$$

4.4.3 A DC Motor Example

Consider the following model of a DC motor:

$$L\frac{di(t)}{dt} = -Ri(t) - k\dot{\theta}(t) + v(t)$$

$$J\frac{d^2\theta(t)}{dt^2} = -B\dot{\theta}(t) + ki(t) \qquad (4.113)$$

where i represents the armature current, θ is given by the rotor angular position, v is the voltage input of the motor. L is the armature induction constant, k denotes the back electromotive force constant, R is the armature resistance, J is the inertia rotor constant, and B is the damping rotor constant.

4.4.3.1 Problem Formulation

Given a reference trajectory θ^*, and the sampled output θ, for the completely uncertain DC motor model (4.113), devise a control law associated with a parameter identification scheme by means of the delta operator to force the global asymptotic convergence of $\theta(t)$ to $\theta^*(t)$ (in spite of measurement additive noises and the absence of any electrical variable measurements).

4.4.3.2 A Certainty Equivalence Control Based on Integral Reconstructors

Let us assume that all system parameters are known and no disturbance inputs are present. Under these circumstances, consider the motor model (4.113). Assuming that the time derivative of the armature current is negligible, we have

$$i(t) = \frac{v(t) - k\dot{\theta}(t)}{R}$$

and then the motor model can be expressed as

$$a_1\ddot{\theta}(t) + a_2\dot{\theta}(t) = v(t) \tag{4.114}$$

where

$$a_1 = \frac{JR}{k}, \quad a_2 = \frac{R}{k}\left(B + \frac{k^2}{R}\right) \tag{4.115}$$

Let us consider the approximate model, derived from a sampling process (with sampling period Δ) using the delta operator:

$$a_1\delta^2\theta(t) + a_2\delta\theta(t) = v(t); \quad t \in \Omega^+(\Delta) \tag{4.116}$$

Consider the system (4.116). An integral reconstructor (Fliess and Sira-Ramírez, 2004) of $\delta\theta$, denoted $\hat{\delta\theta}$, can be proposed as follows. By integrating (4.116), the following relation is obtained:

$$\delta\theta(t) = -\frac{a_2}{a_1}\theta(t) + \frac{1}{a_1}\mathbf{S}_0^t v(\tau)d\tau + \delta\theta(0) \tag{4.117}$$

The integral reconstructor of $\delta\theta$ is then defined as

$$\hat{\delta\theta}(t) = -\frac{a_2}{a_1}\theta(t) + \frac{1}{a_1}\mathbf{S}_0^t v(\tau)d\tau \tag{4.118}$$

We have, purposefully, ignored the influence of the initial conditions in the integral reconstructor expression. Hence, there is a constant error given by

$$\delta\theta - \hat{\delta\theta} = \delta\theta(0)$$

From the problem statement, it is desired to track a given reference of the output signal $\theta^*(t)$. Denote by $v^*(t)$ a feedforward input which, in unperturbed system conditions, generates $\theta^*(t)$. In other words,

$$v^*(t) = a_1\delta^2\theta^*(t) + a_2\delta\theta^*(t)$$

Since there is no knowledge of initial conditions and unexpected perturbations may arise, it is necessary to devise a control law that is robust against these facts and possibilities. In this instance, a GPI control law is proposed which parallels, in the delta-operator domain, that of the discrete-time case in the Z-transform viewpoint. The tracking error $e_\theta(t)$ is defined as follows:

$$e_\theta\Delta = \theta(t) - \theta^*(t) \tag{4.119}$$

We have the following GPI control law with the integral reconstructor (4.88) in terms of the tracking error:

$$v(t) = v^* - a_1[k_2\hat{\delta e}_\theta(t) + k_1 e_\theta(t) + k_0\mathbf{S}_0^t e_\theta(\tau)d\tau]$$
$$\hat{\delta e}_\theta(t) = \hat{\delta\theta}(t) - \delta\theta^*(t) \tag{4.120}$$

Now, using (4.120) in (4.116) we have

$$\delta^2 e_\theta + \left(\frac{a_2}{a_1} + k_2\right)\delta e_\theta + k_1 e_\theta + k_0\mathbf{S}e_\theta(\tau)d\tau = k_2\delta\theta(0) \tag{4.121}$$

Applying the delta operator in (4.121), the closed-loop dynamics is given by

$$\delta^3 e_\theta + \left(\frac{a_2}{a_1} + k_2 \right) \delta^2 e_\theta + k_1 \delta e_\theta + k_0 e_\theta = 0 \tag{4.122}$$

Let λ_1, λ_2, λ_3 be the roots of the characteristic equation associated with (4.122):

$$\gamma^3 + \left(\frac{a_2}{a_1} + k_2 \right) \gamma^2 + k_1 \gamma + k_0 = 0 \tag{4.123}$$

By selecting k_2, k_1, k_0 such that

$$Re\{\lambda_i\} + \frac{\Delta}{2} |\lambda_i|^2 < 0; \ i = 1, 2, 3$$

the asymptotic stability of the tracking error is ensured. It follows that with knowledge of the set of system parameters $\{a_1, a_2\}$, the gains $\{k_2, \quad k_1, \quad k_0\}$ of the controller can be adjusted suitably so that the poles of the closed-loop system coincide with the roots of a desired closed-loop characteristic polynomial $p_d(\gamma)$:

$$p_d(\gamma) = \gamma^3 + \alpha_2 \gamma^2 + \alpha_1 \gamma + \alpha_0 \tag{4.124}$$

Remark 4.4.8 *The proposed controller can be interpreted in frequency-domain terms (transfer function) as follows. Let us define $e_v \Delta = v - v^*$. From (4.120) and the integral reconstructor definition:*

$$e_v(t) = -a_1(k_2 \hat{\delta} e_\theta + k_1 e_\theta + k_0 S_0^t e_\theta(\tau) d\tau)$$

$$\hat{\delta} e_\theta(t) = -\frac{a_2}{a_1} e_\theta(t) + \frac{1}{a_1} S_0^t e_v(\tau) d\tau$$

Applying the delta transform and with some straightforward algebraic manipulations, we obtain

$$\frac{e_v(\gamma)}{e_\theta(\gamma)} = \frac{\gamma(k_2 a_2 - k_1 a_1) - k_0 a_1}{\gamma + k_2}$$

Finally, the control law in terms of a transfer function is given by

$$v(\gamma) = v^*(\gamma) + \frac{\gamma(k_2 a_2 - k_1 a_1) - k_0 a_1}{\gamma + k_2} e_\theta(\gamma) \tag{4.125}$$

The synthesis of the appropriate controller parameters, as well as the generation of the feedforward compensation signal $v^*(t)$, demands acceptable information on the parameter set $\Theta = \{a_1, a_2\}$. A separate online parameter identification process will be proposed to obtain, quite robustly with respect to possible noisy measurements, a parameter estimate set denoted by $\hat{\Theta} = \{\hat{a}_1, \quad \hat{a}_2\}$ which will replace the unknown parameters in a controller identical to the certainty-equivalence controller (4.120) except that it is written in terms of estimated parameter values rather than actual parameters values, as follows:

$$v(t) = v^* - \hat{a}_1[k_2 \delta \hat{e}_\theta(t) + k_1 e_\theta(t) + k_0 S_0^t e_\theta(\tau) d\tau]$$

$$\delta \hat{e}_\theta(t) = \hat{\delta\theta}(t) - \delta\theta^*(t)$$

$$v^*(t) = \hat{a}_1 \delta^2 \theta^*(t) + \hat{a}_2 \delta\theta^*(t) \tag{4.126}$$

with $k_2 = \alpha_2 - (\hat{a}_2/\hat{a}_1)$, $k_1 = \alpha_1$, $k_0 = \alpha_0$.

Thus, it is proposed to use the controller (4.126) as a certainty-equivalence control law. The elements of the unknown parameter set will be estimated online using the delta-operator-based algebraic estimation method to be presented in the following section.

4.4.3.3 Parameter Estimation: The Delta Approach

Consider (4.116). Taking the delta transform results in

$$a_1 \gamma^2 \theta(\gamma) + a_2 \gamma \theta(\gamma) - (1 + \Delta\gamma)[a_1 \gamma \theta(0^-) + a_1 \delta\theta(0^-) + a_2 \theta(0^-)] = v(\gamma) \tag{4.127}$$

To obtain an expression free from the initial conditions, let us take three time derivatives with respect to γ:

$$a_1 \left(6\frac{d\theta}{d\gamma} + 6\gamma \frac{d^2\theta}{d\gamma^2} + \gamma^2 \frac{d^3\theta}{d\gamma^3} \right) + + a_2 \left(3\frac{d^2\theta}{d\gamma^2} + \gamma \frac{d^3\theta}{d\gamma^3} \right) = \frac{d^3 v}{d\gamma^3} \tag{4.128}$$

Now, it is necessary to express (4.128) in the time domain, where the frequency differentiation property will be applied. To use the property, it is necessary to multiply both sides of (4.128) by the term $(1 + \Delta\gamma)^\beta$, where β is the highest derivative with respect to γ, in this case $\beta = 3$. We have

$$(1 + \Delta\gamma)^3 \left[a_1 \left(6\frac{d\theta}{d\gamma} + 6\gamma \frac{d^2\theta}{d\gamma^2} + \gamma^2 \frac{d^3\theta}{d\gamma^3} \right) + a_2 \left(3\frac{d^2\theta}{d\gamma^2} + \gamma \frac{d^3\theta}{d\gamma^3} \right) \right] = (1 + \Delta\gamma)^3 \frac{d^3 v}{d\gamma^3}$$

To avoid having "derivative" terms in the time expression, we multiply by $1/\gamma^2$, which is equivalent to integrating twice with respect to the time in the time domain. Making some algebraic manipulations, we have

$$a_1 \left\{ 6\frac{(1 + \Delta\gamma)^2}{\gamma^2} \left[(1 + \Delta\gamma)\frac{d\theta}{d\gamma} \right] + \frac{6}{\gamma}(1 + \Delta\gamma) \left[(1 + \Delta\gamma)^2 \frac{d^2\theta}{d\gamma^2} \right] + \left[(1 + \Delta\gamma)^3 \frac{d^3\theta}{d\gamma^3} \right] \right\}$$

$$+ a_2 \left\{ 3 \left(\frac{1 + \Delta\gamma}{\gamma^2} \right) \left[(1 + \Delta\gamma)^2 \frac{d^2\theta}{d\gamma^2} \right] + \frac{1}{\gamma} \left[(1 + \Delta\gamma)^3 \frac{d^3\theta}{d\gamma^3} \right] \right\} = \frac{1}{\gamma^2} \left[(1 + \Delta\gamma)^3 \frac{d^3 v}{d\gamma^3} \right] \tag{4.129}$$

Finally, transforming into the time domain and applying the frequency differentiation and integration properties, the following regressor is obtained:

$$p_1(\theta, t)a_1 + p_2(\theta, t)a_2 = q(v, t) \tag{4.130}$$

where

$$p_1(\theta, t) = 6\Delta^2 t\theta - 6\Delta t^2\theta + t^3\theta + S_0^t 12\Delta\tau\theta(\tau) - 6\tau^2\theta(\tau)d\tau + 6S_0^t S_0^\tau \sigma\theta(\sigma)d\sigma d\tau$$

$$p_2(\theta, t) = S_0^t[-3\Delta\tau^2\theta(\tau) + \tau^3\theta(\tau)]d\tau - 3S_0^t S_0^\tau \sigma^2\theta(\sigma)d\sigma d\tau$$

$$q(v, t) = S_0^t S_0^\tau \sigma^3 v(\sigma)d\sigma d\tau$$

The last formula expresses the system in the following regressor form:

$$P(\theta, t)\Theta = q(v, t) \tag{4.131}$$

where $P(\theta, t) = [p_1(\theta, t) \quad p_2(\theta, t)]$, $\Theta = [a_1 \quad a_2]^T$.

Remark 4.4.9 *The last algebraic identification procedure can also be carried out in the time domain: multiplying (4.116) by t^3 and integrating twice with respect to t. This nearly corresponds to the algebraic identification procedure in the continuous-time domain. In the delta-operator approach, however, we have to multiply by an additional power of time (in the continuous-time approach we have to multiply only by t^2). This is due to the effects of the initial conditions when performing a sampling procedure.*

The next step is to obtain a linear equation from which to efficiently find the unknown vector of parameters.

4.4.3.4 Constructing the System of Equations

To obtain the vector of unknown parameters Θ, it is necessary to build a system of two linearly independent equations from (4.130). Traditionally, this is accomplished by considering further integrations of the single linear equation involving the algebraically generated regressor vector. Here we propose, as performed in Chapter 3, the identification of the unknown parameter vector Θ as an optimization problem, by minimizing an integral square-error criterion. The optimization criterion is given in terms of the following cost function:

$$J(\hat{\Theta}, t) = \frac{1}{2}S_0^t \varepsilon^2(\hat{\Theta}, \tau)d\tau \tag{4.132}$$

where

$$\varepsilon(\hat{\Theta}, t) = P(\theta, t)\hat{\Theta} - q(v, t) \tag{4.133}$$

and where $\hat{\Theta}$ is the vector of estimated parameters. Thus, the optimization problem is stated as

$$\min_{\hat{\Theta}} \frac{1}{2}S_0^t(P(\theta, \tau)\hat{\Theta} - q(v, \tau))^2 d\tau \tag{4.134}$$

In order to minimize the cost function (4.132), the gradient method is applied. Differentiating $J(\hat{\Theta}, t)$ with respect to the vector of estimated parameters $\hat{\Theta}$ yields

$$\nabla_{\hat{\Theta}} J(\hat{\Theta}, t) = \frac{\partial}{\partial\hat{\Theta}} \frac{1}{2}S_0^t \varepsilon^2(\hat{\Theta}, \tau)d\tau = S_0^t \frac{\partial\varepsilon}{\partial\hat{\Theta}} \varepsilon(\hat{\Theta}, \sigma)d\tau \tag{4.135}$$

Notice that $\partial(P(\theta, t)\hat{\Theta} - q(v, t))/\partial\hat{\Theta} = P^T(\theta, t)$, thus

$$\nabla_{\hat{\Theta}}J(\hat{\Theta}, t) = \mathbf{S}_0^t P^T(\theta, \tau)[P(\theta, \tau)\hat{\Theta} - q(v, \tau)]d\tau \tag{4.136}$$

Since $J(\hat{\Theta}, t)$ is a convex function, there exists a global minimum that satisfies $\nabla_{\hat{\Theta}}J(\hat{\Theta}, t) = 0$ for all t. We then have

$$[\mathbf{S}_0^t P^T(\theta, \tau)P(\theta, \tau)d\tau]\hat{\Theta} = \mathbf{S}_0^t P^T(\theta, \tau)q(v, \tau)d\tau \tag{4.137}$$

The expression (4.137) constitutes a set of two linearly independent equations for $t > 0$. Therefore, the vector of estimated parameters, $\hat{\Theta}$, can be obtained by means of the formula

$$\hat{\Theta} = [\mathbf{S}_0^t P^T(\theta, \tau)P(\theta, \tau)d\tau]^{-1}\mathbf{S}_0^t P^T(\theta, \tau)q(v, \tau)d\tau \tag{4.138}$$

Remark 4.4.10 *The invertibility of $\mathbf{S}_0^t P^T(\theta, \tau)P(\theta, \tau)d\tau$ is given in terms of the condition $(\mathbf{S}_0^t p_1^2(\theta, \tau)d\tau)(\mathbf{S}_0^t p_2^2(\theta, \tau)d\tau) - (\mathbf{S}_0^t p_1(\theta, \tau)p_2(\theta, \tau)d\tau)^2 \neq 0$. While the signals p_1 and p_2 are independent, we ensure an algebraic conditioning of the system, leading to the invertibility condition.*

Since $\hat{\Theta}$ is independent of time, the above formula is valid for any arbitrary small time integration interval $[0, \epsilon]$, with $\epsilon > 0$. The estimation of Θ can be achieved quite fast.

4.4.3.5 Experimental Results

Experiments were carried out to assess the performance of the proposed delta algebraic identification method and the GPI delta-operator-based output feedback control scheme on a laboratory DC motor plant. A 24-V DC motor with 1000 counts/rev optical encoder was considered for these purposes. The data acquisition system was set up using a National Instruments PCI-6602 data acquisition card, with a sampling time of 0.1 ms. The identification and control scheme was devised in a MATLAB®-xPC Target environment. To carry out the online identification process, the first part of the control input consisted of an open-loop sinusoidal voltage signal of the form $v(t) = 4 + 16\sin(5\pi t)$, $t \in [0, 1)$. Figure 4.20 shows that the actual values of the parameters are achieved in approximately 0.5 s. After 1 s had elapsed, we connected the delta-operator-based GPI output feedback control. A desired characteristic polynomial was furnished by considering $p_d(\gamma) = (\gamma + 50)(\gamma + 60)(\gamma + 70)$ as the desired closed-loop characteristic polynomial. Figure 4.21 shows the tracking of the output reference signal and the corresponding control input.

4.4.3.6 Discussion

We have proposed the use of the delta-operator approach in the context of digital implementation of the discrete-time algebraic parameter identification method for linear systems. We implemented a discrete GPI output feedback control scheme, within the delta-operator methodology, based on online fast identified parameter values. We achieved accurate output

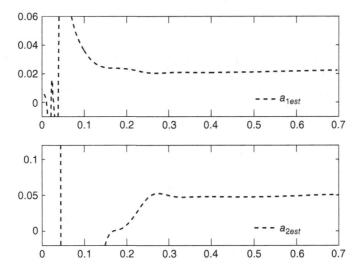

Figure 4.20   Parameter identification results

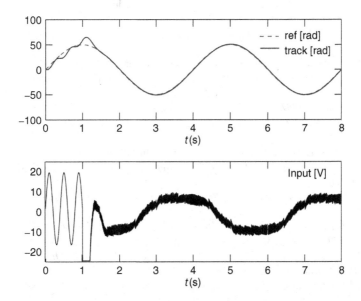

Figure 4.21   Trajectory tracking behavior

reference trajectory-tracking results. The proposed identification and control methods were tested in an experimental framework, and found to be most suitable for purely digital controllers such as DSPs or micro-controllers free from high-level prototyping platforms, such as Matlab or LabView.

4.5 Remarks

- The algebraic identification theory presented here permits a robust online computation of unknown parameters. It may therefore be regarded as a possible alternative to some techniques stemming from adaptive control.
- Extensions to nonlinear systems and delay systems are currently being investigated. Let us also emphasize some possible applications of our methods to other fields, like convolutional error-control codes, whose connection with linear system theory is already well known (see, e.g., Forney-Jr *et al.*, 1995; Rosenthal, 2000 and references cited therein). In Section 4.3.3, the repetition at several time instants of the calculations of the uncertain parameters is reminiscent of some of the *sliding* techniques in convolutional decoding (see, e.g., Blahut, 1983), which is being developed further (see Fliess, 2002 for a first attempt).
- The identification technique can be extended to work with time-series analysis with motivating results (Fliess and Join, 2008).

References

Åstrom, K. and Wittenmark, B. (1995) *Adaptive Control*, 2nd edn. Addison-Wesley, New York.

Blahut, R. (1983) *Theory and Practice of Error Control Codes*. Addison-Wesley, New York.

Bourlès, H. and Fliess, M. (1997) Poles and zeros of linear systems: An intrinsic approach. *International Journal of Control* **68**, 897–922.

Caines, P. (1988) *Linear Stochastic Systems*. John Wiley & Sone, Inc., New York.

Chadwick, M., Kadirkamanathan, V. and Billings, S. (2006) Analysis of fast-sampled non-linear systems: Generalised frequency response functions for δ-operator models. *Signal Processing* **86**, 3246–3257.

Cheng, H. and Chiu, G. (2007) Coupling between sample rate and required wordlength for finite precision controller implementation with delta transform. *Proceedings of the American Control Conference*, pp. 3588–3593, New York, NY.

Cohn, R. (1965) *Difference Algebra*. John Wiley & Sons, Inc., New York.

De-Freitas, J. and De-Freitas, J. (2005) *Digital Filter Design Solutions*. Artech House Microwave Library, London.

Feuer, A. and Goodwin, G. (1996) *Sampling in Digital Signal Processing and Control*. Birkhäuser, Boston, MA.

Fliess, M. (1992) Reversible linear and nonlinear discrete-time dynamics. *IEEE Transactions on Automatic Control* **37**, 1144–1153.

Fliess, M. (2002) On the structure of linear recurrent error-control codes. *ESAIM, Control, Optimization and Calculus of Variations* **8**, 703–713.

Fliess, M. and Glad, S. (1993) *Essays on Control: Perspectives in the Theory and its Applications*, pp. 223–267. Birkhäuser, Boston, MA.

Fliess, M. and Join, C. (2008) Time series technical analysis via new fast estimation methods: A preliminary study in mathematical finance. *23rd IAR Workshop on Advanced Control Diagnosis-IAR-ACD08*. http://hal.inria.fr/inria-00338099/en/.

Fliess, M. and Marquez, R. (2000) Continuous-time linear predictive control and flatness: A module-theoretic setting with examples. *International Journal of Control* **73**, 606–623.

Fliess, M. and Marquez, R. (2001) Une approche intrinsèque de la commande prédictive linéaire discrète. *APII - Journal Européen des Systmes Automatisés* **35**, 127–147.

Fliess, M. and Sira-Ramírez, H. (2003) An algebraic framework for linear identification. *ESAIM, Control, Optimization and Calculus of Variations* **9**(1), 151–168.

Fliess, M. and Sira-Ramírez, H. (2004) Reconstructeurs d'etat. *C.R. Academie des Sciences de Paris, Série I* **338**(1), 91–96.

Fliess, M., Fuchsummer, S., Schöberl, M., Schlacher, K. and Sira-Ramírez, H. (2008) An introduction to algebraic discrete-time linear parametric identification with a concrete application. *Journal Européen des Systèmes Automatisés* **42**(2&3), 210–232.

Fliess, M., Lévine, J., Martin, P. and Rouchon, P. (1995) Flatness and defect of non-linear systems: Introductory theory and applications. *International Journal of Control* **61**, 1327–1361.

Forney-Jr, G., Marcus, B., Sindhushayana, N. and Trott, M. (1995) A multilingual dictionary: System theory, coding theory, symbolic dynamics and automata theory. *Proceedings of the Symposium on Applied Mathematics,* vol. 50, pp. 109–138, AMS.

Fradkov, A. and Pogromsky, AY. (1998) *Introduction to Control of Oscillations and Chaos,* vol. 35. World Scientific Publishing Co., Singapore.

Goodwin, G. and Middleton, R. (1992) High-speed digital signal processing and control. *Proceedings of the IEEE* **80**(2), 240–259.

Goodwin, G. and Sin, K. (1984) *Adaptive Filtering, Prediction and Control.* Prentice-Hall, Englewood Cliffs, NJ.

Goodwin, GC., Graebe, SF. and Salgado, ME. (2001) *Control Systems Design.* Prentice-Hall, Englewood Cliffs, NJ.

Hénon, M. (1976) A two-dimensional mapping with a strange attractor. *Communications in Mathematical Physics* **50**(2), 69–77.

Hsu, L., Lopes Zachi, A. and Lizarralde, F. (2001) Adaptive visual tracking for motions on smooth surfaces. *Proceedings of the 40th IEEE Conference on Decision and Control, 2001,* vol. 3, pp. 2430–2435, Orlando, FL.

Isermann, R. (1987) *Identifikation dynamischer Systeme.* Springer-Verlag, Berlin.

Jacobson, N. (1974) *Basic Algebra*, vols I & II. W. H. Freeman & Co., San Francisco, CA.

Kailath, T. (1979) *Linear Systems.* Prentice-Hall, Englewood Cliffs, NJ.

Landau, I. (1990) *System Identification and Control Design.* Prentice-Hall, Englewood Cliffs, NJ.

Liu, D., Gibaru, O., Perruquetti, W., Fliess, M. and Mboup, M. (2008) An error analysis in the algebraic estimation of a noisy sinusoidal signal. *16th Mediterranean Conference on Control and Automation, 2008,* pp. 1296–1301, Ajaccio Corsica, France.

Ljung, L. (1987) *System Identification: Theory for the User.* Prentice-Hall, Englewood Cliffs, NJ.

McConnell, J. and Robson, J. (1987) *Noncommutative Noetherian Rings.* Pure and Applied Mathematics. Wiley Interscience, New York.

Middleton, R. and Goodwin, G. (1990) *Digital Control and Estimation: A unified approach.* Prentice-Hall, Englewood Cliffs, NJ.

Newman, M. and Holmes, D. (2003) Delta operator digital filters for high performance inverter applications. *IEEE Transactions on Power Electronics* **18**(1), 447–454.

Parks, T. and Burrus, C. (1987) *Digital Filter Design.* Topics in Digital Signal Processing. John Wiley & Sons, Inc., New York.

Polyak, T. (1987) *Introduction to Optimization.* Optimization Software, New York.

Richalet, J. (1998) *Pratique de l'identification*, 2nd edn. Hermès, Paris.

Rosenthal, J. (2000) *Codes, Systems and Graphical Models*, pp. 39–66. Springer-Verlag, New York.

Samuelson, PA. (1939) Interactions between the multiplier analysis and the principle of acceleration. *Review of Economic Statistics* **21**(2), 75–78.

Sira-Ramírez, H. and Agrawal, S. (2004) *Differentially Flat Systems.* Marcel Dekker, New York.

Sira-Ramírez, H. and Fliess, M. (2002) On discrete-time uncertain visual based control of planar manipulators: An online algebraic identification approach. *Proceedings of the 41st IEEE Conference on Decision and Control,* vol. **4**, pp. 4509–4514, Las Vegas, NV.

Söderström, P. and Stoica, P. (1989) *System Identification.* Prentice-Hall, Englewood Cliffs, NJ.

Tesfaye, A. and Tomizuka, M. (1995) Zeros of discretized continuous systems expressed in the Euler operator – an asymptotic analysis. *IEEE Transactions on Automatic Control* **40**(4), 743–747.

Trapero, JR., Sira-Ramírez, H. and Feliu Batlle, V. (2007) An algebraic frequency estimator for a biased and noisy sinusoidal signal. *Signal Processing* **87**(6), 1188–1201.

Venkataramanan, V., Peng, K., Chen, BM. and Lee, T. (2003) Discrete-time composite nonlinear feedback control with an application in design of a hard disk servo system. *IEEE Transactions on Control Systems Technology* **11**(1), 16–23.

5

State and Parameter Estimation in Linear Systems

5.1 Introduction

The theory of asymptotic observers for linear systems originated in the pioneering work of Kalman, Luenberger, as well as many other important contributions that occurred during the second half of the 20th century. The Kalman filter, thanks to its state-space formulation, represented a significant advancement for applications over the so-called *Wiener optimal filter*, specified in terms of complex variables.

The extension of dynamic observers and Kalman filters to the nonlinear case has not achieved the same success, and the general case remains pretty much open. It is clear that for an observable system, represented in state-space, the state estimation problem is intimately related to the computation of time derivatives of the output signals, in sufficient number.

In this chapter we present an algebraic method, of non-asymptotic nature, for the estimation of states by computing a finite number of time derivatives of the outputs. The theoretical framework of this methodology was introduced by M. Fliess *et al.* (see Fliess *et al.*, 2008 for a deeper analysis of the topic). The method is based on elementary algebraic manipulations that lead to specific formulae for the unmeasured states. The method may be combined with the notion of *flatness* to complete the feedback loop, imposing a desired closed-loop dynamics, and based solely on the feeding back of the system outputs.

Sections 5.2, 5.2.2, and 5.2.3 introduce an algebraic approach to fast state estimation in linear systems via rather elementary open-loop and closed-loop controlled system examples. First, we assume that the system is an open-loop system and then we show that for closed-loop systems state estimation can also be accomplished in an elementary manner too. In Section 5.2.4, we present a slightly more complex state estimation problem for a linear third-order controlled system. In Section 5.2.5, we place special attention on the classical sinusoid frequency estimation problem. This classical problem, which has been solved via nonlinear adaptive control methods in recent times, results in a particularly simple approach to simultaneous state and parameter estimation. We show in Section 5.2.6 that more complex problems, such as the gravitational wave parameter identification problem, can, in principle, be dealt

Algebraic Identification and Estimation Methods in Feedback Control Systems, First Edition.
Hebertt Sira-Ramírez, Carlos García-Rodríguez, Alberto Luviano-Juárez, John Alexander Cortés-Romero.
© 2014 John Wiley & Sons, Ltd. Published 2014 by John Wiley & Sons, Ltd.

with using the proposed algebraic approach. In Section 5.2.7, we present a power electronics, online, simultaneous state and parameter estimation example. Sections 5.2.8 and 5.2.9 deal with some mechanical control systems: a hydraulic press and a plotter. Section 5.3 presents a solution for the problem of state estimation in some chaotic oscillators with applications in communication systems. Finally, some remarks are given in Section 5.4. The reader can find some information concerning the mathematical framework of the methodology presented in this chapter in Appendix C.

5.1.1   Signal Time Derivation Through the "Algebraic Derivative Method"

As mentioned in the last chapter, given a smooth signal $y(t)$, it is possible to obtain close approximations of its finite time derivatives by means of the algebraic derivative method which consists of making a local approximation of the signal using a truncated Taylor series and some algebraic manipulations in order to obtain a triangular system of linear equations from which the time derivatives of the approximating signal can be computed, solely in terms of time convolutions of $y(t)$. The idea is then to adopt these obtained signals as local approximations to the actual time derivatives of the original signal $y(t)$. Naturally, one reverts to the time domain the calculations made in the frequency domain in order to obtain explicit formulae for approximating the different time derivatives of $y(t)$.

The importance of having accurate values of the time derivatives of the output has to do with the nonlinear observability criterion, where the observability of a state is achieved if that state can be expressed in terms of the output, the input, and their finite time derivatives. In the next section, the nonlinear observability concept will be discussed.

5.1.2   Observability of Nonlinear Systems

Consider a smooth nonlinear system, characterized by a state vector $x \in R^n$, of the form

$$\dot{x} = f(x)$$
$$y = h(x) \tag{5.1}$$

where y is the output of the system and $h(\cdot)$ is a smooth scalar map taking values on the real line. The output $y = h(x)$ of the system is said to be locally observable if the following map is locally full rank n:

$$\begin{bmatrix} y \\ \dot{y} \\ \vdots \\ y^{(n-1)} \end{bmatrix} = \begin{bmatrix} h(x) \\ L_f h(x) \\ \vdots \\ L_f^{n-1} h(x) \end{bmatrix} \tag{5.2}$$

where $L_f^k h(x)$ stands, in local coordinates, for $\frac{\partial L_f^{k-1} h(x)}{\partial x} f(x)$ with $L_f^0 h(x) = h(x)$.

A well-known result establishes that if the above map is locally full rank n, then the state vector x of the system can be locally expressed as a smooth differential function of y, that is, a smooth function of y and a finite number (in fact $n-1$) of its time derivatives (see Diop and Fliess, 1991 and also the work of Fliess, 1987). We also address this type of function as a *differential parametrization* of the state x in terms of the observable output y. We then have that x can be uniquely expressed as

$$x = \Phi(y, \dot{y}, \ddot{y}, \ldots, y^{(n-1)})$$

for some smooth function Φ.

5.2 Fast State Estimation

5.2.1 An Elementary Second-Order Example

Consider the problem of estimating the initial velocity in

$$\ddot{y} = u + \kappa \mathbf{1}(t) \tag{5.3}$$

Take the Laplace transform to obtain

$$s^2 \hat{y}(s) - sy_0 - \dot{y}_0 = \hat{u}(s) + \frac{\kappa}{s} \tag{5.4}$$

then multiplying out by s we obtain

$$s^3 \hat{y}(s) - s^2 y_0 - s\dot{y}_0 = s\hat{u}(s) + \kappa \tag{5.5}$$

Taking the derivative, with respect to s, we find

$$3s^2 \hat{y}(s) + s^3 \frac{d\hat{y}(s)}{ds} - 2sy_0 - \dot{y}_0 = \hat{u}(s) + s\frac{d\hat{u}(s)}{ds} \tag{5.6}$$

Integrating the previous expression, three times, we get

$$3s^{-1} \hat{y}(s) + \frac{d\hat{y}(s)}{ds} - 2s^{-2} y_0 - s^{-2}\left(\frac{y_0}{s}\right) = s^{-3}\hat{u}(s) + s^{-2}\frac{d\hat{u}(s)}{ds} \tag{5.7}$$

Solving for $\dot{y}_0$ in the time domain yields

$$\dot{y}_0 = \frac{3(\int_0^t y(\sigma)d\sigma) - ty(t) - 2y_0 t - (\int_0^t \int_0^\sigma \int_0^\lambda u(\rho)d\rho d\lambda d\sigma)}{t^2/2}$$

$$+ \frac{(\int_0^t \int_0^\sigma \lambda u(\lambda)d\lambda d\sigma)}{t^2/2} \tag{5.8}$$

which is singular at time $t = 0$ but allows for an accurate evaluation at the end of any time interval of the form $[0, \epsilon]$.

We used the following simulation data:

$$\ddot{y} = u + \kappa \mathbf{1}(t) + v(t) \tag{5.9}$$

$$u = \cos(t) - 2\sin(2t) \tag{5.10}$$

$$y_0 = 1, \quad \dot{y}_0 = -0.5, \quad \kappa = -0.5, \quad \epsilon = 0.1$$

$v(t)$ was taken to be a computer-generated stochastic noise with uniform probability distribution function in the interval $[-0.5, 0.5]$.

The estimation process may be continued by reinitializing (i.e., resetting) the integrations in the previous calculation formula.

The unmeasured state $\dot{y}_0(t_1)$ at any time t_1 may then be computed over a small time interval of the form $[t_1, t_1 + \epsilon]$, as follows (Figure 5.1):

$$\dot{y}_0(t_1) = \frac{3(\int_{t_1}^{t} y(\sigma)d\sigma) - (t - t_1)y(t) - 2y_0(t - t_1) - (\int_{t_1}^{t}\int_{t_1}^{\sigma}\int_{t_1}^{\lambda} u(\rho)d\rho d\lambda d\sigma)}{(t - t_1)^2/2}$$

$$+ \frac{(\int_{t_1}^{t}\int_{t_1}^{\sigma}(\lambda - t_1)u(\lambda)d\lambda d\sigma)}{(t - t_1)^2/2} \tag{5.11}$$

5.2.2 An Elementary Third-Order Example

Consider the third-order system

$$z^{(3)} = u$$

$$y = z \tag{5.12}$$

Using operational calculus and successively taking derivatives with respect to the variable s, we obtain the following description of the system, free from initial conditions:

$$s^3 \frac{d^3 Z(s)}{ds^3} + 9s^2 \frac{d^2 Z(s)}{ds^2} + 18s \frac{dZ(s)}{ds} + 6Z(s) = \frac{d^3 U(s)}{ds^3} \tag{5.13}$$

where $Z(s)$ and $U(s)$ represent, respectively, the transforms of $z(t)$ and $u(t)$ (Figure 5.2).

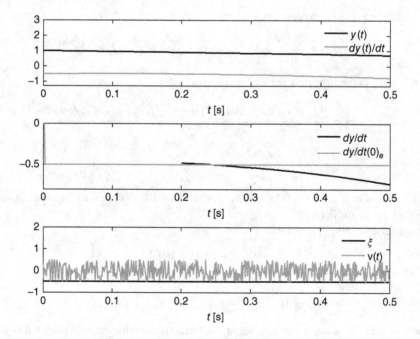

Figure 5.1 Estimation of the initial conditions

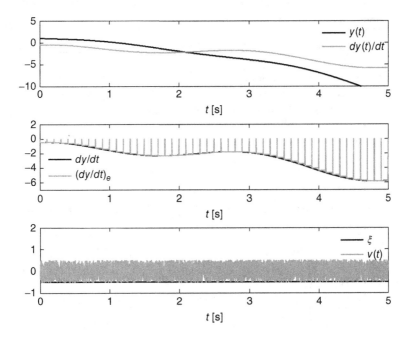

Figure 5.2 Time derivative reconstruction

Note that the preceding expression is also valid in terms of the transform of the measured signal y, that is,

$$s^3\frac{d^3Y(s)}{ds^3} + 9s^2\frac{d^2Y(s)}{ds^2} + 18s\frac{dY(s)}{ds} + 6Y(s) = \frac{d^3U(s)}{ds^3} \tag{5.14}$$

where $Y(s)$ is the transform of $y(t)$.

Integrating *twice* the previous expression yields

$$s\frac{d^3Y(s)}{ds^3} + 9\frac{d^2Y(s)}{ds^2} + 18s^{-1}\frac{dY(s)}{ds} + 6s^{-2}Y(s) = s^{-2}\frac{d^3U(s)}{ds^3} \tag{5.15}$$

In the time domain, we have

$$-\frac{d}{dt}(t^3y) + 9t^2y - 18\int_0^t \sigma y(\sigma)d\sigma + 6\int_0^t\int_0^\sigma y(\lambda)d\lambda d\sigma$$
$$= -\int_0^t\int_0^\sigma \lambda^3 u(\lambda)d\lambda d\sigma \tag{5.16}$$

From here we obtain

$$\frac{dy}{dt} = \frac{1}{t^3}\left[6t^2y - 18\int_0^t \sigma y(\sigma)d\sigma + 6\int_0^t\int_0^\sigma y(\lambda)d\lambda d\sigma\right.$$
$$\left. + \int_0^t\int_0^\sigma \lambda^3 u(\lambda)d\lambda d\sigma\right] \tag{5.17}$$

Integrating now <u>just once</u> the obtained expression, we have

$$s^2 \frac{d^3 Y(s)}{ds^3} + 9s \frac{d^2 Y(s)}{ds^2} + 18 \frac{dY(s)}{ds} + 6s^{-1} Y(s) = s^{-1} \frac{d^3 U(s)}{ds^3} \tag{5.18}$$

which may be written in the time domain as

$$-\frac{d^2}{dt^2}(t^3 y) + 9 \frac{d}{dt}(t^2 y) - 18ty + 6 \int_0^t y(\sigma) d\sigma = - \int_0^t \sigma^3 u(\sigma) d\sigma \tag{5.19}$$

We obtain the following expression for $d^2 y / dt^2$:

$$\frac{d^2 y}{dt^2} = \frac{1}{t^3} \left[3t^2 \left(\frac{dy}{dt} \right) - 6ty + 6 \int_0^t y(\sigma) d\sigma + \int_0^t \sigma^3 u(\sigma) d\sigma \right] \tag{5.20}$$

This expression may now be evaluated with the help of the already-computed estimate of $(\frac{dy}{dt})$. The state estimates are proposed as follows:

$$[\dot{y}]_e = \begin{cases} \text{arbitrary} \quad \forall\, t \in [0, \delta) \\ \frac{1}{t^3} \left[6t^2 y - 18 \int_0^t \sigma y(\sigma) d\sigma + 6 \int_0^t \int_0^\sigma y(\lambda) d\lambda d\sigma \right. \\ \left. \qquad + \int_0^t \int_0^\sigma \lambda^3 u(\lambda) d\lambda d\sigma \right] \\ \forall\, t \geq \delta \end{cases} \tag{5.21}$$

$$[\ddot{y}]_e = \begin{cases} \text{arbitrary} \quad \forall\, t \in [0, \delta) \\ \frac{1}{t^3} \left[3t^2 [\dot{y}]_e - 6ty + 6 \int_0^t y(\sigma) d\sigma + \int_0^t \sigma^3 u(\sigma) d\sigma \right] \\ \forall\, t \geq \delta \end{cases} \tag{5.22}$$

A stabilizing control law, which uses the technique of pole placement, may be obtained with the help of the fast state estimates in the following manner:

$$u = -\omega_n^2 p y - (\omega_n^2 + 2\zeta \omega_n p)[\dot{y}]_e - (2\zeta \omega_n + p)[\ddot{y}]_e \tag{5.23}$$

The characteristic polynomial of the closed-loop system coincides with

$$p_d(s) = (s + p_1)(s + p_2)(s + p_3) \tag{5.24}$$

where we have selected

$$p_1 = 2, \quad p_2 = 3, \quad p_3 = 4$$

5.2.2.1 Simulations

The following numerical simulations were carried out with the selected control parameters. Figure 5.3 depicts the estimated parameters for the time derivatives, the output and input variables.

We now evaluate how the closed-loop system behaves when it is subject to unmodeled significant external input and measurement noise, as in the following case (Figure 5.4):

$$z^{(3)} = u + \xi(t), \quad y = z + v(t) \tag{5.25}$$

with

$$\xi(t) = 0.1(rect(t) - 0.5), \quad v(t) = 0.001(rect(t) - 0.5)$$

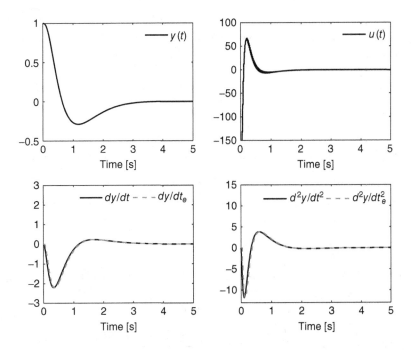

Figure 5.3 Input, output, and estimated values

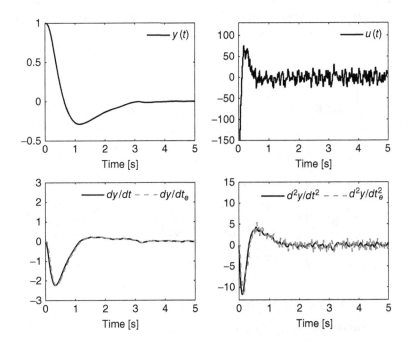

Figure 5.4 Estimation with additive noise

5.2.2.2 Remark

If we know, with certainty, that the system may be affected by a *classical* perturbation input (step, ramp, etc.), whose amplitudes, or slopes, may be totally unknown, we proceed as in the following example:

$$z^{(3)} = u + a\mathbf{1}(t), \quad y = z \tag{5.26}$$

Taking transforms:

$$s^3 Y(s) - s^2 y_0 - s\dot{y}_0 - \ddot{y}_0 = U(s) + \frac{a}{s} \tag{5.27}$$

Multiplying by s, differentiating four times with respect to s, and integrating three times we obtain, in the time domain,

$$24\left(\int^{(3)} y\right) - 96\left(\int^{(2)} ty\right) + 72\left(\int t^2 y\right) - 16t^3 y + \frac{d}{dt}[t^4 y]$$

$$-\left(\int^{(2)} t^4 u\right) + 4\left(\int^{(3)} t^3 u\right) = 0 \tag{5.28}$$

From this expression we obtain the formula for $\dot{y}(t)$, in a manner which is totally independent of the magnitude a of the perturbation.

Indeed, the first-order time-derivative estimate of $y(t)$ is given by

$$[\dot{y}]_e = \frac{-24(\int^{(3)} y) + 96(\int^{(2)} ty) - 72(\int t^2 y) + 12t^3 y + (\int^{(2)} t^4 u) - 4(\int^{(3)} t^3 u)}{t^4} \tag{5.29}$$

The validity of the formulae for the computation of the time derivatives is, in principle, uniform in time. However, due to the limited precision of every arithmetic processor, it is possible to reach a numerical saturation as time increases.

For this reason, it becomes necessary to *reinitialize* or *reset* the calculations at some time $t_i \gg 0$:

$$[\dot{y}]_e(t) = \begin{cases} [\dot{y}]_e(t_i^-) & \forall t \in [t_i, t_i + \delta) \\ \frac{1}{(t-t_i)^3}\left[6(t-t_i)^2 y - 18\int_{t_i}^t (\sigma - t_i)y(\sigma)d\sigma + 6\int_{t_i}^t \int_{t_i}^\sigma y(\lambda)d\lambda d\sigma \right. \\ \left. \quad + \int_{t_i}^t \int_{t_i}^\sigma (\lambda - t_i)^3 u(\lambda)d\lambda d\sigma\right] \\ \forall t \geq t_i + \delta \end{cases} \tag{5.30}$$

$$[\ddot{y}]_e(t) = \begin{cases} [\ddot{y}]_e(t_i^-) & \forall t \in [t_i, t_i + \delta) \\ \frac{1}{(t-t_i)^3}\left[3(t-t_i)^2[\dot{y}]_e - 6(t-t_i)y + 6\int_{t_i}^t y(\sigma)d\sigma \right. \\ \left. \quad + \int_{t_i}^t (\sigma - t_i)^3 u(\sigma)d\sigma\right] \\ \forall t \geq t_i + \delta \end{cases} \tag{5.31}$$

5.2.3 A Control System Example

Consider the following second-order system:

$$\ddot{z} = u, \quad y = z \tag{5.32}$$

In the *operational calculus* notation, we have

$$s^2 Z(s) - sz(0) - \dot{z}(0) = U(s), \quad Y(s) = Z(s) \tag{5.33}$$

Taking derivatives with respect to s yields

$$s^2 \frac{dY(s)}{ds} + 2sY(s) - z(0) = \frac{dU(s)}{ds} \tag{5.34}$$

A second derivation with respect to s results in an expression which is *independent* of the initial conditions:

$$s^2 \frac{d^2Y(s)}{ds^2} + 4s\frac{dY(s)}{ds} + 2Y(s) = \frac{d^2U(s)}{ds^2} \tag{5.35}$$

Integrating once, with respect to time, in the frequency domain we obtain

$$s\frac{d^2Y(s)}{ds^2} + 4\frac{dY(s)}{ds} + 2s^{-1}Y(s) = s^{-1}\frac{d^2U(s)}{ds^2} \tag{5.36}$$

whose expression, in the time domain, reads as follows:

$$\frac{d}{dt}(t^2 y(t)) - 4ty(t) + 2\int_0^t y(\sigma)d\sigma = \int_0^t \sigma^2 u(\sigma)d\sigma \tag{5.37}$$

Evidently, from the last expression, we can solve for the unknown quantity $(\frac{dy}{dt})$. Certainly,

$$\frac{dy}{dt} = \frac{2ty(t) - 2\int_0^t y(\sigma)d\sigma + \int_0^t \sigma^2 u(\sigma)d\sigma}{t^2} \tag{5.38}$$

This expression, which happens to be *indefinite* at time $t = 0$, may, nevertheless, be evaluated for all times $t \geq \delta$ where $\delta > 0$ is an arbitrarily small positive quantity. We propose a derivative estimate $\dot{y}(t)$ of the form

$$[\dot{y}(t)]_e = \begin{cases} \text{arbitrary} & \text{for } t \in [0, \delta) \\ \dfrac{2ty(t) - 2\int_0^t y(\sigma)d\sigma + \int_0^t \sigma^2 u(\sigma)d\sigma}{t^2} & \text{for } t \geq \delta \end{cases} \tag{5.39}$$

5.2.3.1 Simulations

See Figures 5.5 and 5.6:

$$\ddot{z} = u, \quad y = z$$

$$u = -2\zeta\omega_n[\dot{y}]_e - \omega_n^2 y$$

$$\zeta = 0.81, \quad \omega_n = 2$$

$$\ddot{z} = u + \xi(t)$$

$$y = z$$

$$\ddot{z} = u + \xi(t)$$

$$y = z + v(t)$$

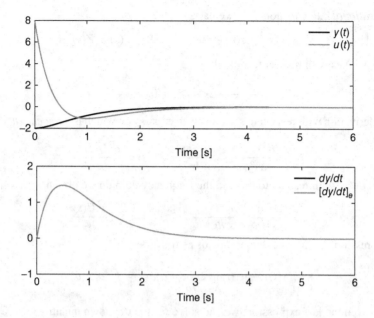

Figure 5.5 System output and time derivative estimation

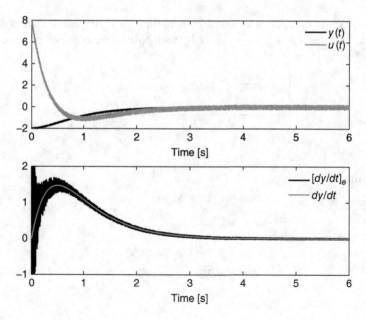

Figure 5.6 System output and time derivative estimation in the presence of additive noise

5.2.4   Control of a Perturbed Third-Order System

Given the following linear perturbed system:

$$y^{(3)} = u + v(t) \tag{5.40}$$

it is desired to make the signal y follow a given signal $y^*(t)$ regardless of the initial conditions and the additive perturbation input $v(t)$ of the form

$$v(t) = b + \gamma(t) \tag{5.41}$$

where b is a known bias input and $\gamma(t)$ is a zero-mean stochastic input. We consider a nominal perturbed system as

$$y^{(3)} = u + b \tag{5.42}$$

A simple feedback controller accomplishing, on average, the desired objective is given by

$$u = [y^*(t)]^{(3)} - k_3(\ddot{y} - \ddot{y}^*(t)) - k_2(\dot{y} - \dot{y}^*(t))$$

$$-k_1(y - y^*(t)) - k_0 \left(\int_0^t (y - y^*(\sigma))d\sigma \right) \tag{5.43}$$

The controller requires knowledge of $y(t)$, which is assumed to be known, and of the signals $\dot{y}(t)$ and $\ddot{y}(t)$, which are unavailable for measurement.

The representation of the system in a state-space framework is given by

$$\dot{x}_1 = x_2, \quad y = x_1$$

$$\dot{x}_2 = x_3 \tag{5.44}$$

$$\dot{x}_3 = u + b$$

The state of the system is thus constituted by $x_1 = y$, $x_2 = \dot{y}$, and $x_3 = \ddot{y}$. If these quantities can be estimated, out control problem is solved.

A full-order observer, with integral control action, which asymptotically and exponentially supplies the unknown states of the system, is of the form

$$\dot{\hat{x}}_1 = \hat{x}_2 + \gamma_3(y - \hat{x}_1)$$

$$\dot{\hat{x}}_2 = \hat{x}_3 + \gamma_2(y - \hat{x}_1)$$

$$\dot{\hat{x}}_3 = u + \gamma_1(y - \hat{x}_1) + \gamma_0 \int_0^t (y - \hat{x}_1)d\sigma \tag{5.45}$$

$$y = x_1$$

The output reconstruction error $e = y - x_1$ satisfies

$$e^{(4)} + \gamma_3 e^{(3)} + \gamma_2 \ddot{e} + \gamma_1 \dot{e} + \gamma_0 e = 0 \tag{5.46}$$

which is independent of the unknown bias perturbation b.

Note that a suitable choice of $\gamma_3, \gamma_2, \gamma_1$, and γ_0 renders $e = y - \hat{x}_1$ globally exponentially and asymptotically stable:

$$e = y - \hat{x}_1 \to 0 \tag{5.47}$$

In contrast:

$$e_2 = \dot{y} - \hat{x}_2 = \dot{e} + \gamma_3 e, \quad e_3 = \ddot{y} - \hat{x}_3 = \ddot{e} + \gamma_3 \dot{e} + \gamma_2 e \tag{5.48}$$

also exponentially and asymptotically tend to zero as t tends to infinity, independently of observer initial conditions.

Exponential state estimation is sometimes not desirable due to transient performance limitations.

In operational calculus notation, we have

$$s^3 y(s) - s^2 y_0 - s \dot{y}_0 - \ddot{y}_0 = u(s) + \frac{b}{s} \tag{5.49}$$

where y_0, $\dot{y}_0$, and $\ddot{y}_0$ form the initial system state at time t_0.

Multiplying out by s, we have

$$s^4 y(s) - s^3 y_0 - s^2 \dot{y}_0 - s \ddot{y}_0 = s u(s) + b \tag{5.50}$$

Derivation with respect to s yields

$$s^4 \frac{dy(s)}{ds} + 4s^3 y(s) - 3s^2 y_0 - 2s \dot{y}_0 - \ddot{y}_0 = s \frac{du(s)}{ds} + u(s) \tag{5.51}$$

and further derivation yields

$$s^4 \frac{d^2 y(s)}{ds^2} + 8s^3 \frac{dy(s)}{ds} + 12 s^2 y(s) - 6 s y_0 - 2 \dot{y}_0$$

$$= s \frac{d^2 u(s)}{ds^2} + 2 \frac{du(s)}{ds} \tag{5.52}$$

By repeated integration we obtain the following system of equations:

$$\begin{bmatrix} s^{-3} & 2s^{-2} \\ 0 & 2s^{-3} \end{bmatrix} \begin{bmatrix} \left(\frac{\ddot{y}_0}{s} \right) \\ \left(\frac{\dot{y}_0}{s} \right) \end{bmatrix}$$

$$= \begin{bmatrix} -s^{-3} \frac{du}{ds} - s^{-4} u + \frac{dy}{ds} + 4s^{-1} y - 3s^{-1} \left(\frac{y_0}{s} \right) \\ -s^{-3} \frac{d^2 u}{ds^2} - 2s^{-4} \frac{du}{ds} + \frac{d^2 y}{ds^2} + 8s^{-1} \frac{dy}{ds} + 12 s^{-2} y - 6s^{-2} \left(\frac{y_0}{s} \right) \end{bmatrix} \tag{5.53}$$

which in the time domain reads

$$\begin{bmatrix} \frac{t^3}{6} & t^2 \\ 0 & \frac{t^3}{3} \end{bmatrix} \begin{bmatrix} \ddot{y}_0 \\ \dot{y}_0 \end{bmatrix}$$

$$= \begin{bmatrix} (\int^{(3)} tu) - (\int^{(4)} u) - ty + 4(\int y) - 3ty_0 \\ 2(\int^{(4)} tu) - (\int^{(3)} t^2 u) + t^2 y - 8(\int ty) + 12(\int^{(2)} y) - 3t^2 y_0 \end{bmatrix} \tag{5.54}$$

5.2.4.1   Simulation Results

We ran some simulations on the perturbed system (5.40) with $\gamma(t)$ a computer-generated stochastic process characterized by a normal probability distribution function at each time t. We also set the offset parameter $b = 0.4$.

Additionally, we also simulated the results of the estimation process under the influence of a measurement noise $v(t)$ given by a zero-mean computer random process. The algebraic observer-based control (5.43) was established to follow a desired trajectory $y^*(t)$ generated by a Genesio–Tesi chaotic system (see Section 3.8). Figure 5.7 depicts the tracking process. Notice that $y(t)$ reaches $y^*(t)$ in a short amount of time and without any overshooting effects. The control input u is also present in this figure. Finally, this figure illustrates the disturbance inputs used in the system and in the measurements given by γ and $v(t)$, respectively.

On the contrary, the state identification process is shown in Figure 5.8. The noise effects are present in the estimation of the second-order time derivative, but the estimation errors do not affect the control process and the first-order time derivative estimate is quite accurate.

5.2.5   A Sinusoid Estimation Problem

Consider the classical problem of determining the amplitude, phase, and frequency of a sinusoid signal generated by a perturbed oscillator system with one unknown initial condition:

$$\ddot{y} = -\omega^2 y + \xi(t)$$
$$y_0 = A \sin \phi \tag{5.55}$$
$$\dot{y}_0 = A\omega \cos \phi$$

This is an example of a simultaneous parameter and state estimation problem. We proceed as before:

$$s^2 \hat{y}(s) - s y_0 - \dot{y}_0 + \omega^2 \hat{y}(s) = \hat{\xi}(s) \tag{5.56}$$

Taking two derivatives with respect to s, multiplying out the resulting expression by s^{-2}, and solving for ω^2, we obtain, after writing everything back in the time domain:

$$\omega^2 = \frac{-2(\int^{(2)} y) + 4(\int ty) - t^2 y + (\int^{(2)} t^2 \xi)}{(\int^{(2)} t^2 y)} \tag{5.57}$$

From the previous algebraic manipulations, we have (Figures 5.7 and 5.8):

$$2sy(s) + s^2 \frac{dy(s)}{ds} - s\left(\frac{y_0}{s}\right) + \omega^2 \frac{dy(s)}{ds} = \frac{d\hat{\xi}(s)}{ds} \tag{5.58}$$

Multiplying out by s^{-2} we obtain, in the time domain,

$$y_0 = \frac{2(\int y) - ty(t) - \omega^2(\int^{(2)} ty(t) + (\int^{(2)} t\xi))}{t} \tag{5.59}$$

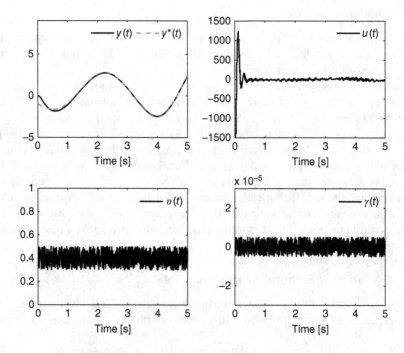

Figure 5.7   Tracking process and disturbance signals

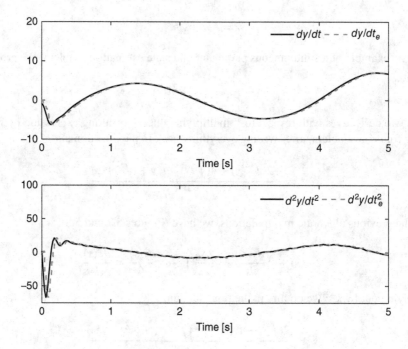

Figure 5.8   Algebraic state estimation

Following a similar procedure, the initial condition $\dot{y}_0$ is determined as

$$\dot{y}_0 = \frac{y(t) - y_0 + \omega^2(\int^{(2)} y(t)) - (\int^{(2)} \xi)}{t} \tag{5.60}$$

The unknown parameters ω^2, y_0, and $\dot{y}_0$ are clearly linearly identifiable since they can be computed (modulo perturbations) from the linear equation

$$\begin{bmatrix} (\int^{(2)} t^2 y) & 0 & 0 \\ (\int^{(2)} ty) & t & 0 \\ -(\int^{(2)} y) & 1 & t \end{bmatrix} \begin{bmatrix} \omega^2 \\ y_0 \\ \dot{y}_0 \end{bmatrix} = \begin{bmatrix} -2(\int^{(2)} y) + 4(\int ty) - t^2 y \\ 2(\int y) - ty \\ y \end{bmatrix} \tag{5.61}$$

Once ω, y_0, and $\dot{y}_0$ are known, we compute the parameters A and ϕ using

$$A = \sqrt{\left(\frac{\dot{y}_0^2}{\omega^2}\right) + y_0^2}, \quad \phi = \arctan\left(\omega \frac{y_0}{\dot{y}_0}\right) \tag{5.62}$$

5.2.5.1 Simulation Results

The last procedure was verified by means of some numerical simulations. The sinusoidal wave parameters were set as follows: peak amplitude $A = 127$, angular frequency $\omega = 120\pi$ s^{-1}. A phase value of $\phi = 0.7$ was chosen. Finally, the output was affected by a zero-mean computer-generated random signal with a peak amplitude of 5% of the nominal peak amplitude of the measured output signal. Figure 5.9 shows the numerical simulation results. The estimation process for both parameters was accurate in spite of the noisy signal. The parameters, and their estimates, are located on the left of the figure while the output and the noisy signals are shown on the right.

5.2.6 Identification of Gravitational Wave Parameters

Consider the perturbed signal

$$y(t) = \sin(at + \sin(bt + c) + d) + v(t) \tag{5.63}$$

where $v(t)$: is a stochastic perturbation signal
  a, b, c, d : are constant, unknown, parameters

5.2.6.1 Problem Formulation

It is desired to accurately compute the unknown parameters in the signal model

$$y(t) = \sin(at + \sin(bt + c) + d) + v(t) \tag{5.64}$$

from measurements performed on the signal y(t) alone.

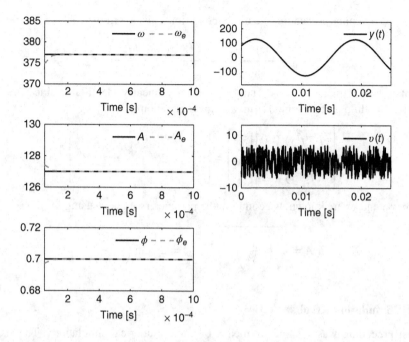

Figure 5.9   Algebraic state estimation

We first consider the nominal, unperturbed signal, which we denote by $z(t)$, without the presence of the measurement noise, and propose an algorithm to compute the unknown parameters in a rather fast manner.

We go ahead and apply the developed algorithm on the perturbed signal to find that the presence of the noise does not prevent us from obtaining a fast and quite accurate estimate of the parameters.

We restrict our considerations to the interval $[-\pi/2, +\pi/2]$.

Let $\theta(t) = \arcsin(z(t))$. Knowledge of θ is equivalent to knowledge of z. It is easy to see that θ satisfies the following linear differential equation with unknown initial conditions:

$$\theta^{(4)}(t) + b^2 \ddot{\theta}(t) = 0 \tag{5.65}$$

$$\theta(0) = \sin(c) + d \tag{5.66}$$

$$\dot{\theta}(0) = b\cos(c) + a \tag{5.67}$$

$$\ddot{\theta}(0) = -b^2 \sin(c) \tag{5.68}$$

$$\theta^{(3)}(0) = -b^3 \cos(c) \tag{5.69}$$

Suppose, for a moment, that b is known then note that the problem is reduced to the accurate computation of the uncertain initial state set $\{\theta(0), \dot{\theta}(0), \ldots, \theta^{(3)}(0)\}$ since, then, the uncertain parameters can readily be computed as

$$a = \dot{\theta}(0) + \frac{\theta^{(3)}(0)}{b^2} \tag{5.70}$$

$$c = \arctan\left(\frac{b\ddot{\theta}(0)}{\theta^{(3)}(0)}\right) \tag{5.71}$$

$$d = \theta(0) + \frac{\ddot{\theta}(0)}{b^2} \tag{5.72}$$

Although it is easy to establish that the parameter b also satisfies the following relationship:

$$b^6 - \ddot{\theta}(0)b^2 - (\theta^{(3)}(0))^2 = 0 \tag{5.73}$$

we shall concentrate on the *simultaneous* computation of b and the unknown initial conditions. Our problem is then a parameter identification problem, coupled to an initial state estimation problem to be solved from corrupted output measurements.

In the usual operational calculus, the differential equation defining θ reads

$$s^4\theta(s) - s^3\theta(0) - s^2\dot{\theta}(0) - s\ddot{\theta}(0) - \theta^{(3)}(0)$$
$$+b^2[s^2\theta(s) - s\theta(0) - \dot{\theta}(0)] = 0 \tag{5.74}$$

Differentiating this quantity with respect to s *four times*, we obtain an expression which is free from all the initial conditions:

$$24\theta(s) + 96s\frac{d\theta(s)}{ds} + 72s^2\frac{d^2\theta(s)}{ds^2} + 16s^3\frac{d^3\theta(s)}{ds^3} + s^4\frac{d^4\theta(s)}{ds^4}$$
$$+b^2\left(12\frac{d^2\theta(s)}{ds^2} + 8s\frac{d^3\theta}{ds^3} + s^2\frac{d^4\theta(s)}{ds^4}\right) = 0 \tag{5.75}$$

Integrating this expression *four times* we obtain, still in the frequency domain,

$$24s^{-4}\theta(s) + 96s^{-3}\frac{d\theta(s)}{ds} + 72s^{-2}\frac{d^2\theta(s)}{ds^2} + 16s^{-1}\frac{d^3\theta(s)}{ds^3} + \frac{d^4\theta(s)}{ds^4}$$
$$+b^2\left(12s^{-4}\frac{d^2\theta(s)}{ds^2} + 8s^{-3}\frac{d^3\theta}{ds^3} + s^{-2}\frac{d^4\theta(s)}{ds^4}\right) = 0$$

Solving for b^2, we readily obtain

$$b^2 = -\frac{24s^{-4}\theta(s) + 96s^{-3}\frac{d\theta(s)}{ds} + 72s^{-2}\frac{d^2\theta(s)}{ds^2} + 16s^{-1}\frac{d^3\theta(s)}{ds^3} + \frac{d^4\theta(s)}{ds^4}}{\left(12s^{-4}\frac{d^2\theta(s)}{ds^2} + 8s^{-3}\frac{d^3\theta}{ds^3} + s^{-2}\frac{d^4\theta(s)}{ds^4}\right)} \tag{5.76}$$

In the time domain the previous formula, which should be evaluated during a small time interval $[0, \epsilon]$, reads

$$b^2 = -\frac{24(\int^{(4)}\theta) - 96(\int^{(3)}t\theta) + 72(\int^{(2)}t^2\theta) - 16(\int t^3\theta) + t^4\theta}{12(\int^{(4)}t^2\theta) - 8(\int^{(3)}t^3\theta) + (\int^{(2)}t^4\theta)} \tag{5.77}$$

where, as usual, we have used the following notation for the iterated convolutions:

$$\left(\int^{(j)}t^p\theta\right) = \int_0^t\int_0^{\sigma_1}\cdots\int_0^{\sigma_{j-1}}\sigma_j^p\theta(\sigma_j)d\sigma_j\cdots d\sigma_1$$

$$\left(\int t^p\theta\right) = \left(\int^{(1)}t^p\theta\right) = \int_0^t\sigma_1^p\theta(\sigma_1)d\sigma_1 \tag{5.78}$$

Note that at time $t = 0$, the previous formula yields an undefined number. At time $t = \epsilon >$ 0, arbitrarily small, the computation of b^2 yields an accurate estimate (see the simulations section).

To compute the required initial conditions, basically we should only concern ourselves with the set $\{\dot{\theta}(0), \ddot{\theta}(0), \theta^{(3)}(0)\}$ since $\theta(0)$ is known from signal measurements. Nevertheless, we proceed to derive, respectively, three, two, and one times the transformed differential equation for θ to consecutively eliminate the initial conditions multiplying the smaller powers of the variable s.

In this manner, we obtain *all* the initial conditions, including $\theta(0)$, from our measurements carried out during a small time interval $[0, \epsilon]$.

After sufficient integrations, and using the fact that transforms of constants exhibit an "s" operator dividing the value of the constant, we obtain the following expressions for the unknown initial conditions:

$$6s^{-3}\frac{\theta(0)}{s} = 24s^{-3}\theta(s) + 36s^{-2}\frac{d\theta(s)}{ds} + 12s^{-1}\frac{d^2\theta(s)}{ds^2} + \frac{d^3\theta(s)}{ds^3}$$

$$+ b^2(6s^{-4}\frac{d\theta(s)}{ds} + 6s^{-3}\frac{d^2\theta(s)}{ds^2} + s^{-2}\frac{d^3\theta(s)}{ds^3}) \tag{5.79}$$

In the time domain this is interpreted as

$$\theta(0) = \frac{1}{t^3}[m_0(t) + b^2 n_0(t)] \tag{5.80}$$

$$m_0(t) = 24\left(\int^{(3)}\theta\right) - 36\left(\int^{(2)}t\theta\right) + 12\left(\int^{(1)}t^2\theta\right) - t^3\theta(t) \tag{5.81}$$

$$n_0(t) = -6\left(\int^{(4)}t\theta\right) + 6\left(\int^{(3)}t^2\theta\right) - \left(\int^{(2)}t^3\theta\right) \tag{5.82}$$

Similarly, we obtain

$$2s^{-3}\frac{\dot{\theta}(0)}{s} = 12s^{-2}\theta(s) + 8s^{-1}\frac{d\theta}{ds} + \frac{d^2\theta}{ds^2} - 6s^{-2}\frac{\theta(0)}{s}$$

$$+ b^2\left(2s^{-4}\theta(s) + 4s^{-3}\frac{d\theta(s)}{ds} + s^{-2}\frac{d^2\theta(s)}{ds^2}\right) \tag{5.83}$$

from were it is immediate that

$$\dot{\theta}(0) = \frac{3}{t^3}[m_1(t) + b^2 n_1(t)] \tag{5.84}$$

$$m_1(t) = 12\left(\int^{(2)}\theta\right) - 8\left(\int^{(1)}t\theta\right) + t^2\theta(t) - 3t^2\theta(0) \tag{5.85}$$

$$n_1(t) = 2\left(\int^{(4)}\theta\right) - 4\left(\int^{(3)}t\theta\right) + \left(\int^{(2)}t^2\theta\right) \tag{5.86}$$

We also have

$$s^{-3}\frac{\ddot{\theta}(0)}{s} = 4s^{-1}\theta(s) + \frac{d\theta(s)}{ds} - 3s^{-1}\frac{\theta(0)}{s} - 2s^{-2}\frac{\dot{\theta}(0)}{s}$$

$$+ b^2\left(2s^{-3}\theta(s) + s^{-2}\frac{d\theta(s)}{ds} - s^{-3}\frac{\theta(0)}{s}\right) \tag{5.87}$$

from where

$$\ddot{\theta}(0) = \frac{6}{t^3}[m_2(t) + b^2 n_2(t)] \tag{5.88}$$

$$m_2(t) = 4\left(\int^{(1)}\theta\right) - t\theta(t) - 3t\theta(0) - t^2\dot{\theta}(0) \tag{5.89}$$

$$n_2(t) = \left[2\left(\int^{(3)}\theta\right) - \left(\int^{(2)}t\theta\right) - \frac{t^3}{6}\theta(0)\right] \tag{5.90}$$

Finally,

$$s^{-3}\frac{\theta^{(3)}(0)}{s} = \theta(s) - \frac{\theta(0)}{s} - s^{-1}\frac{\dot{\theta}(0)}{s} - s^{-2}\frac{\ddot{\theta}(0)}{s}$$

$$+ b^2\left(s^{-2}\theta(s) - s^{-2}\frac{\theta(0)}{s} - s^{-3}\frac{\dot{\theta}(0)}{s}\right) \tag{5.91}$$

Hence,

$$\theta^{(3)}(0) = \frac{6}{t^3}[m_3(t) + b^2 n_3(t)] \tag{5.92}$$

$$m_3(t) = \theta(t) - \theta(0) - t\dot{\theta}(0) - \frac{t^2}{2}\ddot{\theta}(0) \tag{5.93}$$

$$n_3(t) = \left(\int^{(2)}\theta\right) - \frac{t^2}{2}\theta(0) - \frac{t^3}{6}\dot{\theta}(0) \tag{5.94}$$

5.2.6.2 Simulations

Here we will present the simulation results of the proposed parameter identification algorithm for the gravitational wave, including in the simulations the fact that $y(t)$ is a noise-corrupted signal.

We used the following perturbed wave-generating system and perturbed measurements:

$$\theta^{(4)} + b^2\ddot{\theta} = \eta(t) \tag{5.95}$$

$$y(t) = \sin(\theta(t)) + \xi(t) = \sin(at + \sin(bt + c) + d) + \xi(t) \tag{5.96}$$

The actual parameters used for the simulation of the wave generator were

$$a = 1, \quad b = 1.5, \quad c = 0.4, \quad d = 0.1$$

with $\eta(t) = 1.5 \, \text{NORM}(t)$, that is, a stochastic process constituted by a normally distributed quasi-random variable at each instant of time t, with amplitude 1.5.

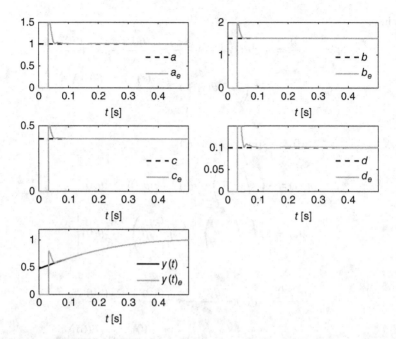

Figure 5.10 Result of unknown parameter calculation processed with generating system input noise and output measurement noise

The measurement noise $\xi(t)$ was set to be $\xi(t) = 1.0 \times 10^{-8}(rect(t) - 0.5)$, where $rect(t)$ is a computer-generated quasi-random variable uniformly distributed in the interval [0, 1] of the real line.

Figure 5.10 shows the outcome of the proposed algorithm for the estimated values of the parameters.

5.2.7   A Power Electronics Example

$$\begin{cases} u = 1 : & S_1 = ON, S_2 = ON, S_3 = OFF, S_4 = OFF \\ u = 0 : & S_1 = OFF, S_2 = OFF, S_3 = ON, S_4 = ON \end{cases}$$

Consider the average dynamics of a DC-to-DC double-bridge buck converter (Figure 5.11):

$$L\dot{x}_1 = -x_2 + (2v - 1)E$$

$$C\dot{x}_2 = x_1 - \frac{x_2}{R}$$

$$\eta = x_2 \tag{5.97}$$

x_1 : inductor current
x_2 : capacitor voltage
$v \in [0, 1]$ represents the *duty ratio* of the switch position function
L, C, E are assumed to be known while R is an unknown constant

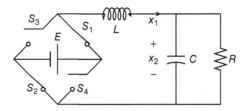

Figure 5.11 The double-bridge buck converter

The average normalized model of the system is given by

$$\dot{z}_1 = -z_2 + (2v - 1)$$
$$\dot{z}_2 = z_1 - \frac{z_2}{Q} \tag{5.98}$$
$$y = z_2$$

where

$$z_1 = \left(\frac{1}{E}\sqrt{\frac{L}{C}}\right)x_1, \quad z_2 = \frac{x_2}{E}, \quad \frac{1}{Q} = \frac{1}{R}\sqrt{\frac{L}{C}}, \quad \tau = \frac{t}{\sqrt{LC}}$$

We denote by μ the expression $2v - 1$, $\mu \in [-1, 1]$.

5.2.7.1 Identification of the Load

$$\ddot{y} + \frac{1}{Q}\dot{y} + y = \mu \tag{5.99}$$

To obtain $\frac{1}{Q}$ in terms of y, μ, we proceed as follows.

1. Multiply the input–output relation by t^2:

$$t^2\ddot{y} + t^2\frac{1}{Q}\dot{y} + t^2 y = t^2\mu \tag{5.100}$$

2. Integrate by parts:

$$t^2\dot{y} - 2\left[ty - \left(\int y\right)\right] + \frac{1}{Q}\left[t^2 y - 2\left(\int ty\right)\right] + \left(\int t^2 y\right) = \left(\int t^2\mu\right) \tag{5.101}$$

3. Integrate by parts once more:

$$t^2 y - 4\left(\int ty\right) + 2\left(\int^{(2)} y\right) + \frac{1}{Q}\left[\left(\int t^2 y\right) - 2\left(\int^{(2)} ty\right)\right]$$
$$+ \left(\int^{(2)} t^2 y\right) = \left(\int^{(2)} t^2\mu\right) \tag{5.102}$$

We can now solve for the unknown constant $\frac{1}{Q}$:

$$\frac{1}{Q} = \frac{\int_0^t \int_0^\sigma \{\lambda^2[\mu(\lambda) - y(\lambda)] - 2y(\lambda)\}d\lambda d\sigma - t^2 y(t) + 4\int_0^t \sigma y(\sigma)d\sigma}{\int_0^t \sigma^2 y(\sigma)d\sigma - 2\int_0^t \int_0^\sigma \lambda y(\lambda)d\lambda d\sigma} \quad (5.103)$$

5.2.7.2 State Estimation

Note that if y, $\dot{y}$ are known then, from the average dynamics equations of the system, we may compute the average states once the parameter $\frac{1}{Q}$ has been determined (Figure 5.12).

Indeed, from the system equations we have

$$z_1 = \dot{y} + \frac{1}{Q}y, \quad z_2 = y \quad (5.104)$$

To determine the two state variables z_1, z_2, we only need to determine $\dot{y}$ since y is known.

From the first integration by parts, we have

$$t^2\dot{y} - 2\left[ty - \left(\int y\right)\right] + \frac{1}{Q}\left[t^2 y - 2\left(\int ty\right)\right] + \left(\int t^2 y\right) = \left(\int t^2 \mu\right) \quad (5.105)$$

that is, the quantity $\dot{y}$ may be computed once $\frac{1}{Q}$ has been determined:

$$\dot{y} = \frac{2(ty - (\int y)) - \frac{1}{Q}[t^2 y - 2(\int ty)] - (\int t^2 y) + (\int t^2 \mu)}{t^2} \quad (5.106)$$

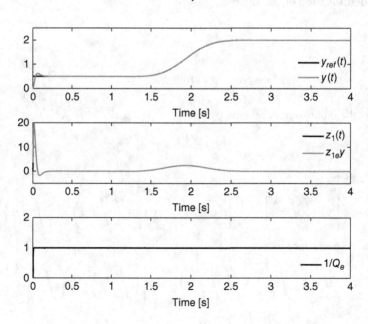

Figure 5.12 Load estimation and state estimation for the double-bridge converter

It then follows that

$$z_1 = \frac{2(ty - (\int y)) + \frac{2}{Q}(\int ty) - (\int t^2 y) + (\int t^2 \mu)}{t^2} \qquad (5.107)$$

$$z_2 = y \qquad (5.108)$$

5.2.7.3 Simulations

We considered an output trajectory-tracking problem $y \to y^*(t)$ with simultaneous identification of the unknown parameter and estimation of the unmeasured state.

5.2.8 A Hydraulic Press

Figure 5.13 represents a hydraulic press. A plunger is actuated by means of a control signal $u(t)$. The plunger's mass M has some reaction forces due to a linear viscous friction component with a damping parameter B. The pressed body has negligible mass and behaves as an ideal spring with an elastic parameter K_1. The body to be pressed is over a platform which is isolated by four other springs of elastic parameter K_2.

We have the following equations:

$$M\ddot{x}_1 + B\dot{x}_1 + K_1(x_1(t) - x_2(t)) = u \qquad (5.109)$$

$$K_1(x_1(t) - x_2(t)) = 4K_2 x_2(t) \qquad (5.110)$$

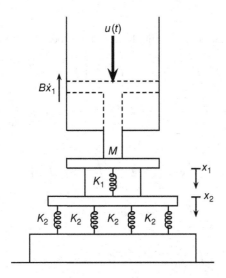

Figure 5.13 The hydraulic press

Let us define $x = x_1 - x_2$, which represents the compression range of the body. From the last equations:

$$K_1 x(t) = 4K_2(x_1(t) - x) \Rightarrow x_1 = \frac{K_1 + 4K_2}{4K_2}x \tag{5.111}$$

We obtain the following simplified equation:

$$\ddot{x} + a_1\dot{x} + a_0 x = bu(t) \tag{5.112}$$

where

$$a_1 = \frac{B}{M}$$

$$a_0 = \frac{4K_1 K_2}{M(K_1 + 4K_2)}$$

$$b = \frac{4K_2}{M(K_1 + 4K_2)}$$

The main problem consists of finding a control law u to force x to track a predefined trajectory x^* using the estimates of the parameters a_1, a_0, and b but also the estimation of the state $\dot{x}$.

5.2.8.1   Parameter Estimation

We take the Laplace transform in (5.112) to obtain

$$s^2 x(s) - sx_0 - \dot{x}_0 + a_1(sx(s) - x_0) + a_0 x(s) = bu(s) \tag{5.113}$$

Taking two iterated derivatives with respect to s:

$$\left(s\frac{d^2x}{ds^2} + 2\frac{dx}{ds}\right)a_1 + \left(\frac{d^2x}{ds^2}\right)a_0 - \left(\frac{d^2u}{ds}\right)b = -s^2\frac{d^2x}{ds^2} - 4s\frac{dx}{ds} - 2x \tag{5.114}$$

Multiplying out by s^{-2} and transforming back to the time domain:

$$\left[\int_0^t \tau^2 x(\tau)d\tau - 2\int_0^t\int_0^\tau \sigma x(\sigma)d\sigma d\tau\right]a_1 + \left[\int_0^t\int_0^\tau \sigma^2 x(\sigma)d\sigma d\tau\right]a_0 -$$
$$\left[\int_0^t\int_0^\tau \sigma^2 u(\sigma)d\sigma d\tau\right]b = -t^2 x(t) + 4\int_0^t \tau x(\tau)d\tau - 2\int_0^t\int_0^\tau x(\sigma)d\sigma d\tau \tag{5.115}$$

Now, define

$$p_1(t) = \int_0^t \tau^2 x(\tau)d\tau - 2\int_0^t\int_0^\tau \sigma x(\sigma)d\sigma d\tau$$

$$p_2(t) = \int_0^t\int_0^\tau \sigma^2 x(\sigma)d\sigma d\tau$$

$$p_3(t) = -\int_0^t\int_0^\tau \sigma^2 u(\sigma)d\sigma d\tau$$

$$q(t) = -t^2 x(t) + 4\int_0^t \tau x(\tau)d\tau - 2\int_0^t\int_0^\tau x(\sigma)d\sigma d\tau$$

and let us define the vector $P(t) = \begin{bmatrix} p_1(t) & p_2(t) & p_3(t) \end{bmatrix}$, $\theta = \begin{bmatrix} a_1 & a_0 & b \end{bmatrix}^T$. The parameter estimation is then given by the following expression:

$$\hat{\theta}(t) = \left[\int_0^t P^T(\tau)P(\tau)d\tau \right]^{-1} \left[\int_0^t P^T(\tau)q(\tau)d\tau \right] \tag{5.116}$$

5.2.8.2   State Estimation and Controller Design

Since the parameter estimation vector reaches the actual values in a short time $\epsilon > 0$, the state $\dot{x}$ can be obtained as follows.

Multiplying (5.112) by t^2 and integrating with respect to time, we have

$$t^2 \dot{x}(t) - 2tx(t) + 2 \int_0^t x(\tau)d\tau + a_1 \left(t^2 x(t) - 2 \int_0^t \tau x(\tau)d\tau \right)$$

$$+ a_0 \int_0^t \tau^2 x(\tau)d\tau = b \int_0^t \tau^2 u(\tau) \tag{5.117}$$

and finally, using the parameter identified values we obtain the following expression for the estimation of $\dot{x}$:

$$\hat{\dot{x}} = \frac{\hat{b} \int_0^t \tau^2 u(\tau) - a_1(t^2 x(t) - 2 \int_0^t \tau x(\tau)d\tau) + a_0 \int_0^t \tau^2 x(\tau)d\tau + 2tx(t) - 2 \int_0^t x(\tau)d\tau}{t^2}$$

$$\tag{5.118}$$

We propose the following feedback control law:

$$u = \frac{1}{\hat{b}} \left(\ddot{x}^* + \hat{a}_1 \hat{\dot{x}} + \hat{a}_1 x - k_2(\hat{\dot{x}} - \dot{x}^*) - k_1(x - x^*) - k_0 \int (x - x^*) \right) \tag{5.119}$$

which leads to a closed-loop error dynamics given by

$$e^{(3)} + k_2 \ddot{e} + k_1 \dot{e} + k_0 e = 0$$

$$\text{with } e = x - x^*$$

5.2.8.3   Simulation Results

We ran some simulations for the press system under the described control strategy. The plant parameters were $K_1 = 500\,\text{N/m}$, $K_1 = 700\,\text{N/m}$, a damping parameter $B = 15\,\text{N/m}^2$, and a mass parameter $m = 2\,\text{kg}$. The control parameters were selected such that the poles of the closed-loop error dynamics were -4, -5, and -10, respectively. Two different simulations were carried out: the first with a lack of noisy measurements; the second in the presence of a zero-mean computer-generated noise of 1% of the nominal peak output value. Figures 5.14 and 5.15 show the tracking process and the state and parameter estimation for the free from noise case while Figures 5.16 and 5.17 depict the effects of the additive noise in the estimation process.

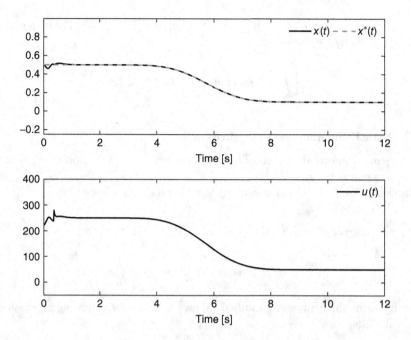

Figure 5.14 Trajectory tracking and control signal

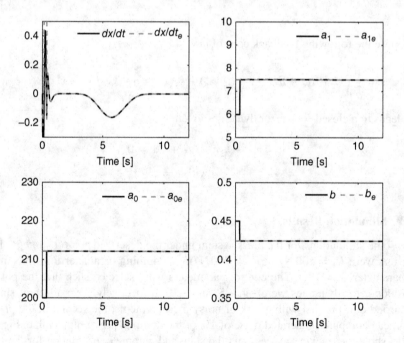

Figure 5.15 Parameter and state estimation

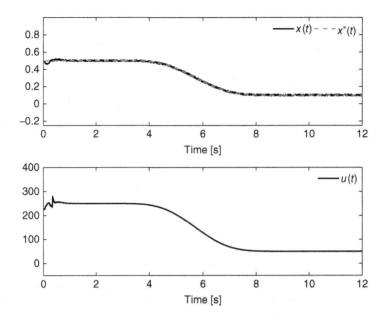

Figure 5.16 Trajectory tracking and control signal under noisy measurements

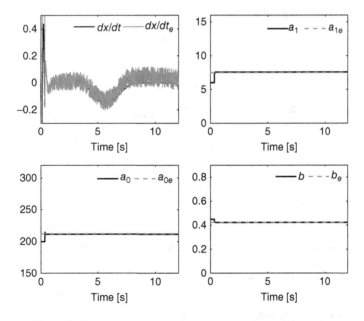

Figure 5.17 Parameter and state estimation with additive noise

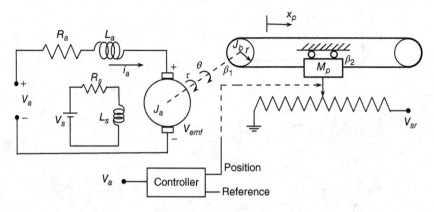

Figure 5.18   Simplified diagram of a plotter

5.2.9   Identification and Control of a Plotter

Figure 5.18 shows a plotter system, where R_a, L_a, R_s, and L_s are electric parameters of a DC motor, J_a, J_b, and β_1 are inertia and viscous damping parameters of the motor. The parameter M_c corresponds to the mass of the writing arm and pen holder and β_2 is a viscous damping coefficient of the bearing system. The position of the pen is given by a linear potentiometer.

5.2.9.1   Mathematical Model

Using the current Kirchoff law, the electrical part of the system can be characterized by the equation

$$L_a \frac{di_a}{dt} = V_a - V_{emf} - R_a i_a \tag{5.120}$$

The sum of pairs and velocity relations describe the mechanical subsystem

$$\frac{dx_p}{dt} = r\frac{d\theta}{dt} \tag{5.121}$$

$$(J_a + J_b)\frac{d^2\theta}{dt^2} = \tau - \beta_1 \frac{d\theta}{dt} - r\left(M_p \frac{d^2 x_p}{dt} + \beta_2 \frac{dx_p}{dt} \right) \tag{5.122}$$

$$T_{torque} = rF_t \tag{5.123}$$

From these equations, we can make some algebraic manipulations to obtain

$$\left(M_p + \frac{J_m + J_b}{r^2} \right) \frac{d^2 x_p}{dt^2} = \frac{1}{r}\tau - \left(\beta_2 + \frac{\beta_1}{r^2} \right) \frac{dx_p}{dt} \tag{5.124}$$

The relations between both subsystems are given by

$$V_{emf} = K_{emf} \frac{d\theta}{dt} \tag{5.125}$$

$$\tau = k_e i_a \tag{5.126}$$

where k_e is the electrical constant which indicates how much armature current the motor converts into mechanical torque applied to the shaft. The back emf voltage, V_{emf}, is the voltage generated across the motor's terminals as the windings move through the motor's magnetic field.

So, we can write the equations of the system as follows.

Electrical part:

$$L_a \frac{di_a}{dt} = V_a - \frac{K_{emf}}{r} \frac{dx_p}{dt} - R_a i_a \tag{5.127}$$

Mechanical part:

$$\left(M_p + \frac{J_m + J_b}{r^2} \right) \frac{d^2 x_p}{dt^2} = \frac{k_e}{r} i_a - \left(\beta_2 + \frac{\beta_1}{r^2} \right) \frac{dx_p}{dt} \tag{5.128}$$

Notice that the armature current can be written as

$$i_a = \frac{r}{k_e} \left(\left(M_p + \frac{J_m + J_b}{r^2} \right) \frac{d^2 x_p}{dt^2} + \left(\beta_2 + \frac{\beta_1}{r^2} \right) \frac{dx_p}{dt} \right) \tag{5.129}$$

Now, differentiating (5.128) once with respect to time and using (5.127) and (5.129), the plotter system is only written in terms of the position of the pen:

$$\left(M_p + \frac{J_m + J_b}{r^2} \right) \frac{d^3 x_p}{dt^3} + \left(\beta_2 + \frac{\beta_1}{r^2} + \frac{R_a}{L_a} \left(M_p + \frac{J_m + J_b}{r^2} \right) \right) \frac{d^2 x_p}{dt^2}$$

$$+ \frac{1}{L_a} \left(\frac{k_e K_{emf}}{r^2} + R_a \left(\beta_2 + \frac{\beta_1}{r^2} \right) \right) \frac{dx_p}{dt} = \frac{k_e}{r L_a} V_a \tag{5.130}$$

5.2.9.2 Control Law Design

The model (5.130) is written as

$$\frac{d^3 y}{dt^3} + \gamma_2 \frac{d^2 y}{dt^2} + \gamma_1 \frac{dy}{dt} = \gamma_0 u \tag{5.131}$$

where

$$\gamma_2 = \frac{\beta_2 + \frac{\beta_1}{r^2} + \frac{R_a}{L_a} \left(M_p + \frac{J_m + J_b}{r^2} \right)}{M_p + \frac{J_m + J_b}{r^2}}$$

$$\gamma_1 = \frac{\frac{k_e K_{emf}}{r^2} + R_a \left(\beta_2 + \frac{\beta_1}{r^2} \right)}{L_a \left(M_p + \frac{J_m + J_b}{r^2} \right)}$$

$$\gamma_0 = \frac{k_e}{r L_a \left(M_p + \frac{J_m + J_b}{r^2} \right)}$$

$$u = V_a$$

$$y = x_p$$

Assuming that the parameters γ_i and all the states are known, the following controller can be proposed such that the pen plotter tracks a reference trajectory:

$$u = \frac{1}{\gamma_0}\left(\gamma_2\ddot{y} + \gamma_1\dot{y} + [y^*]^{(3)} - \alpha_3(\ddot{y} - \ddot{y}^*) - \alpha_2(\dot{y} - \dot{y}^*)\right. \tag{5.132}$$

$$\left.-\alpha_1(y - y^*) - \alpha_0\int_0^t (y(\lambda) - y^*(\lambda))d\lambda\right) \tag{5.133}$$

With the coefficients α_1 corresponding to a desired Hurwitz polynomial of the form

$$s^4 + \alpha_3 s^3 + \alpha_2 s^2 + \alpha_1 s + \alpha_0 \tag{5.134}$$

the tracking error converges asymptotically to zero.

5.2.9.3 Algebraic Estimation of Parameters and States

The system (5.131) in the frequency domain is described by

$$[s^3 Y(s) - s^2 Y(0) - s\dot{Y}(0) - \ddot{Y}(0)] + \gamma_2[s^2 Y(s) - sY(0) - \dot{Y}(0)]$$

$$+\gamma_1[sY(s) - Y(0)] = \gamma_0 U(s) \tag{5.135}$$

This expression is differentiated three times with respect to the complex variable s, and next multiplied by s^{-3} to obtain a regressor form free from any time derivative of the output or input:

$$\gamma_2\left[6s^{-3}\frac{dY}{ds} + 6s^{-2}\frac{d^2Y}{ds^2} + s^{-1}\frac{d^3Y}{ds^3}\right] + \gamma_1\left[3s^{-3}\frac{d^2Y}{ds^2} + s^{-2}\frac{d^3Y}{ds^3}\right]$$

$$-\gamma_0 s^{-3}\frac{d^3U}{ds^3} = -\left[6s^{-3}Y(s) + 18s^{-2}\frac{dY}{ds} + 9s^{-1}\frac{d^2Y}{ds} + \frac{d^3Y}{ds^3}\right] \tag{5.136}$$

Returning to the time domain and rewriting the regressor in matrix form, we have

$$P(t)\Theta = Q(t) \tag{5.137}$$

where

$$P(t) = \left[-6\int^{(3)}ty + 6\int^{(2)}t^2 y - \int t^3 y \quad 3\int^{(3)}t^2 y - \int^{(2)}t^3 y \quad \int^{(3)}t^3 u\right]$$

$$\Theta = \begin{bmatrix}\gamma_2 \\ \gamma_1 \\ \gamma_0\end{bmatrix}$$

$$Q(t) = -6\int^{(3)} y + 18\int^{(2)} ty - 9\int t^2 y + t^3 y$$

The algebraic estimator for the unknown parameters is given by

$$\hat{\Theta} = \left[\int P^T P\right]^{-1}\int P^T Q \tag{5.138}$$

The estimator of the first derivative of y is obtained by differentiating (5.135) three times with respect to s and multiplying the result by s^{-2}:

$$\gamma_2 \left[6s^{-2}\frac{dY}{ds} + 6s^{-1}\frac{d^2Y}{ds^2} + \frac{d^3Y}{ds^3} \right] + \gamma_1 \left[3s^{-2}\frac{d^2Y}{ds^2} + s^{-1}\frac{d^3Y}{ds^3} \right]$$
$$-\gamma_0 s^{-2}\frac{d^3U}{ds^3} = -\left[6s^{-2}Y(s) + 18s^{-1}\frac{dY}{ds} + 9\frac{d^2Y}{ds} + s\frac{d^3Y}{ds^3} \right] \tag{5.139}$$

Returning the expression to the time domain and after some algebraic manipulations, the formula for estimating $\dot{y}$ is

$$\hat{\dot{y}} = \frac{1}{t^3}\left(6\int^{(2)} y - 18\int ty + 6t^2y + \hat{\gamma}_2\left(-6\int^{(2)} ty + 6\int t^2y - t^3y \right) \right.$$
$$\left. +\hat{\gamma}_1\left(3\int^{(2)} t^2y - \int t^3y \right) + \hat{\gamma}_0\int^{(2)} t^3u \right) \tag{5.140}$$

The expression for the algebraic estimator of $\ddot{y}$ is deduced by differentiating (5.135) three times with respect to s, multiplying immediately after by s^{-1}, and returning to the time domain:

$$\hat{\ddot{y}} = \frac{1}{t^3}\left(3t^2\hat{\dot{y}} - 6ty + 6\int y + \hat{\gamma}_2\left(-6\int ty + 3t^2y - t^3\hat{\dot{y}} \right) \right.$$
$$\left. +\hat{\gamma}_1\left(3\int t^2y - t^3y \right) + \hat{\gamma}_0\int t^3u \right) \tag{5.141}$$

Rewriting both estimators in matrix form, we have

$$\begin{bmatrix} 1 & 0 \\ \frac{-3t^2 + \hat{\gamma}_2 t^3}{t^3} & 1 \end{bmatrix}\begin{bmatrix} \hat{\dot{y}} \\ \hat{\ddot{y}} \end{bmatrix} = \frac{1}{t^3}\begin{bmatrix} 6\int^{(2)} y - 18\int ty + 6t^2y \\ -6ty + 6\int y \end{bmatrix}$$
$$+ \frac{1}{t^3}\begin{bmatrix} -6\int^{(2)} ty + 6\int t^2y - t^3y & 3\int^{(2)} t^2y - \int t^3y & \int^{(2)} t^3u \\ -6\int ty + 3t^2y & 3\int t^2y - t^3y & \int t^3u \end{bmatrix}\begin{bmatrix} \hat{\gamma}_2 \\ \hat{\gamma}_1 \\ \hat{\gamma}_0 \end{bmatrix} \tag{5.142}$$

Simplifying,

$$\begin{bmatrix} \hat{\dot{y}} \\ \hat{\ddot{y}} \end{bmatrix} = \frac{1}{t^3}\begin{bmatrix} 1 & 0 \\ \frac{3}{t} - \hat{\gamma}_2 & 1 \end{bmatrix}\left(\begin{bmatrix} 6\int^{(2)} y - 18\int ty + 6t^2y \\ -6ty + 6\int y \end{bmatrix}\right.$$
$$\left. + \begin{bmatrix} -6\int^{(2)} ty + 6\int t^2y - t^3y & 3\int^{(2)} t^2y - \int t^3y & \int^{(2)} t^3u \\ -6\int ty + 3t^2y & 3\int t^2y - t^3y & \int t^3u \end{bmatrix}\begin{bmatrix} \hat{\gamma}_2 \\ \hat{\gamma}_1 \\ \hat{\gamma}_0 \end{bmatrix} \right) \tag{5.143}$$

5.2.9.4 Simulation Results

The parameters used in the simulations were

Parameter	Value	Units
R_a	0.5	Ω
L_a	1.5e − 3	H
k_e	0.1	N m/A
J_a	1e − 5	kg m^2
β_1	0.2e − 3	N m s/rad
K_{emf}	0.09	V s/rad
r	0.01	m
J_b	2.5e − 5	kg m^2
M_p	0.45	kg
β_2	2e − 3	N s/m

The initial conditions were

$$\left[i_0 \ \left(\frac{di}{dt} \right)_0 \ x_0 \ \left(\frac{dx}{dt} \right)_0 \right]^T = \begin{bmatrix} 0 & 0 & 0.05 & 0 \end{bmatrix}^T$$

Figures 5.19, 5.20, and 5.21 show the closed-loop behavior of the system when both parameter and states are being estimated online simultaneously by means of algebraic estimators. Figures 5.22, 5.23, and 5.24 show the closed-loop behavior under noisy conditions. The signal noise was set to

$$y(t) = \mu(t) + x(t)$$

where $\mu(t) = 1 \times 10^{-8}(rect(t) - 0.5)$.

5.3 Recovering Chaotically Encrypted Signals

The field of chaotic systems has undergone considerable development, with a fairly good understanding of the phenomenon and its many implications in applied mathematics, physics, engineering, and other scientific research areas. The many interesting developments are due to mathematicians, physicists, computer scientists, control engineers, and biologists. The state-of-the-art has been summarized in several special issues of known journals which have been devoted to the problem of chaos, in general, and to chaotic systems' synchronization and control, in particular (see, for instance, Special Issue, 1993, 1997a,b, 1999, 2000, 2001).

The reader may look at the enormous collection of references about chaotic systems, and related fields, gathered by Chen (1997). A number of books already exist on the subject (see, for instance, Holden, 1986; Mira, 1987; Afraimovitch *et al.*, 1994; Ott *et al.*, 1994; Fradkov and Pogromsky, 1998; Chen, 1999; and many others). The interest in the topic of synchronization and chaotic system state estimation arises from the possibilities of encoding, or masking, messages using as analog "carrier" a signal representing a state, or an output, of a given chaotic

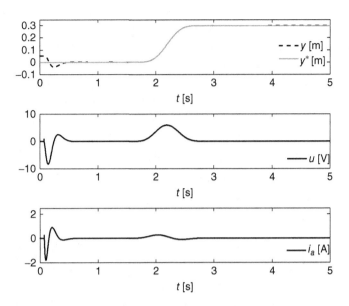

Figure 5.19 Closed-loop behavior of the plotter system with simultaneous state and parameter estimation

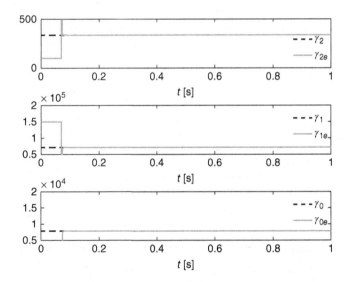

Figure 5.20 Online parameter identification of the plotter system

system. The effectively random nature of the carrier signal additively, or multiplicatively, modulated by the masked message signal makes it "difficult" to attempt decoding of the message from an intercepted transmission (see Cuomo *et al.*, 1993). The problem is then one of effectively recovering the hidden, or encrypted, message at the receiving end by means of an estimator system, or an algorithm, which uses one or several of the transmitted signals.

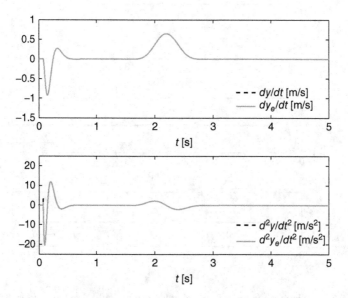

Figure 5.21   Estimation of $\dot{y}$ and $\ddot{y}$ in the plotter system

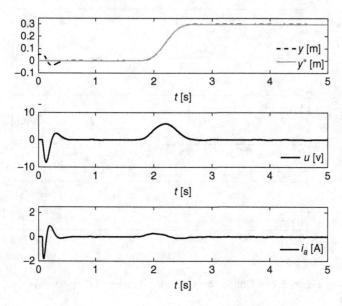

Figure 5.22   Closed-loop behavior of the plotter system with simultaneous state and parameter estimation under noisy conditions

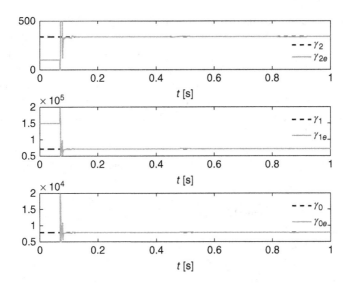

Figure 5.23 Online parameter identification of the plotter system under noisy conditions

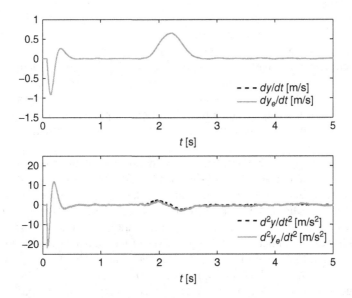

Figure 5.24 Estimation of $\dot{y}$ and $\ddot{y}$ in the plotter system under noisy conditions

The chaotic system synchronization problem is, therefore, intimately related to the design of a nonlinear *state observer* for the chaotic encoding system (see Nijmeijer and Mareels, 1997). In fact, a possible decoding process is based on the remote generation of the state estimates of the coding system, from a transmitted chaotic output signal, and a suitable comparison of such generated state estimates with the transmitted signals containing the message-modulated states. However, in contradistinction to observer design, important limitations and freedoms must be taken into account in the decoding system design problem. Traditionally, asymptotic tracking of the actual transmitter's state is demanded by exciting the designed receiving system with a message-free output. The receiving end system should asymptotically track (synchronize) the states of the transmitting system. This approach, however, generates a need to robustly sustain the "unmodeled" addition of a masked signal input after synchronization has taken place (see Pecora and Carroll, 1991). This insensitivity, or robustness, property is questionable and difficult to achieve in practice. Several research articles deal with some of these important robustness issues. For a passivity-based adaptive approach to synchronization the reader is referred to the interesting articles by Fradkov and Markov (1997) and Pogromsky (1998). In contrast a purely state estimation-based approach entitles transmission of the chaotic system output for the purpose of remotely generating the unperturbed states of the transmitting system via a properly designed asymptotic observer. The secret messages are then coded in the chaotic states and these are transmitted for comparison with the unperturbed estimated states. The message signal recovery is then immediate. The Hamiltonian structure of a collection of well-known examples of chaotic continuous-time systems is exploited in Sira-Ramírez and Cruz-Hernández (2001) to obtain asymptotic state observers requiring the message-free output signal. One important feature is that this observer design merely requires linear-based output injection techniques. A similar approach for signal encryption strategies, dealing with the exact state estimation of discrete-time nonlinear chaotic systems, was presented in Sira-Ramírez *et al.* (2002).

In this section, we take an *algebraic viewpoint* for the state estimation problem associated with the chaotic encryption-decoding problem. Here we emphasize the use of the "algebraic derivative method" for efficient and fast computation of accurate approximations to the successive time derivatives of the transmitted observable output signal received at the decoding end.

The observability of the system output allows one to establish a map constituted by a *differential function* of the output (i.e., a function of the output and a finite number of its time derivatives) from which the state can be computed immediately in a static manner (see Diop and Fliess, 1991).

Instead of attempting the construction of an asymptotic nonlinear observer for the transmitter or coding system, a set of model-independent formulae is developed for the required approximate computation of the time derivatives of an observable transmitted output signal. From these locally valid output time derivatives, the related transmitter system state vector can easily be computed using the static differential parametrization of the states in terms of the observable chaotic output. The time derivatives of the output signal are computed on the basis of a sufficiently accurate truncated Taylor series approximation in combination with the "algebraic derivative method" (see Fliess and Sira-Ramírez, 2004a,b) for state estimation. The key issue here is to initially view the transmitted chaotic system output as a time signal, with no other systems-oriented view of its possible functional dependence upon the system state. As a result, a non-asymptotic, fast state estimation scheme is obtained. The result of our algebraic estimation approach is a set of accurate piecewise continuous approximations to the actual chaotic system state vector components. The calculation method also provides an online updating mechanism that allows for automatic resetting of the involved computations when the validity of the adopted truncated Taylor series approximation ceases to be valid. Incidentally, our formulae for the online generation of the output signal time derivatives consist solely of terms involving integrations and time convolutions of the original observable output signal. The problem of efficiently calculating time derivatives of a given output signal is not entirely new, and some rather interesting approaches have been proposed (see Diop *et al.*, 1994, 2000; Plestan and Grizzle, 1999). Some of the presented information is based on the work of Sira-Ramírez and Fliess (2006).

In this section, several state estimation examples are provided, such as the well-known Lorenz system, Chen's system, Chua's chaotic circuit, Rossler's system, and the hysteretic chaotic system. Owing to a local observability property found in the first two examples, the output time derivative-based state estimation leads to a singularity problem in the reconstruction of one of the state variables. This singularity would invalidate the use of that particular state as a coding signal. Section 5.3.5 explains the coding–decoding process based on the algebraic approach to state estimation and provides some simulation examples of a secret message signal extraction which includes dealing with computer-generated additive transmission noise. The singularity problem encountered for the Lorenz and Chen systems is circumvented by using as coding signal a nonlinear function of the chaotic observable output and the involved singular state. A simulation example is also furnished, depicting the proposed singularity-free coding–decoding scheme.

5.3.1   State Estimation for a Lorenz System

Consider the model of the popular Lorenz system (see Lorenz, 1963):

$$\dot{x}_1 = \sigma(x_2 - x_1)$$

$$\dot{x}_2 = rx_1 - x_2 - x_1 x_3$$

$$\dot{x}_3 = x_1 x_2 - bx_3 \tag{5.144}$$

where $y = x_1$ is the measured output variable. The parameters σ, r, and b are assumed to be known parameters. The system is observable from the output y in all of R^3 except on the line $y = x_1 = 0$.

A local differential parametrization of the system states in terms of the measured output y is given by

$$x_1 = y$$

$$x_2 = \frac{1}{\sigma}\dot{y} + y$$

$$x_3 = -\frac{1}{y}\left[\frac{1}{\sigma}\ddot{y} + \left(1 + \frac{1}{\sigma}\right)\dot{y} + (1 - r)y\right] \tag{5.145}$$

For the generation of the time derivatives of the measured output $y(t) = x_1(t)$, we may propose a seventh-order truncated Taylor series expansion around the reinitialization time t_r of the form

$$y(t) = \sum_{i=1}^{7} \frac{y^{(i-1)}(t_r)}{(i-1)!}(t - t_r)^{(i-1)}$$

which leads, modulo a fixed time translation, to the identity

$$\frac{d^7}{ds^7}[s^7 y(s)] = 0$$

Based on this, we use the specific formulae

$$n_1(t) = 42(t - t_r)^6 y(t) - 882 \left(\int_{t_r} (t - t_r)^5 y \right) + 7350 \left(\int_{t_r}^{(2)} (t - t_r)^4 y \right)$$

$$-29400 \left(\int_{t_r}^{(3)} (t - t_r)^3 y \right) + 52920 \left(\int_{t_r}^{(4)} (t - t_r)^2 y \right)$$

$$-35280 \left(\int_{t_r}^{(5)} (t - t_r) y \right) + 5040 \left(\int_{t_r}^{(6)} y \right)$$

$$d(t) = (t - t_r)^7$$

$$(\dot{y}(t))_e = \begin{cases} (\dot{y}(t_r^-))_e + (t - t_r)(\ddot{y}(t_r^-))_e & \text{for } t \in [t_r, t_r + \epsilon) \\ \dfrac{n_1(t)}{d(t)} & \text{for } t \geq t_r + \epsilon \end{cases}$$

$$n_2(t) = -630(t - t_r)^5 y(t) + 35(t - t_r)^6 (\dot{y}(t))_e + 7350 \left(\int_{t_r} (t - t_r)^4 y \right)$$

$$-29400 \left(\int_{t_r}^{(2)} (t - t_r)^3 y \right) + 52920 \left(\int_{t_r}^{(3)} (t - t_r)^2 y \right)$$

$$-35280 \left(\int_{t_r}^{(4)} (t - t_r) y \right) + 5040 \left(\int_{t_r}^{(5)} y \right)$$

$$d(t) = (t - t_r)^7$$

$$(\ddot{y}(t))_e = \begin{cases} \ddot{y}(t_r^-) & \text{for } t \in [t_r, t_r + \epsilon) \\ \dfrac{n_2(t)}{d(t)} & \text{for } t \geq t_r + \epsilon \end{cases}$$

The differential parametrization (5.145) allows one to propose the following state estimates for the unmeasured states x_2 and x_3:

$$x_{2e} = \frac{1}{\sigma}(\dot{y})_e + y$$

$$x_{3e} = -\frac{1}{y} \left[\frac{1}{\sigma}(\ddot{y})_e + \left(1 + \frac{1}{\sigma} \right)(\dot{y})_e + (1 - r)y \right] \tag{5.146}$$

Note, however, as clarified before, that the estimate of the state variable x_3 undergoes a singularity every time the signal $y = x_1$ goes through the value of 0. We will propose a singularity-free coding–decoding process which allows us to also use x_3 as part of a chaotic coding signal.

5.3.1.1 Simulations

For the computer simulations we have taken the following parameter values:

$$\sigma = 10, \quad r = 28, \quad b = \frac{8}{3}$$

Figure 5.25 shows the computer simulation of the Lorenz system actual state trajectories along with the estimated values of the states x_2 and x_3. The computation of the first and second time

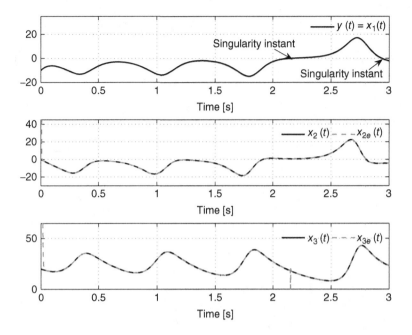

Figure 5.25   State estimates of a Lorenz system

derivatives of the measured output allows us, of course, to estimate the state variables x_2 and x_3 using the static state estimation formula (5.146). The calculation intervals were chosen to be fixed of value 0.15 s, while the small interval of time right after the calculation resettings was set to be defined by $\epsilon = 0.01$ s. Note that the calculation resetting interval was taken to be rather "large." Nevertheless, the accuracy of the estimations and the performance of the algorithm are quite remarkable.

From Figure 5.25, it is evident that when $y(t) = x_1(t)$ goes through zero, a singularity centers around this time instant for the estimation of x_3 (modulo the effects of the finite-step integration algorithm). Naturally, this fact makes the state x_3 a questionable candidate for coding message signals that need to be secretly transmitted.

In order to give an idea of the speed of the fast non-asymptotic convergence as well as the accuracy of the state calculations around the resetting times, we show in Figure 5.26 an inset of the previous simulations around the initial time $t = 0$. The calculation intervals of 0.15 s and the calculation accuracy holding time of 0.01 s are clearly depicted in this figure.

5.3.2   State Estimation for Chen's System

Consider now Chen's system (see Chen, 1993):

$$\dot{x}_1 = a(x_2 - x_1)$$
$$\dot{x}_2 = (c - a)x_1 + cx_2 - x_1 x_3$$
$$\dot{x}_3 = x_1 x_2 - bx_3 \tag{5.147}$$

where $y = x_1$ is the output variable. The parameters a, b, and c are assumed to be known. The system is observable from the output y, except at the line $y = x_1 = 0$.

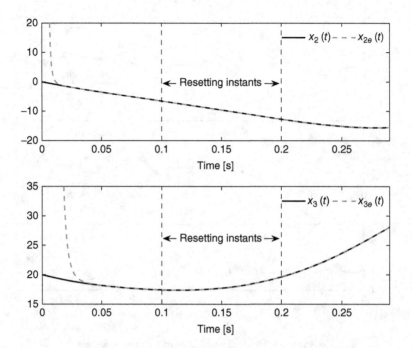

Figure 5.26   An inset for the Lorenz state estimation

A local differential parametrization of the system states, in terms of the measured output y, is given by

$$x_1 = y$$

$$x_2 = \frac{1}{a}\dot{y} + y$$

$$x_3 = -\frac{1}{y}\left[\frac{1}{a}\ddot{y} + \left(1 - \frac{c}{a}\right)\dot{y} + (a - 2c)y\right] \tag{5.148}$$

For the generation of the time derivatives of the output $y(t) = x_1(t)$, we again propose the same seventh-order truncated Taylor series expansion around the reinitialization time t_r used in the previous example. Therefore, we used the same derivative calculation formulae presented in the Lorenz system example.

5.3.2.1 Simulations

For the computer simulations we have taken the following parameter values:

$$a = 35, \quad b = 3, \quad c = 28$$

Figure 5.27 shows the computer simulation of Chen's system actual state trajectories and the estimated trajectories of the states x_2 and x_3. This time, the calculation interval was chosen to be defined by $t_r = 0.1$ s, while the small interval of time after the calculation resetting was set to be defined by $\epsilon = 0.01$ s.

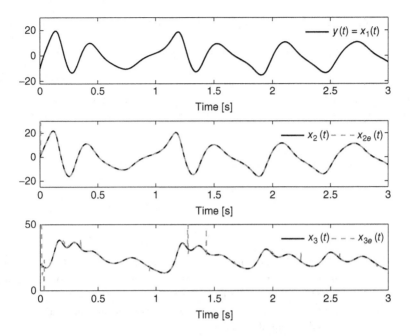

Figure 5.27   State estimates of Chen's system

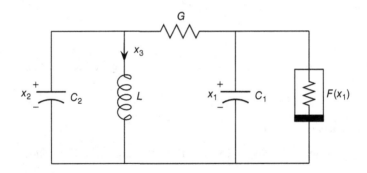

Figure 5.28   Chua's circuit

5.3.3   State Estimation for Chua's Circuit

Consider Chua's circuit (see Wu and Chua, 1993), as shown in Figure 5.28. This circuit is described by the following set of nonlinear differential equations:

$$C_1 \dot{x}_1 = G(x_2 - x_1) - F(x_1)$$

$$C_2 \dot{x}_2 = G(x_1 - x_2) + x_3$$

$$L \dot{x}_3 = -x_2 \tag{5.149}$$

where $F(x_1)$ is a voltage-dependent nonlinear function of the form

$$F(x_1) = ax_1 + \frac{1}{2}(b - a)(|1 + x_1| - |1 - x_1|), \quad a, \ b < 0$$

clearly playing the role of a *negative* resistor.

In order to facilitate the exposition, we adopt a normalized form of the above circuit (see Huijberts *et al.*, 1998):

$$\dot{z}_1 = \beta(-z_1 + z_2 - \phi(z_1))$$

$$\dot{z}_2 = z_1 - z_2 + z_3$$

$$\dot{z}_3 = -\gamma z_2 \tag{5.150}$$

with

$$\phi(z_1) = az_1 + \frac{1}{2}(b - a)\{|\ 1 + z_1\ | - |\ 1 - z_1\ |\}$$

The system is clearly non-differentiable due to the presence of the term $\phi(z_1)$. This makes the output $y = z_1$ not suitable for our state estimation technique, since the corresponding differential parametrization of z_3 requires the time derivative of the function $\phi(z_1)$. Nevertheless, the output $y = x_3$ is *globally observable* and the state of the normalized system enjoys a singularity-free (linear) differential parametrization. Indeed,

$$z_1 = -\frac{1}{\gamma}\ddot{y} - \frac{1}{\gamma}\dot{y} - y$$

$$z_2 = -\frac{1}{\gamma}\dot{y}$$

$$z_3 = y \tag{5.151}$$

The time derivatives of the measured output $y(t) = z_3(t)$ may be generated exactly in the same form as before.

5.3.3.1 Simulations

For the computer simulations, we have taken the following parameter values:

$$a = -\frac{5}{7}, \quad b = -\frac{8}{7}, \quad \beta = 15.6, \quad \gamma = 27$$

Figure 5.29 shows the computer simulation of the normalized Chua's circuit actual state trajectories and the estimated values of the normalized states z_2 and z_3. This time, the calculation interval was chosen to be defined by $t_r = 0.3$ s, while the small interval of time, after the calculation resetting was set to be defined by $\epsilon = 0.02$ s.

5.3.4 State Estimation for Rossler's System

Consider now Rossler's system, described by Pecora and Carroll (1991):

$$\dot{x}_1 = -(x_2 + x_3)$$

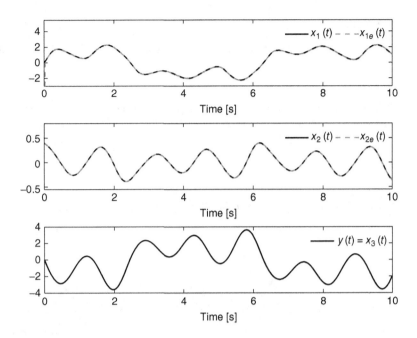

Figure 5.29 State estimates of normalized Chua's chaotic circuit

$$\dot{x}_2 = x_1 + ax_2$$

$$\dot{x}_3 = b + x_1 x_3 - cx_3 \qquad (5.152)$$

where $y = x_2$ is the output variable. The parameters a, b, and c are known quantities. The system is globally observable from the output $y = x_2$.

A (linear) differential parametrization of the system states, in terms of the measured output y, is given by

$$x_1 = \dot{y} - ay$$

$$x_2 = y$$

$$x_3 = -\ddot{y} - a\dot{y} - y \qquad (5.153)$$

As in the previous examples, we used a seventh-order Taylor series expansion around the reinitialization time t_r for the output signal $y(t) = x_2(t)$. The derivative calculation formulae presented in the first example (Lorenz) are still the same in this example.

Note that Rossler's system also exhibits a lack of global observability when the system output is chosen to be $y = x_3$. Indeed, in such a case we have the following differential parametrization of the system states:

$$x_1 = \frac{\dot{y} + cy - b}{y}$$

$$x_2 = -\frac{(\ddot{y} + c\dot{y})y - (\dot{y} + cy - b)\dot{y}}{y^2} - y$$

$$x_3 = y \qquad (5.154)$$

5.3.4.1 Simulations

For the computer simulations, we have taken the following parameter values for Rossler's chaotic system:

$$a = b = 0.2, \quad c = 5$$

Figure 5.30 shows the computer simulation of Rossler's system actual state trajectories and the estimated values of the states x_1 and x_3. This time, the calculation interval was chosen to be 0.1 s, while the small interval of time after the calculation resetting was set to be defined by $\epsilon = 0.01$ s.

5.3.5 State Estimation for the Hysteretic Circuit

Consider now the following chaotic circuit treated by Carroll and Pecora (1991):

$$\dot{x}_1 = x_2 + \gamma x_1 + c x_3$$
$$\dot{x}_2 = -\omega x_1 - \delta x_2$$
$$\epsilon \dot{x}_3 = (1 - x_3^2)(s x_1 + x_3) - \beta x_3 \tag{5.155}$$

where $y = x_2$ is the output variable. The parameters γ, c, ω, β, and ϵ are all perfectly known quantities. The system is globally observable from the output $y = x_2$.

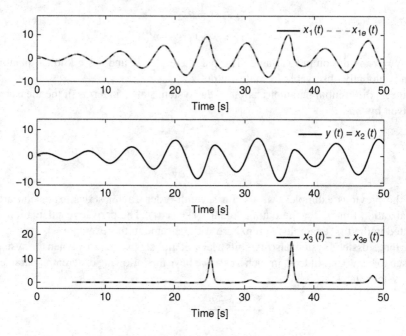

Figure 5.30 State estimates for Rossler's system

A differential parametrization of the system states, in terms of the measured output y, is given by

$$x_1 = -\frac{1}{\omega}[\dot{y} + \delta y]$$

$$x_2 = y$$

$$x_3 = \frac{1}{c}\left[-\frac{1}{\omega}(\ddot{y} + \delta\dot{y}) - y + \frac{\gamma}{\omega}(\dot{y} + \delta y)\right] \quad (5.156)$$

For the computer simulations, we have taken the following parameter values:

$$\gamma = 0.2, \quad c = 2, \quad \omega = 10, \quad \delta = 0.001, \quad s = 1.667$$

$$\beta = 0.001, \quad \epsilon = 0.3$$

Figure 5.31 shows the computer simulation of the hysteretic circuit actual state trajectories $x_1(t)$ and $x_3(t)$ along with the estimated trajectories of those states $x_{1e}(t)$ and $x_{3e}(t)$. This time, the calculation interval was chosen to be 0.25 s, while the small interval of time after the calculation resetting was set to be defined by $\epsilon = 0.04$ s.

Note that the hysteretic circuit also exhibits a lack of global observability when the system output is chosen to be $y = x_3$. Indeed, in such a case we have the following differential

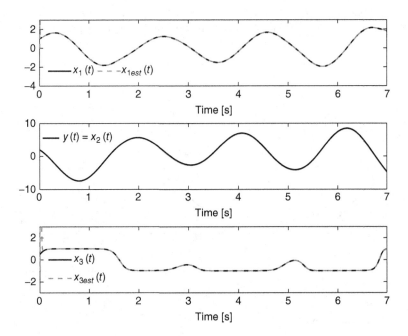

Figure 5.31   State estimates for the hysteretic circuit

parametrization of the system states:

$$x_1 = \frac{1}{s}\left[\frac{\epsilon\dot{y} + \beta y}{1 - y^2} - y\right]$$

$$x_2 = \frac{1}{s}\left[\frac{(\epsilon\ddot{y} + \beta\dot{y})(1 - y^2) + 2(\epsilon\dot{y} + \beta y)y\dot{y}}{(1 - y^2)^2} - \dot{y}\right]$$

$$- \frac{\gamma}{s}\left[\frac{\epsilon\dot{y} + \beta y}{1 - y^2} - y\right] - cy$$

$$x_3 = y \tag{5.157}$$

Clearly, there is a lack of observability at the values $y = \pm 1$. In fact, the hysteretic circuit state variable $y = x_3$ exhibits open intervals of time when y is rather close to either 1 or -1, and it actually achieves these extreme singular values at certain instants of time within those time intervals. The rather singular behavior of the state estimates x_1 and x_2, for this case, is shown in Figure 5.32.

5.3.5.1 Coding–Decoding Process

The previous examples point to the fact that in our algebraic state reconstruction approach, the state variables are accurately reconstructed from the output signal alone. In the particular case of the Lorenz and Chen systems, a hidden signal transmission is possible through at least one of the chaotic states (x_2) of these systems. The subsequent message decoding is performed with the help of the proposed state estimation process at the receiving end, as explained below.

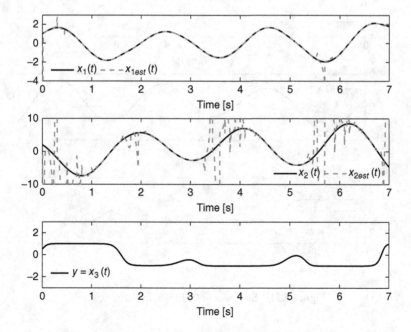

Figure 5.32 Singular state estimates for the hysteretic circuit

Suppose a secret message $m(t)$ is to be sent over a certain communication channel, possibly of analog nature. For the encoding process, we add the secret message signal $m(t)$, say, to the masking state signal $x_2(t)$ of the chaotic system. The obtained signal $z(t) = x_2(t) + m(t)$ is sent toward the receiving end along with the output signal $y(t)$. At the receiving end, the transmitted output signal $y(t)$ is used in the algebraic state estimation scheme for the accurate reconstruction of the chaotic system state $x_2(t)$. This process results in the estimated signal $x_{2e}(t)$. A reconstruction of the hidden message is immediately obtained by forming the estimated secret message signal: $m_e(t) = z(t) - x_{2e}(t)$. The coding–decoding process is depicted in Figure 5.33.

In order to ensure that the addition of the message signal $m(t)$ to the transmitted state does not become evident, one usually scales down the message amplitude so that its maximum amplitude represents only a fraction of the maximum chaotic masking signal amplitude. As a rule of thumb, we use message amplitudes which are roughly 5% of the masking state signal amplitude.

5.3.5.2   A Simulation Example

Using the previously described coding–decoding process, we used Chen's chaotic system for the secret signal encoding–transmission and subsequent decoding process through the algebraic state estimator already discussed at length in the previous section. Figure 5.34 depicts the transmitted signals, $y = x_1(t)$ and $z(t) = x_2(t) + m(t)$, as well as the recovered message $m_e(t)$ compared with the actual message signal $m(t)$. The signal used as $m(t)$ was set to be given by

$$m(t) = \cos[\sqrt{2.15t} - 20\sin(2.02t)]\sin(25t), \quad t \in [1, 2]$$

In order to assess the behavior of our coding–decoding scheme with respect to transmission noises, we used a noisy output signal $y(t) = x_1(t) + \xi(t)$ and a noisy coding state transmission $z(t) = x_2(t) + m(t) + 10\xi(t)$, with $\xi(t)$ being a computer-generated noisy perturbation process taking values in the interval $[-0.0025, 0.0025]$. This computer-generated noise is synthesized on the basis of a rectangular (uniform) probability density function for the corresponding digital computer random number generation comprising the piecewise constant values of the perturbation signal. The simulations are depicted in Figure 5.35. In this instance, we used as secret signal the signal $m(t)$, given in (5.3.5), amplified by a factor of 2.

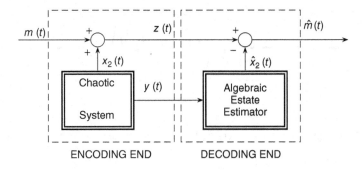

Figure 5.33   Coding–decoding process

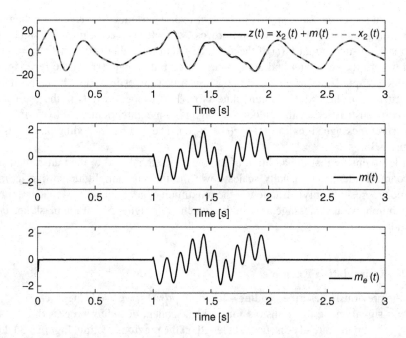

Figure 5.34   A simulation example of encrypted message recovery

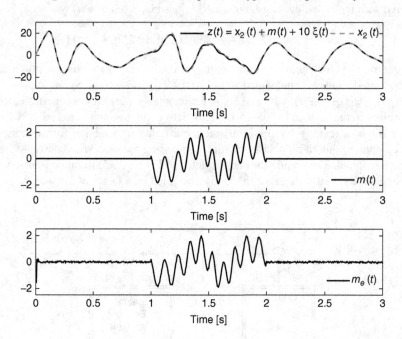

Figure 5.35   Encrypted message recovery from a noisy output and a noisy encoding state transmission

5.3.6 Simultaneous Chaotic Encoding–Decoding with Singularity Avoidance

The singularities present in the estimation of the state x_3 in the Lorenz and Chen system examples make the variable x_3 a useless state for coded signal transmission. Similarly, in the Rossler example and in the hysteretic circuit example, with the output variable taken to be $y = x_3$, both variables x_1 and x_2 would be rather inconvenient for encryption and transmission purposes.

A rather direct way to evade these singularities is suggested by the differential parametrization of the states themselves. For instance, in the Lorenz and Chen examples, rather than using x_3 for coding purposes, we used the product signal $x_3(t)y(t)$; this masking signal could be used to transmit and recover messages without any singularities. Indeed, let $w(t) = x_3(t)y(t)$ and transmit the signal $\zeta(t) = w(t) + n(t)$, where $n(t)$ is a message to be sent toward the receiving end. In Chen's system with $y = x_1$, the estimation of the signal $w(t) = x_3(t)y(t)$, denoted by $\widehat{w}(t)$, is simply obtained from (5.148) as

$$\widehat{w}(t) = -\left[\frac{1}{a}(\ddot{y})_e + \left(1 - \frac{c}{a}\right)(\dot{y})_e + (a - 2c)y\right] \tag{5.158}$$

The message signal estimate $n_e(t)$ is immediately recovered from the simple subtraction operation

$$n_e(t) = \zeta(t) - \widehat{w}(t) \tag{5.159}$$

The fading of $x_3(t)y(t)$ near a zero crossing of $y(t)$ does not affect the encryption, nor the decoding processes. Figure 5.36 depicts the proposed singularity-free encryption process.

Evidently, a similar procedure involving the product signals $x_1(t)y(t)$ and $x_2(t)y^2(t)$ can be proposed to evade the singularities in Rossler's system, when the output is taken to be $y = x_3$ (see (5.154)). In the hysteretic circuit, when $y = x_3$, one must take the nonlinear signals $x_1(t)(1 - y^2(t))$ and $x_2(t)(1 - y^2(t))^2$ for coding–decoding purposes (see (5.157)).

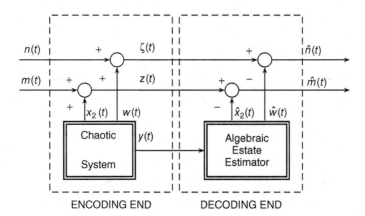

Figure 5.36 Simultaneous chaotic encoding–decoding with singularity avoidance

5.3.6.1 Simulations

Using the previously described coding–decoding process with singularity avoidance, we used Chen's chaotic system for signal encoding–transmission of two secret messages $m(t)$ and $n(t)$. The subsequent decoding process for the coding chaotic signal $w(t) = x_3(t)y(t)$ was carried out through the algebraic state estimator as already discussed above. Figure 5.37 depicts the singularity-free signal $w(t) = x_3(t)y(t)$, along with the transmitted signal $\zeta(t) = x_3(t)y(t) + n(t)$. The figure also shows the recovered message $n_e(t)$ and the actual message $n(t)$. In order to keep the amplitude of the message, roughly speaking, at 5% of the value of the carrier signal amplitude, the signal to be transmitted was amplified by a factor of 40. Evidently, this scaling has no bearing whatsoever on the recovery of the actual signal once its multiple value is safely received at the decoding end and the scaling factor is known. In order to send the message signal $\sin[\sqrt{512t} + 5\sin(10t)]\sin(15t)$, we used in this instance the signal

$$n(t) = 40\sin[\sqrt{512t} + 5\sin(10t)]\sin(15t), \quad t \in [0.5, 1.5]$$

5.3.7 Discussion

In this section we introduced, in the context of well-known chaotic system examples, a fast non-model-based successive time derivative calculation of a measured observable output signal. This procedure is readily used as a tool for the state estimation process to be carried out at the receiving end in a chaotic system state-based modulation, and transmission of encrypted secret message signals. At the receiving end, the state variables of the unperturbed transmitting chaotic system are accurately, locally calculated from formulae using the transmitted

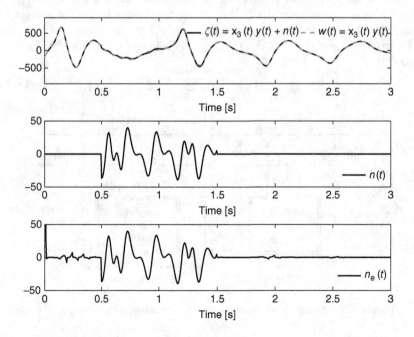

Figure 5.37 Chaotic encoding–decoding with singularity avoidance

observable output alone and some time convolutions. The process enables the online local computation of a sufficient number of its time derivatives and the use of a static map, guaranteed by the local observability of the output variable, relating these output time derivatives to the system states (in fact, in the examples presented here, only two time derivatives of such outputs are required). The generation, at the receiving end, of the required coding system state estimates, or reconstructions, is carried out using the (static) model-based differential parametrization of the encrypting system states in terms of the measured output variable.

An efficient computational method is proposed for the piecewise continuous online computation of the first few time derivatives of the chaotic output signal along with, possibly, an automatic resetting calculation mechanism based on an online evaluated integral quadratic error criterion. In practice, one can also use a fixed calculation interval of sufficiently small length. The successive time derivative generation method is based on a combination of a truncated (polynomial) approximating Taylor series expansion of the output signal and use of the *algebraic derivative method* on a time-invariant homogeneous linear system of sufficiently high order. The estimation of the unperturbed carrier states, at the receiving end, is then used in the traditional message decoding scheme. The masking and recovery of the transmitted message naturally require the transmission of the chaotic system output signal and of the chaotic states additively perturbed by the secret message signal. Several simulation examples were presented which depict the effectiveness of the proposed approach. The proposed estimation scheme for one of the chaotic states, in the Lorenz, Chen, Rossler, and hysteretic circuit examples, suffers from the presence of singularities at each zero crossing of the chosen system output. This is caused by an instantaneous loss of the required output observability (a common phenomenon in nonlinear systems, where observability is definitely a local concept). A method which evades such singularities in the calculations and still allows one to use the troublesome state in the encrypting–decoding process is also proposed. Interestingly enough, some chaotic systems were shown to have global observability properties with linear differential parametrizations of the states.

The method naturally derives from the framework of module theory and the implications of non-commutative algebra in linear systems theory.

Potential areas of application of the proposed output signal derivative calculation scheme can be the extension to hyper-chaotic signal encoding–decoding schemes.

5.4 Remarks

- An algebraic method for fast online estimation of unknown parameters and unmeasured observable states, using only inputs and outputs, has been introduced for the class of linear time-invariant systems subject to "classical" unknown perturbation inputs.
- The technique, which still rests on the concept of an "algebraic derivative," was also shown to be robust with respect to zero-mean high-frequency noises.
- This approach has been applied in other areas, such as vehicle dynamics estimation (Villagra *et al.*, 2008), traffic models (Abouaïssa *et al.*, 2008), DC motor state estimation (Tian *et al.*, 2008), fault-tolerant control (Fliess *et al.*, 2004), among others.
- Identification problems in multi-variable linear time-invariant, continuous, and discrete systems can also be treated similarly. The technique seems to work well in some nonlinear systems.

References

Abouaïssa, H., Fliess, M. and Join, C. (2008) Fast parametric estimation for macroscopic traffic flow model. *Proceedings of the 17th IFAC World Congress*, Seoul, Korea.

Afraimovitch, V., Nekorkin, V., Osipov, G. and Shalfeev, V. (1994) *Stability, Structures and Chaos in Synchronization Networks*. World Scientific Publishing Co., Singapore.

Carroll, T. and Pecora, L. (1991) Synchronizing chaotic circuits. *IEEE Transactions on Circuits and Systems* **38**(4), 453–456.

Chen, G. (1993) From chaos to order. *International Journal of Bifurcation and Chaos* **3**(6), 1363–1409.

Chen, G. (1997) *Control and synchronization of chaotic systems*. ECE Department, University of Houston, TX. Available from ftp.egr.uh.edu/pub/TeX/chaos.tex (login name: "anonymous," password: your email address).

Chen, G. (1999) *Controlling Chaos and Bifurcations in Engineering Systems*. CRC Press, Boca Raton, FL.

Cuomo, K., Oppenheim, A. and Strogatz, S. (1993) Synchronization of Lorenz-based chaotic circuits with applications to communications. *IEEE Transactions on Circuits and Systems II: Analog and Digital Signal Processing* **40**(10), 626–633.

Diop, S. and Fliess, M. (1991) Nonlinear observability, identifiability and persistent trajectories. *Proceedings of the 36th IEEE Conference on Decision and Control*, pp. 714–719, Brighton, UK.

Diop, S., Grizzle, J. and Chaplais, F. (2000) On numerical differentiation algorithms for nonlinear estimation. *Proceedings of the 39th IEEE Conference on Decision and Control*, pp. 1133–1138, Sydney, Australia.

Diop, S., Grizzle, J., Moraal, P. and Stefanopoulou, A. (1994) Interpolation and numerical differentiation for observer design. *Proceedings of the American Control Conference*, pp. 1329–1333, Baltimore, MD.

Fliess, M. (1987) Quelques remarques sur les observateurs non lineaires. *11ᵉᵐᵉ Colloque GRETSI sur le Traitement du Signal et des Images*, Nice, France.

Fliess, M. and Sira-Ramírez, H. (2004a) Control via state estimations of flat systems. *IFAC Nolcos Conference*, Stuttgart, Germany.

Fliess, M. and Sira-Ramírez, H. (2004b) Reconstructeurs d'etat. *C.R. Academie des Sciences de Paris, Série I* **338**(1), 91–96.

Fliess, M., Join, C. and Sira-Ramírez, H. (2004) Robust residual generation for linear fault diagnosis: An algebraic setting with examples. *International Journal of Control* **77**(14), 1223–1242.

Fliess, M., Join, C. and Sira-Ramírez, H. (2008) Non-linear estimation is easy. *International Journal of Modelling, Identification and Control* **4**(1), 12–27.

Fradkov, A. and Markov, AY. (1997) Adaptive synchronization of chaotic systems based on speed gradient method and passification. *IEEE Transactions on Circuit and Systems I: Fundamental Theory and Applications* **44**(10), 905–917.

Fradkov, A. and Pogromsky, AY. (1998) *Introduction to Control of Oscillations and Chaos*, vol. 35. World Scientific Publishing Co., Singapore.

Holden, A. (1986) *Chaos*. Princeton University Press, Princeton, NJ.

Huijberts, H., Nijmeijer, H. and Willems, R. (1998) A control perspective on communication using chaotic systems. *Proceedings of the 37th IEEE Conference on Decision and Control, 1998*, vol. 2, pp. 1957–1962, Tampa, Florida, FL.

Lorenz, EN. (1963) Deterministic non-periodic flow. *Journal of Athmospheric Science* **20**(2), 130–141.

Mira, C. (1987) *Chaotic Dynamics*. World Scientific Publishing Co., Singapore.

Nijmeijer, H. and Mareels, MY. (1997) An observer looks at synchronization. *IEEE Transactions on Circuits and Systems I: Fundamental Theory and Applications* **44**(10), 882–890.

Ott, E., Sauer, T. and Yorke, J. (eds) (1994) *Analysis of Chaotic Data and the Exploitation of Chaotic Systems*. Wiley-Interscience, New York.

Pecora, L. and Carroll, T. (1991) Driving systems with chaotic signals. *Physics Review A* **44**(4), 2374–2383.

Plestan, F. and Grizzle, J. (1999) Synthesis of nonlinear observers via structural analysis and numerical differentiation. *Proceedings of the European Control Conference*, Karlsruhe, Germany.

Pogromsky, AY. (1998) Passivity-based design of synchronizing systems. *International Journal of Bifurcation and Chaos* **8**(2), 295–319.

Sira-Ramírez, H., Aguilar-Ibáñez, C. and Suárez-Castañón, M. (2002) Exact state reconstruction in the recovery of messages encrypted by the states of nonlinear discrete-time chaotic systems. *International Journal of Bifurcation and Chaos* **12**(1), 169–177.

Sira-Ramírez, H. and Cruz-Hernández, C. (2001) Synchronization of chaotic systems: A hamiltonian systems approach. *International Journal of Bifurcation and Chaos* **11**(5), 1381–1395.

Sira-Ramírez, H. and Fliess, M. (2006) An algebraic state estimation approach for the recovery of chaotically encrypted messages. *International Journal of Bifurcation and Chaos* **16**(2), 295–309.

Special Issue (1993) Chaos synchronization and control: Theory and applications. *IEEE Transactions on Circuits and Systems I: Fundamental Theory and Applications*.

Special Issue (1997a) Chaos synchronization and control: Theory and applications. *IEEE Transactions on Circuits and Systems I: Fundamental Theory and Applications*.

Special Issue (1997b) Control of chaos and synchronization. *System Control Letters*.

Special Issue (1999) Communications, information processing and control using chaos. *International Journal of Circuit Theory and Applications*.

Special Issue (2000) Control and synchronization of chaos. *International Journal of Bifurcation and Chaos*.

Special Issue (2001) Application of chaos in modern communication systems. *IEEE Transactions on Circuits and Systems I: Fundamental Theory and Applications*.

Tian, Y., Floquet, T. and Perruquetti, W. (2008) Fast state estimation in linear time-invariant systems: An algebraic approach. *16th Mediterranean Conference on Control and Automation,2008*, pp. 350–355, Ajaccio Corsica, France.

Villagra, J., d'Andrea Novel, B., Fliess, M. and Mounier, H. (2008) Estimation of longitudinal and lateral vehicle velocities: An algebraic approach. *Proceedings of the American Control Conference, 2008*, pp. 3941–3946, Seattle, WA.

Wu, C. and Chua, L. (1993) A simple way to synchronize chaotic systems with applications to secure communication systems. *International Journal of Bifurcation and Chaos* **3**(6), 1619–1627.

6

Control of Nonlinear Systems via Output Feedback

6.1 Introduction

In an observable system, the state estimation problem is intimately related to the problem of computing the successive time derivatives of the output and input signals in a sufficiently large number (see Diop and Fliess, 1991).

In this chapter, we propose a non-asymptotic algebraic procedure for the approximate estimation of the system states from the calculation of a finite number of time derivatives of the output signal. The method is based on results from differential algebra and it furnishes some general formulae for the time derivatives of a measurable signal.

The method proposed may be combined with the notion of *differential flatness* aimed at completing a feedback loop, with desirable closed loop dynamics, based on the feedback of the flat output and some of its time derivatives. There are some other interesting contributions that propose non-asymptotic approaches to state estimation in dynamical systems, which are also based on time derivative calculations in the presence of noise (see, for example, the works by Diop *et al.*, 1994, 2000; Levant, 1998).

Section 6.2 introduces the algebraic approach to time derivative calculation as a natural extension of the linear-state identifiability problem exercised now on an approximate linear time-invariant homogeneous model of the sufficiently smooth signal. This model is based on the truncated Taylor series expansion (i.e., the Taylor polynomial) of the output signal values around a given point, and therefore it is just a local polynomial approximation model of the output variable time signal.

The nonlinear state estimation problem is then solved by proposing the efficient computation of a finite number of time derivatives of the output signal. In this respect, we no longer regard the nonlinear model of the output as a function of the state, but as a linear time-invariant homogeneous model describing a local finite-order approximation to the time realization of the output signal. The rest of the scheme is based on the necessary reinitializations (or resettings) of the local model in a periodic manner.

Algebraic Identification and Estimation Methods in Feedback Control Systems, First Edition.
Hebertt Sira-Ramírez, Carlos García-Rodríguez, Alberto Luviano-Juárez, John Alexander Cortés-Romero.
© 2014 John Wiley & Sons, Ltd. Published 2014 by John Wiley & Sons, Ltd.

In this chapter we present some illustrative application examples. We first deal with the flatness-based control of a rather popular synchronous generator model. We proceed with a more complex example of output feedback controlling a nonlinear multi-variable system representing the dynamic model of a mono-cycle. We also include, in this chapter, experimental results obtained from a laboratory prototype system.

6.2   Time-Derivative Calculations

Given a signal $y(t)$, we want to online compute its time derivatives with the following restrictions:

- The information on $y(t)$ is gathered "online."
- The signal $y(t)$ is corrupted by measurement noises whose statistics are unknown.
- The estimation process should not depend on the model of the system that generates the output signal $y(t)$.
- It is assumed that no other signal is available from the system.

The most popular method for obtaining the first-order time derivative of a signal $y(t)$ is the so-called finite difference method. It consists of the following approximation to the derivative $\dot{y}(t)$:

$$\dot{y}_e(t_i) \approx \frac{y(t) - y(t_i)}{t - t_i} \tag{6.1}$$

where $t - t_i = \epsilon > 0$ is a known scalar and t is an arbitrary instant of time close to t_i. We note:

- The method is quite general and it does not depend on the system model.
- The quality of the approximation depends on $\epsilon = t - t_i$.
- The estimation process is *not asymptotic* and the estimate is available "instantaneously," right after the instant t.
- The method is quite sensitive to the presence of noise perturbations in the signal to be processed.

The finite difference approximation of the derivative is evidently based on a truncated expansion of the Taylor series of the underlying signal:

$$\tilde{y}(t) = y(t_i) + [\dot{y}(t_i)](t - t_i) \tag{6.2}$$

We insist upon the fact that the model is that of the signal and not of the system producing it. Taking one time derivative on the approximation formula (6.2), we obtain the following linear homogeneous model for the approximation:

$$\frac{d^2 \tilde{y}}{dt^2} = 0 \tag{6.3}$$

The local problem is reduced to computing the unmeasured state of a homogeneous second-order linear time-invariant system.

We can generalize the above procedure slightly by adopting a higher-order model for the approximation of the output signal $y(t)$:

$$\tilde{y}(t) = \sum_{k=0}^{N-1} \frac{1}{k!} y^{(k)}(t_i)(t - t_i)^k \tag{6.4}$$

This approximation satisfies the homogeneous linear time-invariant differential equation

$$\frac{d^N \tilde{y}}{dt^N} = 0 \tag{6.5}$$

The local problem of computing time derivatives of $y(t)$ is then reduced to computing the states of the linear time-invariant homogeneous system of order N.

The linear approximation adopted, $y^{(N)}(t) = 0$, satisfies, in terms of operational transforms, the following relation:

$$\frac{d^N}{ds^N}[s^N Y(s)] = 0 \tag{6.6}$$

The expressions given by

$$s^{-k} \frac{d^N}{ds^N}[s^N Y(s)] = 0, \qquad k = N - 1, N - 2, \ldots, N - k \tag{6.7}$$

contain, respectively, implicit information on the first, second, ..., kth derivatives of $y(t)$ in an approximate manner.

6.2.1   An Introductory Example

Consider a fifth-order approximation in $t = 0$ of a sufficiently differentiable signal, $y(t)$:

$$\hat{y}(t) = \sum_{j=0}^{5} \frac{1}{j!} t^j y^{(j)}(0) \tag{6.8}$$

The proposed formula is written as follows:

$$s^{-k} \frac{d^6}{ds^6}[s^6 \hat{y}(s)] = 0, \qquad k = 5, 4, 3 \tag{6.9}$$

For $k = 5$, we have

$$\left[720s^{-5} y(s) + 4320s^{-4} \frac{dy(s)}{ds} + 5400s^{-3} \frac{d^2 y(s)}{ds^2} \right.$$
$$\left. + 2400s^{-2} \frac{d^3 y(s)}{ds^3} + 450s^{-1} \frac{d^4 y(s)}{ds} + 36 \frac{d^5 y(s)}{ds^5} + s \frac{d^6 y(s)}{ds^6} \right] = 0 \tag{6.10}$$

This expression relates the term

$$s \frac{d^6 y(s)}{ds^6}$$

which, in the time domain, is just given by

$$\frac{d}{dt}[t^6 y(t)] = 6t^5 y(t) + t^6 \dot{y}(t) \tag{6.11}$$

with a finite sum of convolutions of the signal $y(t)$ with powers of t:

$$\left[720 \left(\int^{(5)} y(t) \right) - 4320 \left(\int^{(4)} ty(t) \right) + 5400 \left(\int^{(3)} t^2 y(t) \right) \right.$$

$$-2400 \left(\int^{(2)} t^3 y(t) \right) + 450 \int t^4 y(t) - 36 t^5 y(t)$$

$$\left. + \frac{d}{dt} t^6 y(t) \right] = 0 \tag{6.12}$$

where

$$\dot{y}(t) = \frac{1}{t^6} \left[-720 \left(\int^{(5)} y(t) \right) + 4320 \left(\int^{(4)} ty(t) \right) \right.$$

$$-5400 \left(\int^{(3)} t^2 y(t) \right) + 2400 \left(\int^{(2)} t^3 y(t) \right) - 450 \int t^4 y(t)$$

$$\left. + 30 t^5 y(t) \right] \tag{6.13}$$

The formula obtained presents a singularity at $t = 0$, which disappears for any $t = \epsilon > 0$. We propose the following estimate of the first-order time derivative of $y(t)$:

$$\dot{y}(t) = \begin{cases} \text{arbitrary constant} & 0 \leq t < \epsilon \\ \frac{1}{t^6} \left[-720(\int^{(5)} y(t)) + 4320(\int^{(4)} ty(t)) \right. \\ -5400(\int^{(3)} t^2 y(t)) + 2400(\int^{(2)} t^3 y(t)) \\ \left. -450 \int t^4 y(t) + 30 t^5 y(t) \right] & t \geq \epsilon \end{cases} \tag{6.14}$$

This computation can be expressed in the form of a time-varying linear filter:

$$\dot{y}(t) = \begin{cases} \text{arbitrary constant} & 0 \leq t < \epsilon \\ \frac{1}{t^6} \left[30 t^5 y(t) + z_1 \right] & t \geq \epsilon \end{cases} \tag{6.15}$$

where

$$\dot{z}_1 = z_2 - 450 t^4 y(t)$$

$$\dot{z}_2 = z_3 + 2400 t^3 y(t)$$

$$\dot{z}_3 = z_4 - 5400 t^2 y(t)$$

$$\dot{z}_4 = z_5 + 4320 t y(t)$$

$$\dot{z}_5 = -720 y(t)$$

To compute the second-order time derivative of $y(t)$, *we integrate one less time* the expression originally proposed; that is, letting $k = 4$ in (6.9), we have

$$\left[720s^{-4}y(s) + 4320s^{-3}\frac{dy(s)}{ds} + 5400s^{-2}\frac{d^2y(s)}{ds^2}\right.$$
$$+2400s^{-1}\frac{d^3y(s)}{ds^3} + 450\frac{d^4y(s)}{ds} + 36s\frac{d^5y(s)}{ds^5}$$
$$\left.+s^2\frac{d^6y(s)}{ds^6}\right] = 0 \tag{6.16}$$

The last two terms in the sum are written as $-36\frac{d}{dt}[t^5y(t)] + \frac{d^2}{dt^2}[t^6y(t)]$, which allows us to compute the second-order time derivative of $y(t)$ in terms of the first-order time derivative of $y(t)$ and a finite sum of convolutions of $y(t)$ with powers of t.

In the time domain, we obtain

$$\ddot{y} = \frac{1}{t^6}(150t^4y(t) + 24t^5\dot{y}(t) + z_2) \tag{6.17}$$

$$\dot{z}_1 = z_2 - 450t^4y(t)$$

$$\dot{z}_2 = z_3 + 2400t^3y(t)$$

$$\dot{z}_3 = z_4 - 5400t^2y(t)$$

$$\dot{z}_4 = z_5 + 4320ty(t)$$

$$\dot{z}_5 = -720y(t)$$

In an analogous manner, the procedure obtains the third time derivative of $y(t)$.

A numerical simulation was carried out to show the time derivative estimation given in the last procedure. Figure 6.1 shows the performance of the first three time derivative estimates of a signal $y(t) = e^{\sin(\omega t)}$, $\omega = 3$. Notice that the first time derivative has good accuracy for a longer time interval in relation to the second and third time derivative estimations. Thus, the precision of the estimator and the reinitialization must be in terms of the higher-order derivative and the number of approximation terms of the truncated time series used.

6.2.1.1 Calculation Resettings

The validity of the formulae for the calculation of the time derivatives is limited in the time horizon. For this reason, it becomes necessary to *reinitialize* the computations at some time $t_r > 0$.

As the derivatives drift from their actual values, so will the estimated signal computed on the basis of the truncated Taylor series approximation and the estimated values of the signal's time derivative.

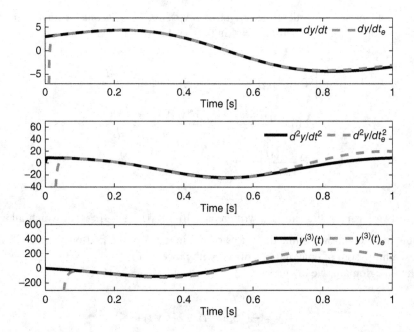

Figure 6.1   Time derivative estimations

An automatic resetting of the calculations can be devised on the basis of the integral square error of the reconstructed signal deviation and a prespecified threshold value for such a reconstruction error:

$$e = \int_{t_r+\epsilon}^{t} |y(\sigma) - \hat{y}(\sigma)|^2 d\sigma \tag{6.18}$$

$$\hat{y}(t) = y(t_r + \epsilon) + [\dot{y}(t_r + \epsilon)](t - t_r - \epsilon) + \frac{1}{2}[\ddot{y}(t_r + \epsilon)](t - t_r - \epsilon)^2 + \cdots$$

or

$$\hat{y}(t) = y(t_r + \epsilon) + \int_{t_r+\epsilon}^{t} [\dot{y}]_e(\sigma) d\sigma$$

It is not difficult to see that for any resetting time t_r, we also have the following approximation formulae valid:

$$\left(\frac{dy}{dt}\right)_e = \frac{1}{(t - t_r)^6}\left[-720\left(\int^{(5)} y(t)\right) + 4320\left(\int^{(4)}(t - t_r)y(t)\right)\right.$$

$$- 5400\left(\int^{(3)}(t - t_r)^2 y(t)\right) + 2400\left(\int^{(2)}(t - t_r)^3 y(t)\right)$$

$$\left. -450\int (t - t_r)^4 y(t) + 30(t - t_r)^5 y(t)\right] \tag{6.19}$$

$$\left(\frac{d^2y}{dt^2}\right)_e = \frac{1}{(t-t_r)^6}\left[-720\left(\int^{(4)}y(t)\right) + 4320\left(\int^{(3)}(t-t_r)y(t)\right)\right.$$

$$-5400\left(\int^{(2)}(t-t_r)^2y(t)\right) + 2400\left(\int(t-t_r)^3y(t)\right)$$

$$+24(t-t_r)^5\left(\frac{dy}{dt}\right)_e + 150(t-t_r)^4y\right] \tag{6.20}$$

As before, the above formulae for the estimates of $\dot{y}$ and $\ddot{y}$ are valid after a small time interval, of duration ϵ, has elapsed from the instant $t = t_r$, that is, during the interval $[t_r + \epsilon, t)$. A new resetting is to be carried out when the validity of the approximation becomes questionable.

We remark that during the time interval $[t_r, t_r + \epsilon]$, we may adopt as temporary values for the time derivative estimates $(\dot{y}(t))_e$ and $(\ddot{y}(t))_e$ either constant values of the form $\dot{y}(t_r^-)$ and $\ddot{y}(t_r^-)$ (i.e., the last computed values of the time derivatives of the interval $[t_{r-1} + \epsilon, t_r]$) or, alternatively, polynomial splines whose parameters are determined on the basis of the last values of the previously computed time derivatives at time t_r. The last strategy may cause singularity problems when the temporary values are not close enough to the actual estimation, immediately after $t_r + \epsilon$.

Assuming that

$$\epsilon \ll t_r \tag{6.21}$$

an alternative strategy which allows us to reinitialize without singularities consists of using two identifiers in a resetting mode configuration such that the reinitialization of each identifier is separate at time $t_r/2$. This can be performed by defining two *time lines* t_1, t_2 as follows:

$$t_1 = t \bmod T$$

$$t_2 = t - t_r/2 \bmod T$$

where the time line for the first identifier is defined as t_1 and, similarly, t_2 for the second identifier. Figure 6.2 depicts the time lines for both identifiers.

The first identifier is reinitialized when $t_1 = 0$ and the second identifier when $t_2 = 0$. Denote the first and second estimations for the first time derivative as $\dot{y}_{1e}$ and $\dot{y}_{2e}$. The first time derivative estimation based on the two identifiers is then given by

$$\dot{y}_e(t) = \begin{cases} \dot{y}_{2e} & \text{for } 0 \leq t \bmod t_r < t_r/2 \\ \dot{y}_{1e} & \text{for } t_r/2 \leq t \bmod t_r < t_r \end{cases} \tag{6.22}$$

and for the second time derivative

$$\ddot{y}_e(t) = \begin{cases} \ddot{y}_{e2} & \text{for } 0 \leq t \bmod t_r < t_r/2 \\ \ddot{y}_{e1} & \text{for } t_r/2 \leq t \bmod t_r < t_r \end{cases} \tag{6.23}$$

Figure 6.3 shows the first time derivative identification of $y(t)$ for this example based on two different identifiers. The t_r parameter was set to be 0.2 s, which is significantly larger than the parameter ϵ. The figure shows the first and second identifiers in which the effects due to the singularity of the resetting are present. The bottom graph shows the time derivative estimation including the switching of the identifiers. Notice that the inclusion of the second identifier eliminates the periodic singularity at each resetting.

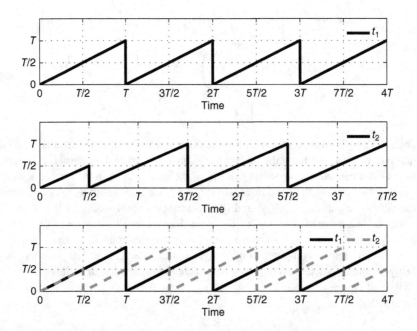

Figure 6.2   Time setting for each identifier

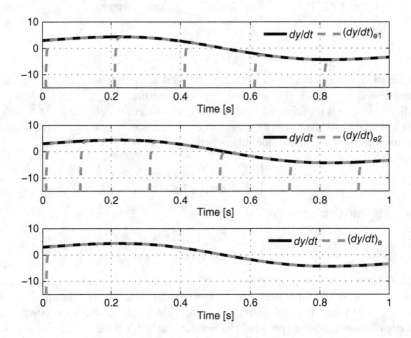

Figure 6.3   Response of the two-identifiers approach

6.2.2 Identifying a Switching Input

In many instances it is desired to identify the input sequence of a closed-loop system based on observations of the output and knowledge of the model of the system. Of particular interest is the case of switched systems which do not undergo sliding motions. Here we briefly present a couple of examples of a linear system controlled by a hard nonlinearity. The first example is a relay controller with a dead zone and the second a relay with hysteresis.

6.2.2.1 Identifying a Dead-Zone Nonlinearity

Consider the system described in the book by Vukić *et al.* (2003) and depicted in Figure 6.4.

The system is a linear time-invariant system in a closed loop with a relay controller with dead zone:

$$y = \frac{K}{s(1 + T_1 s)(1 + T_2 s)} u, \qquad u = -\varphi(y) \tag{6.24}$$

where $\varphi(y)$ describes the nonlinearity:

$$\varphi(y) = \begin{cases} 0 & \text{for} \quad |y| \le b \\ c\,\text{sign}(y) & \text{for} \quad |y| > b \end{cases} \tag{6.25}$$

It is desired to identify the sequence of control inputs given to the system, with b and c being completely unknown. The gain K is assumed to be known and is taken here to be 1.

In the time domain, the system is expressed by means of the following differential equation:

$$T_1 T_2 y^{(3)} + (T_1 + T_2)\ddot{y} + \dot{y} = -\varphi(y) \tag{6.26}$$

Clearly, the computation of the time derivatives of y allows one to reconstruct the control input. Indeed, an estimate of the signal u would be obtained using

$$u = T_1 T_2 y_e^{(3)} + (T_1 + T_2)\ddot{y}_e + \dot{y}_e \tag{6.27}$$

where $\dot{y}_e$, $\ddot{y}_e$, and $y_e^{(3)}$ are the time derivative estimates of the measured output signal y.

Figure 6.5 shows the performance of the input estimator for this hard nonlinearity case. We used the following parameters:

$$b = 0.3, \qquad c = 3, \qquad T_1 = 0.2, \qquad T_2 = 0.1$$

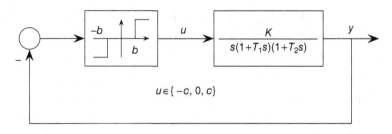

Figure 6.4 A closed-loop control system with hard nonlinearity

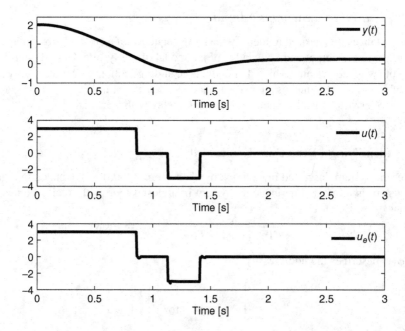

Figure 6.5   Performance of the hard nonlinearity estimator using time derivative calculations

Since we do not know the instants at which the input changes value, we must reinitialize quite often the derivative calculation process so that the new value of the control input is obtained as close as possible to the time when it underwent the discontinuous change.

In this case, we reinitialized every 0.1 s and allowed an ϵ interval of 0.015 s after reinitialization in order to settle the calculation precision.

6.2.2.2   Identifying a Relay with Hysteresis

Consider again the linear system treated in the previous example, but this time in a closed loop with a relay exhibiting hysteretic behavior shown in Figure 6.6.

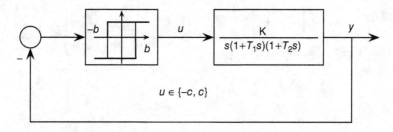

Figure 6.6   A closed-loop control system with hysteretic relay

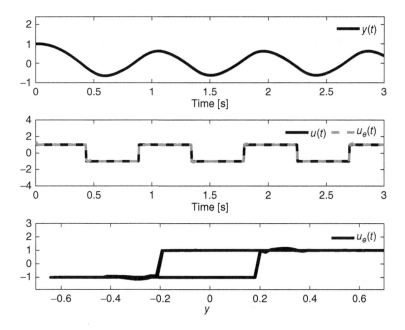

Figure 6.7 Performance of the hysteretic relay estimator using output time-derivative calculations

An estimate of the switching signal u would be obtained by using the derivative-based estimator

$$u_e = T_1 T_2 y_e^{(3)} + (T_1 + T_2)\ddot{y}_e + \dot{y}_e \tag{6.28}$$

where $\dot{y}_e$, $\ddot{y}_e$, and $y_e^{(3)}$ are the time derivative estimates of the measured output signal y.

Figure 6.7 shows the performance of the input estimator for this hard nonlinearity case. We used the following parameters:

$$b = 0.2, \qquad k = 5, \qquad c = 1, \qquad T_1 = 0.1, \qquad T_2 = 0.1$$

We reinitialized the derivative calculation process every 0.1 s and allowed an ϵ interval of 0.015 s after reinitialization in order to settle the calculation precision.

6.3   The Nonlinear Systems Case

The asymptotic estimation of the states in a nonlinear system (whether for its own sake or in order to complete a state feedback control law based on the available input and output signals) has led to the understanding of a fundamental limitation of nonlinear control theory. In most of the formal mathematical approaches proposed so far, it becomes clear that there is not much hope of finding a general solution to the problem of state estimation. This need has resulted in a shift of attention to other methods, such as numerical methods and brute force methods (genetic algorithms, fuzzy). Here we explore the algebraic methods in state estimation for feedback purposes. The approach is based on knowledge of the model and a sufficient number of time derivatives of inputs and output signals.

6.3.1 Control of a Synchronous Generator

Consider the following model of a synchronous generator (Sira-Ramírez and Fliess, 2004):

$$\dot{x}_1 = x_2$$
$$\dot{x}_2 = -b_1 x_3 \sin x_1 - b_2 x_2 + P$$
$$\dot{x}_3 = b_3 \cos x_1 - b_4 x_3 + E - u \tag{6.29}$$

where x_1 is the load angle, x_2 is the angular velocity of the rotor axis, x_3 is the internal voltage, and u is the control input. The parameters b_1, b_2, b_3, b_4, P, and E are assumed to be strictly positive constants which are temporarily assumed to be known in order to derive a certainty equivalence controller.

The system (6.29) is differentially flat, with flat output given by the load angle $F = x_1$. Indeed:

$$x_1 = F$$
$$x_2 = \dot{F}$$
$$x_3 = \frac{P - b_2 \dot{F} - \ddot{F}}{b_1 \sin F}$$
$$u = E - b_4 \left(\frac{P - b_2 \dot{F} - \ddot{F}}{b_1 \sin F} \right) + b_3 \cos F$$
$$-\frac{[-b_2 \ddot{F} - F^{(3)}] \sin F - [P - b_2 \dot{F} - \ddot{F}] \dot{F} \cos F}{b_1 \sin^2 F} \tag{6.30}$$

Given a reference trajectory for the load angle, represented by $F^*(t)$, a control law which achieves trajectory tracking is given by

$$u = E - b_4 \left(\frac{P - b_2 \dot{F} - \ddot{F}}{b_1 \sin F} \right) + b_3 \cos F$$
$$-\frac{[-b_2 \ddot{F} - v] \sin F - [P - b_2 \dot{F} - \ddot{F}] \dot{F} \cos F}{b_1 \sin^2 F}$$
$$v = [F^*(t)]^{(3)} - \lambda_2 (\ddot{F} - \ddot{F}^*(t)) - \lambda_1 (\dot{F} - \dot{F}^*(t))$$
$$-\lambda_0 (F - F^*(t)) \tag{6.31}$$

with $\{\lambda_2, \lambda_1, \lambda_0\}$ being design parameters. The closed-loop characteristic polynomial for the load angle tracking error response is just $p_{x_1}(s) = s^3 + \lambda_2 s^2 + \lambda_1 s + \lambda_0$.

The proposed controller is difficult to synthesize due to the need to feedback the measured output $x_1 = F$, and its two first time derivatives. $\dot{F}$ and $\ddot{F}$. See Figure 6.8.

We demonstrate, by resorting to digital computer simulations, that an algebraic method for the calculation of time derivatives may be used to synthesize the required nonlinear state feedback control law and still obtain a robust performance with respect to noisy plant perturbations and output measurement noises, synthesized by computer generated random processes.

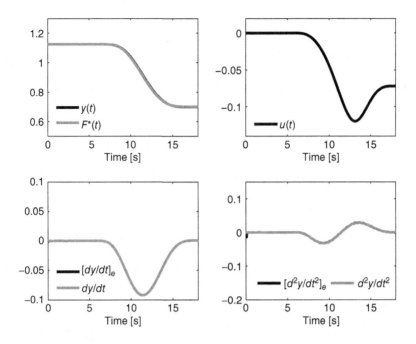

Figure 6.8   Closed-loop response of output feedback-controlled synchronous generator

We propose, for the generation of the time derivatives of the measured output $y(t) = x_1(t)$, a truncated Taylor series expansion of $y(t)$ up to sixth order around the re-initialization time t_r of the form

$$y(t) = \sum_{i=0}^{6} \frac{y^{(i)}(t_r)}{(i)!}(t - t_r)^{(i)} \tag{6.32}$$

which, in the frequency domain, leads to the identity

$$\frac{d^7}{ds^7}[s^7 y(s)] = 0 \tag{6.33}$$

Based on these developments, we write

$$n_1(t) = 42(t - t_r)^6 y(t) - 882 \left(\int_{t_i} (t - t_r)^5 y \right)$$

$$+ 7350 \left(\int_{t_i}^{(2)} (t - t_r)^4 y \right) - 29400 \left(\int_{t_i}^{(3)} (t - t_r)^3 y \right)$$

$$+ 52920 \left(\int_{t_i}^{(4)} (t - t_r)^2 y \right) - 35280 \left(\int_{t_i}^{(5)} (t - t_r) y \right)$$

$$+ 5040 \left(\int_{t_i}^{(6)} y \right) \tag{6.34}$$

$$d(t) = (t - t_r)^7 \tag{6.35}$$

$$\dot{y}(t) = \begin{cases} \dot{y}(t_r^-) & \text{for} \quad t \in [t_r, t_r + \epsilon) \\ \frac{n_1(t)}{d(t)} & \text{for} \quad t \geq t_r + \epsilon \end{cases} \tag{6.36}$$

$$n_2(t) = -630(t - t_r)^5 y(t) + 35(t - t_r)^6 \dot{y}$$

$$+7350 \left(\int_{t_r} (t - t_r)^4 y \right) - 29400 \left(\int_{t_r}^{(2)} (t - t_r)^3 y \right)$$

$$+52920 \left(\int_{t_i}^{(3)} (t - t_r)^2 y \right) - 35280 \left(\int_{t_i}^{(4)} (t - t_r) y \right)$$

$$+5040 \left(\int_{t_i}^{(5)} y \right) \tag{6.37}$$

$$d(t) = (t - t_r)^7 \tag{6.38}$$

$$\ddot{y}(t) = \begin{cases} \ddot{y}(t_r^-) & \text{for} \quad t \in [t_r, t_r + \epsilon) \\ \frac{n_2(t)}{d(t)} & \text{for} \quad t \geq t_r + \epsilon \end{cases} \tag{6.39}$$

The derivative calculations may again be realized in terms of a time varying linear filter as follows:

$$\dot{y}(t) = \begin{cases} \dot{y}(t_r^-) & \text{for} \quad t \in [t_r, t_r + \epsilon) \\ \frac{1}{(t-t_r)^7} \left[42(t - t_r)^6 y(t) + z_1 \right] & \text{for} \quad t \geq t_r + \epsilon \end{cases} \tag{6.40}$$

$$\ddot{y}(t) = \begin{cases} \ddot{y}(t_r^-) & \text{for} \quad t \in [t_r, t_r + \epsilon) \\ \frac{1}{(t-t_r)^7} \left[-630(t - t_r)^5 y(t) + 35(t - t_r)^6 \dot{y}(t) + z_2 \right] \\ \quad \text{for} \quad t \geq t_r + \epsilon \end{cases} \tag{6.41}$$

where

$$\dot{z}_1 = z_2 - 882(t - t_r)^5 y(t)$$

$$\dot{z}_2 = z_3 + 7350(t - t_r)^4 y(t)$$

$$\dot{z}_3 = z_4 - 29400(t - t_r)^3 y(t)$$

$$\dot{z}_4 = z_5 + 52920(t - t_r)^2 y(t)$$

$$\dot{z}_5 = z_6 - 35280(t - t_r) y(t)$$

$$\dot{z}_6 = 5040 y(t)$$

We take the following numerical data, after Bazanella et al. (1997):

$$b_1 = 34.29, \qquad b_2 = 0, \qquad b_3 = 0.1490, \qquad b_4 = 0.3341$$

$$P = 28.220, \qquad E = 0.2405$$

An equilibrium point for the system is given by

$$x_{1e} = 1.12 \quad \text{rad}, \qquad x_{2e} = 0, \qquad x_{3e} = 0.914 \quad \text{pu},$$
$$u = 0 \quad \text{pu}$$

It is desired to transfer the load angle $x_1 = F$ from an initial equilibrium value $x_1(t_0)$ with $t_0 = 0$ to a new equilibrium characterized by $x_{1e}(T)$ in the instant $T > t_0$.

We propose, as a reference trajectory, the following Bézier interpolating polynomial:

$$F^*(t) = x_1(t_0) + (x_1(T) - x_1(t_0))\varphi(t, t_0, T) \tag{6.42}$$

where $\varphi(t, t_0, T)$ is a polynomial taking the initial value 0 at time t_0 and the value 1 at time T, that is,

$$\varphi(t_0, t_0, T) = 0, \qquad \varphi(T, t_0, T) = 1$$

We specifically propose

$$\varphi(t, t_0, T) = \Delta^5(t)[r_1 - r_2\Delta(t) + \cdots - r_6\Delta^5(t)] \tag{6.43}$$

with $\Delta = [t - t_0]/(T - t_0)$ and

$$r_1 = 252, \quad r_2 = 1050, \quad r_3 = 1800, \quad r_4 = 1575, r_5 = 700, \quad r_6 = 126$$

Consider the following perturbed model of a synchronous generator:

$$\dot{x}_1 = x_2$$
$$\dot{x}_2 = -b_1 x_3 \sin x_1 - b_2 x_2 + P - \eta(t)$$
$$\dot{x}_3 = b_3 \cos x_1 - b_4 x_3 + E - u + \eta(t)$$
$$y = x_1 + \xi(t) \tag{6.44}$$

The signals $\eta(t)$ and $\xi(t)$ are computer-generated stochastic processes based on a sequence of random variables, piecewise discrete, uniformly distributed over a symmetric interval around the origin of the real line (Figure 6.9). We select an amplitude for η of 0.03 and for the measurement noise $\xi(t)$ an amplitude of 10^{-4}. The closed-loop characteristic polynomial is set to be $(s^2 + 2\zeta\omega_n s + \omega_n^2)(s + p)$, with $\omega_n = 1.5$, $\zeta = 0.81$, and $p = 2$.

In particular for this problem, two estimation processes are launched with a specified shift time, so that we are assured of finding a better final estimation than that obtained with only one identifier subject to a reinitialization mechanism. For the first time derivative estimation we reinitialize every 2 s within a time delay between estimators of 1 s. In the case of the second time derivative estimation, the estimators are reinitialized every 1 s with a time delay of 0.5 s.

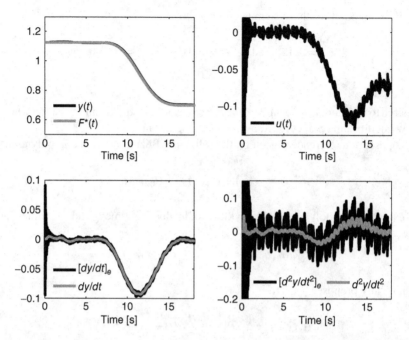

Figure 6.9 Closed-loop response of the perturbed synchronous generator. Actual and estimated state variables and control input

6.3.1.1 Control of an Uncertain Generator

In order to illustrate the possibilities of *simultaneously* estimating the unmeasured states of the system, through output derivations and the estimation of unknown parameters, we assume now that the parameter b_3 in the system model is constant but unknown.

We recall that the unmeasured state variable x_3 was estimated with the help of time derivatives of the noisy output y as

$$x_{3e} = \frac{P - b_2 \ddot{y}_e - \ddot{y}_e}{b_1 \sin y} \tag{6.45}$$

Multiply the state equation by t and integrate the left-hand side by parts, then solve for b_3 as $\hat{b}_3$ to obtain.

$$b_{3e} = \frac{t x_{3e}(t) + \int_0^t [x_{3e}(\sigma)(b_4 \sigma - 1) - \sigma(E - u(\sigma))]d\sigma}{\int_0^t \sigma \cos(y(\sigma))d\sigma} \tag{6.46}$$

We use this parameter estimate in the c.e. state feedback controller (Figure 6.10):

$$u = E - b_4 \left(\frac{P - b_2 \ddot{y}_e - \ddot{y}_e}{b_1 \sin y} \right) + b_{3e} \cos y$$

$$- \frac{[-b_2 \ddot{y}_e - v] \sin y - [P - b_2 \ddot{y}_e - \ddot{y}_e] \dot{y}_e \cos y}{b_1 \sin^2 y}$$

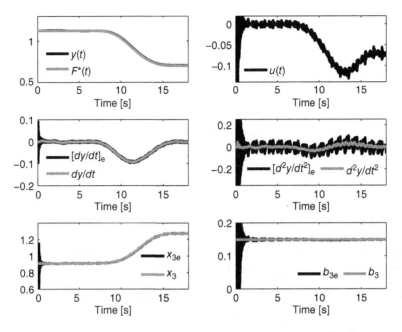

Figure 6.10   Closed-loop response of the perturbed synchronous generator. Simultaneous estimation of parameter and states

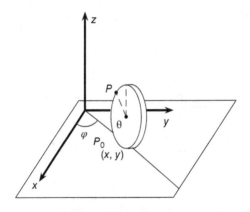

Figure 6.11   Rolling disk

$$v = [y^*(t)]^{(3)} - \lambda_3(\ddot{y}_e - \ddot{y}^*(t)) - \lambda_2(\dot{y}_e - \dot{y}^*(t))$$
$$- \lambda_1(y - y^*(t)) \tag{6.47}$$

6.3.2   Control of a Multi-variable Nonlinear System

Consider a vertically rolling homogeneous disk on the horizontal plane, rolling without slipping or tilting, as in Figure 6.11.

The dynamic model of the system is written in the following manner, using the Euler–Lagrange formulation with non-holonomic restrictions:

$$\dot{x} = R(\cos \varphi)\dot{\theta}$$

$$\dot{y} = R(\sin \varphi)\dot{\theta}$$

$$J\ddot{\varphi} = u_\varphi$$

$$(I + mR^2)\ddot{\theta} = u_\theta$$

The first two equations represent the *kinematics* of the system and the last two equations the *dynamics* of the system. Note that θ does not intervene in the system equations, only $\dot{\theta}$ (cyclic variable).

The typical problem consists of controlling the system from the ports u_φ, u_θ in such a manner that a specified trajectory in parametric form $x^*(t)$, $y^*(t)$ on the coordinate plane is tracked. The *independent* variables x and y play a crucial role in the understanding of the structure of the non-holonomic system and in the design of a feedback control law for the tracking of smooth trajectories specified in the plane of coordinates x, y. Indeed, all variables in the system are expressible in terms of x, y, and a finite number of their time derivatives:

$$\dot{\theta} = \frac{1}{R^2}\sqrt{\dot{x}^2 + \dot{y}^2}$$

$$\varphi = \arctan\left(\frac{\dot{y}}{\dot{x}}\right)$$

$$u_\theta = (m + \frac{I}{R^2})\left[\frac{\ddot{x}\dot{x} + \ddot{y}\dot{y}}{\sqrt{\dot{x}^2 + \dot{y}^2}}\right]$$

$$u_\varphi = J\frac{[\dot{x}y^{(3)} - \dot{y}x^{(3)}](\dot{x}^2 + \dot{y}^2) - 2(\ddot{y}\dot{x} - \ddot{x}\dot{y})(\ddot{x}\dot{x} + \ddot{y}\dot{y})}{(\dot{x}^2 + \dot{y}^2)^2}$$

$$\dot{u}_\theta = (m + \frac{I}{R^2})\left[\frac{(\dot{x}x^{(3)} + \dot{y}y^{(3)} + \ddot{x}^2 + \ddot{y}^2)(\dot{x}^2 + \dot{y}^2) - (\ddot{x}\dot{x} + \ddot{y}\dot{y})^2}{(\dot{x}^2 + \dot{y}^2)^{\frac{3}{2}}}\right]$$

Evidently, it was necessary to perform a first-order extension of the control input variable u_θ.

The state-dependent change of coordinates in the input space

$$u_\varphi = J\frac{[\dot{x}v_y - \dot{y}v_x](\dot{x}^2 + \dot{y}^2) - 2(\ddot{y}\dot{x} - \ddot{x}\dot{y})(\ddot{x}\dot{x} + \ddot{y}\dot{y})}{(\dot{x}^2 + \dot{y}^2)^2}$$

$$\dot{u}_\theta = (m + \frac{I}{R^2})\left[\frac{(\dot{x}v_x + \dot{y}v_y + \ddot{x}^2 + \ddot{y}^2)(\dot{x}^2 + \dot{y}^2) - (\ddot{x}\dot{x} + \ddot{y}\dot{y})^2}{(\dot{x}^2 + \dot{y}^2)^{\frac{3}{2}}}\right]$$

reduces the control problem to the regulation of two independent chains of integrators:

$$x^{(3)} = v_x, \qquad y^{(3)} = v_y$$

The trajectory-tracking controller is given by

$$u_\varphi = J \frac{[\dot{x}v_y - \dot{y}v_x](\dot{x}^2 + \dot{y}^2) - 2(\dot{y}\dot{x} - \ddot{x}\dot{x})(\ddot{x}\dot{x} + \ddot{y}\dot{y})}{(\dot{x}^2 + \dot{y}^2)^2}$$

$$\dot{u}_\theta = (m + \frac{I}{R^2}) \left[\frac{(\dot{x}v_x + \dot{y}v_y + \dot{x}^2 + \dot{y}^2)(\dot{x}^2 + \dot{y}^2) - (\ddot{x}\dot{x} + \ddot{y}\dot{y})^2}{(\dot{x}^2 + \dot{y}^2)^{\frac{3}{2}}} \right]$$

$$v_x = [x^*]^{(3)} - k_{2x}(\ddot{x} - \ddot{x}^*) - k_{1x}(\dot{x} - \dot{x}^*) - k_{0x}(x - x^*)$$

$$v_y = [y^*]^{(3)} - k_{2y}(\ddot{y} - \ddot{y}^*) - k_{1y}(\dot{y} - \dot{y}^*) - k_{0y}(y - y^*)$$

The implementation of this controller requires the generation of the derivative signals $\dot{x}, \dot{y}, \ddot{x}, \ddot{y}$.

Consider an output signal $x(t)$, which is sufficiently smooth and defined on the interval $[t_0, \infty]$. The well-known *truncated Taylor series* expansion around $t = t_0$ is written

$$x(t) \approx \sum_{j=0}^{K-1} \frac{1}{j!}[x^{(j)}(t_0)](t - t_0)^j$$

If we accept this polynomial representation for $x(t)$ as valid, then it is true that $x(t)$ is also approximated by a linear time-invariant homogeneous system of the form

$$\frac{d^K x(t)}{dt^K} = 0$$

with unknown initial conditions at time t_0, represented by the value of the derivatives of $x(t)$ from order 0 to order $K - 1$, both inclusive.

For the nonlinear system under study, we propose the following models (see Figures 6.12 and 6.13), piecewise linear, of the system outputs $x(t), y(t)$, comprising linear homogeneous systems (polynomials or splines) with unknown initial conditions:

$$x^{(6)} = 0, \qquad y^{(6)} = 0$$

Utilizing operational calculus, we obtain

$$s^6 X(s) - s^5 x_0 - s^4 \dot{x}_0 - s^3 \ddot{x}_0 - s^2 x_0^{(3)} - s x_0^{(4)} - x_0^{(5)} = 0$$

$$s^6 Y(s) - s^5 y_0 - s^4 \dot{y}_0 - s^3 \ddot{y}_0 - s^2 y_0^{(3)} - s y_0^{(4)} - y_0^{(5)} = 0$$

Taking derivatives with respect to s *six times*, we have the following identities:

$$\frac{d^6[s^6 X(s)]}{ds^6} = 0, \qquad \frac{d^6[s^6 Y(s)]}{ds^6} = 0$$

and, as a consequence,

$$s^{-j}\frac{d^6[s^6 X(s)]}{ds^6} = 0, \qquad s^{-j}\frac{d^6[s^6 Y(s)]}{ds^6} = 0, j = 5, 4$$

The crucial observation is that, in the time domain, these expressions contain, in a hidden manner, all the time derivatives of order less than or equal to 2 for both signals.

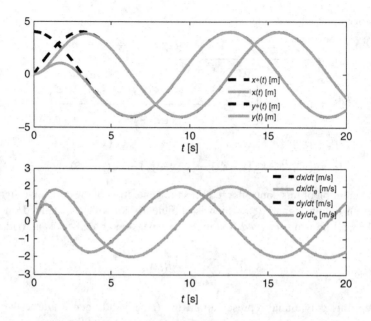

Figure 6.12   Closed-loop behavior with velocity estimation in a multi-variable nonlinear system

Indeed, the expression

$$s^{-5}\frac{d^6[s^6 Y(s)]}{ds^6} = 0$$

results, in the time domain, in the equality

$$720\left(\int^{(5)} y\right) - 4320\left(\int^{(4)} ty\right) + 5400\left(\int^{(3)} t^2 y\right)$$

$$-2400\left(\int^{(2)} t^3 y\right) + 450\left(\int t^4 y\right) - 36t^5 y + \frac{d}{dt}[t^6 y] = 0$$

but, since

$$\frac{d}{dt}[t^6 y] = 6t^5 y + t^6 \dot{y}$$

we have the following estimator:

$$[\dot{y}]_e = \begin{cases} \text{arbitrary} & \forall\, t \in [0, \epsilon) \\[2mm] \dfrac{1}{t^6}\left[-720(\int^{(5)} y) + 4320(\int^{(4)} ty) \right. \\[1mm] \quad -5400(\int^{(3)} t^2 y) + 2400(\int^{(2)} t^3 y) \\[1mm] \quad \left. -450(\int t^4 y) + 30t^5 y \right] & \forall\, t > \epsilon \end{cases}$$

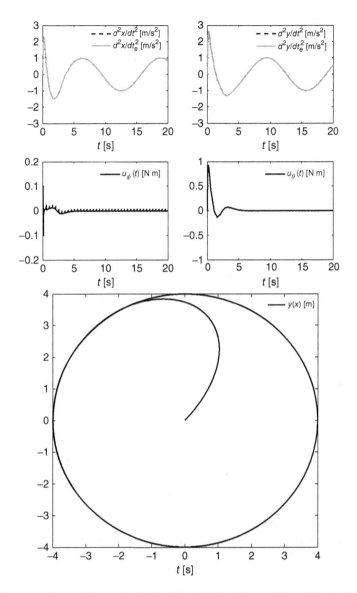

Figure 6.13 Trajectory tracking and estimated accelerations in a multi-variable nonlinear system

and, in a similar form:

$$
[\dot{x}]_e =
\begin{cases}
\text{arbitrary} & \forall\, t \in [0, \epsilon) \\[2mm]
\dfrac{1}{t^6}\Big[-720(\int^{(5)} x) + 4320(\int^{(4)} tx) \\
\quad -5400(\int^{(3)} t^2 x) + 2400(\int^{(2)} t^3 x) \\
\quad -450(\int t^4 x) + 30 t^5 x\Big] & \forall\, t > \epsilon
\end{cases}
$$

The expression

$$s^{-4} \frac{d^6[s^6 Y(s)]}{ds^6} = 0$$

results, in the time domain, in the equality

$$720 \left(\int^{(4)} y \right) - 4320 \left(\int^{(3)} ty \right) + 5400 \left(\int^{(2)} t^2 y \right)$$

$$-2400 \left(\int t^3 y \right) + 450 t^4 y - 36 \frac{d}{dt}[t^5 y] + \frac{d^2}{dt^2}[t^6 y] = 0$$

but, since

$$450 t^4 y - 36 \frac{d}{dt}[t^5 y] + \frac{d^2}{dt^2}[t^6 y] = 300 t^4 y - 24 t^5[\dot{y}] + t^6 \ddot{y}$$

we propose the following (non-asymptotic) state estimator:

$$[\ddot{y}]_e(t) = \begin{cases} \text{arbitrary} & \forall t \in [0, \epsilon) \\ \frac{1}{t^6} \left[-720 (\int^{(4)} y) + 4320 (\int^{(3)} ty) \right. & \\ \left. -5400 (\int^{(2)} t^2 y) + 2400 (\int t^3 y) \right. & \\ \left. -300 t^4 y + 24 t^5 [\dot{y}]_e \right] & \forall t > \epsilon \end{cases}$$

and, in a similar form:

$$[\ddot{x}]_e(t) = \begin{cases} \text{arbitrary} & \forall t \in [0, \epsilon) \\ \frac{1}{t^6} \left[-720 (\int^{(4)} x) + 4320 (\int^{(3)} tx) \right. & \\ \left. -5400 (\int^{(2)} t^2 x) + 2400 (\int t^3 x) \right. & \\ \left. -300 t^4 x + 24 t^5 [\dot{x}]_e \right] & \forall t > \epsilon \end{cases}$$

The previous estimators must be *re-initialized* so as to prevent lack of validity of the used polynomial approximation of the signal. For this, we recall the form of a polynomial approximation from an arbitrary time t_i:

$$x(t) = \sum_{j=0}^{6} \frac{1}{j!} [x^{(j)}(t_i)](t - t_i)^j$$

This leads to similar formulae as obtained at time $t_0 = 0$. Indeed,

$$[\dot{x}]_e(t) = \begin{cases} [\dot{x}]_e(t_i^-) & \forall t \in [t_i, t_i + \epsilon) \\ \frac{1}{(t-t_i)^6} \left[-720 (\int_{t_i}^{(5)} x) + 4320 (\int_{t_i}^{(4)} (t - t_i)x) \right. & \\ \left. -5400 (\int_{t_i}^{(3)} (t - t_i)^2 x) + 2400 (\int_{t_i}^{(2)} (t - t_i)^3 x) \right. & \\ \left. -450 (\int_{t_i} (t - t_i)^4 x) + 30(t - t_i)^5 x \right] & \forall t > t_i + \epsilon \end{cases}$$

where we have used the following notation:

$$\left(\int_{t_i}^{(k)} (t - t_i)^j x \right) = \int_{t_i}^{t} \int_{t_i}^{\sigma_1} \cdots \int_{t_i}^{\sigma_{k-1}} (\sigma_k - t_i)^j x(\sigma_k) d\sigma_k \cdots d\sigma_1$$

The reinitialization of the estimator may be carried out in an *automatic* fashion, simply evaluating an integral quadratic error of the output signal reconstruction and fixing a threshold for the allowable error. We may use, for instance,

$$e_x(t) = \int_{t_i}^{t} (x(\sigma) - \hat{x}(\sigma))^2 d\sigma$$

with

$$\hat{x}(t) = \begin{cases} x(t_i) + (t - t_i)[\dot{x}]_e(t_i) + \frac{1}{2}(t - t_i)^2[\ddot{x}]_e(t_i) + \cdots \\ \hspace{4cm} \forall t \in [t_i, t_i + \epsilon] \\ x(t_i + \epsilon) + \int_{t_i+\epsilon}^{t} [\dot{x}]_e(\sigma) d\sigma \quad \forall t \in [t_i + \epsilon, t_{i+1}) \end{cases}$$

We tried the algorithm when the system is subject to computer-generated noise perturbations (Figure 6.14):

$$\dot{x} = R(\cos \varphi)\dot{\theta}$$

$$\dot{y} = R(\sin \varphi)\dot{\theta}$$

$$J\ddot{\varphi} = u_\varphi + \xi(t)$$

$$(I + mR^2)\ddot{\theta} = u_\theta + \eta(t)$$

with $\xi(t)$ and $\eta(t)$ synthesized as random processes constituted by sequences of piecewise constant uniformly distributed random variables in the interval $[-0.5, 0.5]$ and known in most computer packages as the functions $rect(t)$.

6.3.3 Experimental Results on a Mechanical System

The laboratory prototype of Figure 6.15 is a mass–spring–damper system in a 2-DOF configuration. These devices are commonly used to study vibration phenomena in vehicle suspension systems, shakers, or vibration generators, to attenuate, or increase, these vibrations. Another control application is the high-precision positioning task in modern computer numerical controlled (CNC) machine tools, where the nonlinear friction effects are more significant. We have used this plant to show, in an experimental manner, the reliability of state estimation via the algebraic methodology. In this case, the goal is to track a reference trajectory with precision and robustness using estimated states supplied by algebraic estimators.

The schematic diagram of Figure 6.16 shows the variables that describe the dynamical behavior of this system. The plant is equipped with an actuator to apply a horizontal input force and is provided with encoders to measure the displacement of each mass. However, the corresponding velocities are not available and must be estimated from these measurements.

The mathematical model of the mechanical plant is given by

$$m_1\ddot{x}_1 + (k_1 + k_2)x_1 - k_2x_2 + c_1\dot{x}_1 = F$$

$$m_2\ddot{x}_2 + k_2(x_2 - x_1) + c_2\dot{x}_2 = 0 \tag{6.48}$$

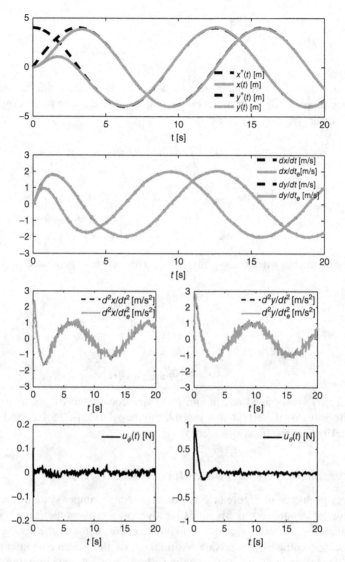

Figure 6.14 Trajectory tracking and estimated accelerations with input noise in the vertically rolling homogeneous disk

6.3.3.1 Formulation of the Problem

The control of the position of mass, m_2, is by no means trivial because this mass motion corresponds to the underactuated degree of freedom. The formulation of the problem is as follows.

Given the plant dynamics (6.48), it is desired to asymptotically track an arbitrary smooth reference trajectory $x_2^*(t)$ for the output variable x_2 under the assumption that the plant parameters $m_1, m_2, c_1, c_2, k_1, k_2$ are known and the mass positions are measurable. The velocity variables, however, are not available for measurement; they are to be algebraically estimated on-line to complete the feedback loop.

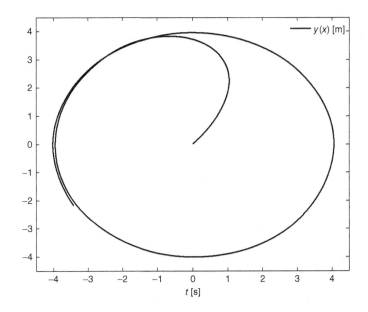

Figure 6.14   (*continued*)

Figure 6.15   Electromechanical 2-DOF control plant

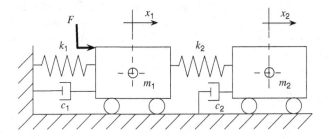

Figure 6.16   A 2-DOF mass−spring−damper system

6.3.3.2 Controller Design

Rewriting (6.48) in state-space form, with $z_1 = x_1, z_2 = \dot{x}_1, z_3 = x_2, z_4 = \dot{x}_2, u = F$, we obtain

$$\dot{z} = Az + bu$$
$$y = Cz \tag{6.49}$$

where

$$A = \begin{bmatrix} 0 & 1 & 0 & 0 \\ -\dfrac{(k_1+k_2)}{m_1} & -\dfrac{c_1}{m_1} & \dfrac{k_2}{m_1} & 0 \\ 0 & 0 & 0 & 1 \\ \dfrac{k_2}{m_2} & 0 & -\dfrac{k_2}{m_1} & \dfrac{c_2}{m_1} \end{bmatrix} \quad b = \begin{bmatrix} 0 \\ \dfrac{1}{m_1} \\ 0 \\ 0 \end{bmatrix}$$

$$z = \begin{bmatrix} z_1 & z_2 & z_3 & z_4 \end{bmatrix}^T \quad C = \begin{bmatrix} 1 & 0 & 1 & 0 \end{bmatrix}$$

The controllability matrix of the system is

$$C = \begin{bmatrix} 0 & \dfrac{1}{m_1} & -\dfrac{c_1}{m_1^2} & \alpha_1 \\ \dfrac{1}{m_1} & -\dfrac{c_1}{m_1^2} & \alpha_1 & \alpha_2 \\ 0 & 0 & 0 & \dfrac{k_2}{m_1 m_2} \\ 0 & 0 & \dfrac{k_2}{m_1 m_2} & \alpha_3 \end{bmatrix} \tag{6.50}$$

with

$$\alpha_1 = \frac{1}{m_1}\left(-\frac{1}{m_1}(k_1 + k_2) + \frac{c_1^2}{m_1^2}\right)$$

$$\alpha_2 = \frac{1}{m_1}\left(\frac{c_1}{m_1^2}(k_1 + k_2) - \frac{c_1}{m_1}\left(-\frac{1}{m_1}(k_1 + k_2) + \frac{c_1^2}{m_1^2}\right)\right)$$

$$\alpha_3 = \frac{1}{m_1}\left(-c_2\frac{k_2}{m_2^2} - c_1\frac{k_2}{m_1 m_2}\right)$$

The determinant of (6.50) is $\det C = k_2^2/(m_1^4 m_2^2)$. For physical reasons, $k_2, m_1, m_2 \neq 0$, which indicates that (6.49) is controllable and consequently also differentially flat. Therefore, there exists a linear function of the state vector z, called the *flat output*, which completely differentially parameterizes all the variables of the system. This output is given by (Sira-Ramírez and Agrawal, 2004)

$$\vartheta = \begin{bmatrix} 0 & 0 & 0 & 1 \end{bmatrix} C^{-1} z \tag{6.51}$$

So, using (6.51), the flat output of the system is

$$\vartheta = \frac{m_1 m_2}{k_2} z_3 \tag{6.52}$$

Note that ϑ is proportional to the variable which describes the unactuated degree of freedom, the position of the second mass. For simplicity, $\vartheta = x_2 = z_3$ is taken to be the flat output of the system. All the original system variables are expressible as functions of ϑ and a finite number of its time derivatives, that is,

$$x_2 = \vartheta$$

$$\dot{x}_2 = \dot{\vartheta}$$

$$x_1 = \frac{1}{k_2}[m_2\ddot{\vartheta} + c_2\dot{\vartheta} + k_2\vartheta]$$

$$\dot{x}_1 = \frac{1}{k_2}[m_2\vartheta^{(3)} + c_2\ddot{\vartheta} + k_2\dot{\vartheta}]$$

$$F = \frac{m_1 m_2}{k_2}\vartheta^{(4)} + \left(\frac{m_1 c_2 + c_1 m_2}{k_2}\right)\vartheta^{(3)} + \left(m_1 + m_2 + \frac{m_2 k_1 + c_1 c_2}{k_2}\right)\vartheta^{(2)}$$

$$+ \left(c_1 + c_2 + \frac{c_2 k_1}{k_2}\right)\dot{\vartheta} + k_1\vartheta$$

So, we can propose the following flatness-based control law:

$$u = F = \frac{m_1 m_2}{k_2}v + \left(\frac{m_1 c_2 + c_2 m_2}{k_2}\right)\vartheta^{(3)} + \left(m_1 + m_2 + \frac{m_2 k_1 + c_1 c_2}{k_2}\right)\vartheta^{(2)}$$

$$+ \left(c_1 + c_2 + \frac{c_2 k_1}{k_2}\right)\dot{\vartheta} + k_1\vartheta \tag{6.53}$$

with

$$v = [\vartheta^*(t)]^{(4)} - \alpha_4(\vartheta^{(3)} - [\vartheta^*(t)]^{(3)}) - \alpha_3(\ddot{\vartheta} - \ddot{\vartheta}^*(t)) - \alpha_2(\dot{\vartheta} - \dot{\vartheta}^*(t))$$

$$- \alpha_1(\vartheta - \vartheta^*(t)) - \alpha_0 \int_0^t (\vartheta - \vartheta^*(\sigma))d\sigma$$

The coefficients α_i are chosen to set the poles of the closed-loop system in the stable portion of the complex plane.

6.3.3.3 Algebraic Velocity Estimation

The implementation of the flatness-based controller can be achieved through the computation of the time derivatives of the measured outputs using a third-degree polynomial approximation for the time signal representing the state. Take $\phi(t)$ to be such a polynomial. We have

$$\frac{d^4}{ds^4}(s^4\phi(s)) = 0 \tag{6.54}$$

Following the algebraic methodology, we obtain from (6.54) the following estimator for the first time derivative of $\phi(t)$:

$$[\dot{\phi}]_e(t) = \begin{cases} \text{arbitrary} & \forall\, t \in [0, \epsilon) \\ \frac{1}{t^4}\left[12t^3\phi - 72\int t^2\phi + 96\int^{(2)}t\phi - 24\int^{(3)}\phi\right] & \forall t > \epsilon \end{cases}$$

Since this estimated value is an approximation, the algebraic estimator formula is adapted to a reinitialization of the online calculations:

$$[\dot{\phi}]_e(t) = \begin{cases} [\dot{\phi}]_e(t_i^-) & \forall\, t \in [t_i, t_i + \epsilon) \\ \frac{1}{(t-t_i)^4}\left[12(t-t_i)^3\phi - 72\int_{t_i}(t-t_i)^2\phi \right. \\ \left. + 96\int_{t_i}^{(2)}(t-t_i)\phi - 24\int_{t_i}^{(3)}\phi\right] & \forall t > t_i + \epsilon \end{cases}$$

A time-varying linear filter representation of this algebraic estimator is

$$[\dot{\phi}]_e(t) = \begin{cases} [\dot{\phi}]_e(t_i^-) & \forall\, t \in [t_i, t_i + \epsilon) \\ \frac{12(t-t_i)^3\phi(t)+\gamma_1}{(t-t_i)^4} & \forall t > t_i + \epsilon \end{cases}$$

$$\dot{\gamma}_1 = \gamma_2 - 72(t - t_i)^2\phi$$

$$\dot{\gamma}_2 = \gamma_3 + 96(t - t_i)\phi$$

$$\dot{\gamma}_3 = -24\phi$$

$$\gamma_1(t_i) = 0$$

$$\gamma_2(t_i) = 0$$

$$\gamma_3(t_i) = 0 \tag{6.55}$$

All time derivatives of the flat output needed for implementing (6.53) can be calculated using the measured mass positions and the estimated velocities provided by (6.55):

$$\vartheta = x_2$$

$$\dot{\vartheta} = [\dot{x}_2]_e$$

$$\vartheta^{(2)} = \frac{1}{m_2}(k_2x_1 - k_2x_2 - c_2[\dot{x}_2]_e)$$

$$\vartheta^{(3)} = \frac{1}{m_2^2}(-c_2k_2x_1 + k_2m_2[\dot{x}_1]_e + c_2k_2x_2 + (-k_2m_2 + c_2^2)[\dot{x}_2]_e) \tag{6.56}$$

6.3.3.4 Simulation and Experimental Results

All the parameters of the 2-DOF mass–spring–damper system were properly identified by traditional methods:

$$m_1 = 2.7945 \quad \text{kg}$$
$$m_2 = 2.5434 \quad \text{kg}$$
$$k_1 = 360.1603 \text{ N/m}$$
$$k_2 = 723.07 \quad \text{N/m}$$
$$c_1 = 2.05 \quad \text{N/(m/s)}$$
$$c_2 = 1.8283 \text{ N/(m/s)}$$

The first test over the experimental plant was the stabilization of the state trajectory to a desired equilibrium, $(\vartheta, \dot{\vartheta}, \vartheta^{(2)}, \vartheta^{(3)}) = (\vartheta^*, 0, 0, 0)$. Simulation and experimental results, Figures 6.17 and 6.18, show the stabilization of the system to different equilibriums using the algebraic velocity estimation. The estimators were implemented with $\epsilon = 0.035$ s and reinitialization of the online calculations every 0.15 s.

For robustness testing purposes, three different reference trajectories based on interpolating the Bézier polynomial were proposed. The first, $\vartheta_1^*(t)$, is a smooth rest-to-rest trajectory defined by

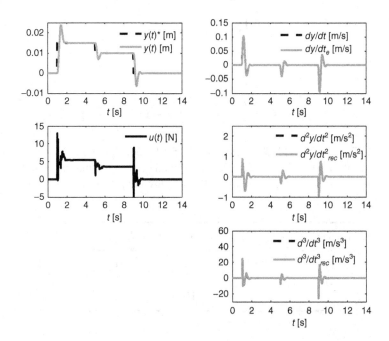

Figure 6.17 Simulation results of a feedback control using derivatives reconstructed from estimated velocities in a 2-DOF mechanical system. Stabilization of the flat output

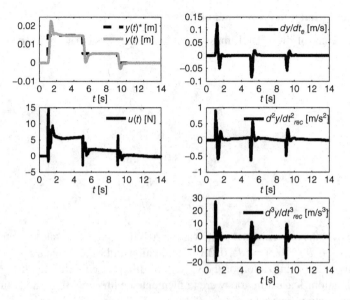

Figure 6.18 Experimental results of a feedback control using derivatives reconstructed from estimated velocities in a 2-DOF mechanical system. Stabilization of the flat output

$$
\vartheta_1^*(t) = \begin{cases} \hbar_0 & \text{for } t < t_0 \\[2mm] \hbar_0 + (\hbar_f - \hbar_0)\phi(t) & \forall t \in [t_0, t_f] \\[2mm] \hbar_f & \text{for } t > t_f \end{cases} \tag{6.57}
$$

$$
\phi(t, t_0, t_f) = \left(\frac{t - t_0}{t_f - t_0} \right)^r \sum_{i=0}^{r} (-1)^i \beta_i \left(\frac{t - t_0}{t_f - t_0} \right)^i \tag{6.58}
$$

with $\beta_0 = 12870$, $\beta_1 = 91520$, $\beta_2 = 288288$, $\beta_3 = 524160$, $\beta_4 = 600600$, $\beta_5 = 443520$, $\beta_6 = 205920$, $\beta_7 = 54912$, and $\beta_8 = 6435$. Figures 6.19 and 6.20 show simulation and experimental results of this tracking task for intervals with the following parameters:

Interval	t_0(s)	t_f(s)	$\hbar_0$(m)	$\hbar_f$(m)
(1)	1	2	0	0.015
(2)	5	6	0.015	0.01
(3)	9	10	0.01	0

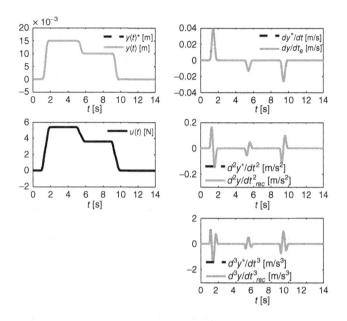

Figure 6.19   Simulation results of rest-to-rest trajectory tracking

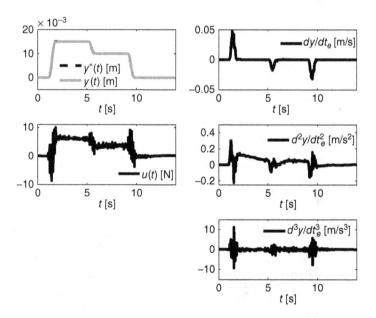

Figure 6.20   Experimental results of rest-to-rest trajectory tracking

The second trajectory is a time-varying amplitude sinusoidal, where the amplitude $A(t)$ is described by the first reference trajectory $\vartheta_1^*(t)$ with the same parameter list as above:

$$\vartheta_2^*(t) = \vartheta_1^*(t)\sin(\omega t) \tag{6.59}$$

The tracking results of the second trajectory are presented in Figures 6.21 and 6.22. The third trajectory is a time-varying frequency sinusoidal, defined as

$$\vartheta_3^*(t) = 0.01\sin(\omega(t)t) \tag{6.60}$$

where $\omega(t)$ is given by (6.58), with the following parameters:

Interval	t_0(s)	t_f(s)	$\hbar_0$(rad/s)	$\hbar_f$(rad/s)
(1)	8	12	1	2
(2)	25	35	2	1

Finally, the simulation and experimental results of the third trajectory are displayed in Figures 6.23 and 6.24. In all experimental cases, the control parameters were fixed to $\omega_n = 20$, $\xi = 0.65$, and $p = 15$.

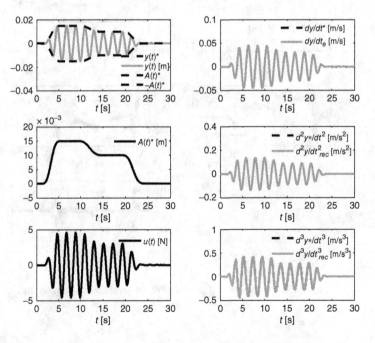

Figure 6.21   Tracking simulation results of a time-variant amplitude sinusoidal trajectory

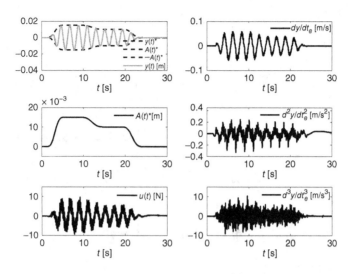

Figure 6.22 Tracking experimental results of a time-variant amplitude sinusoidal trajectory

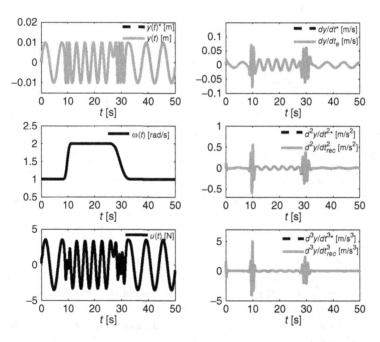

Figure 6.23 Tracking simulation results of a time-variant frequency sinusoidal trajectory

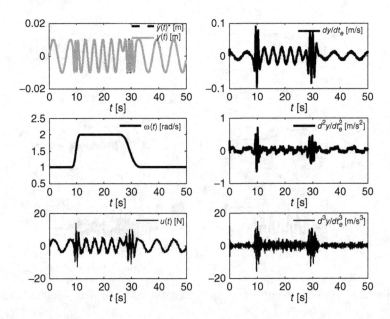

Figure 6.24   Tracking experimental results of a time-variant frequency sinusoidal trajectory

6.4   Remarks

- We have seen the possibilities of an algebraic approach for state estimation in linear and nonlinear systems, whether mono or multi-variable. The approach is characterized by non-asymptotic formulae for the fast calculation of the time derivatives of the output.
- We have also demonstrated the possibility of using such an algebraic method for approximate state estimation in the control of a real-life linear system. In the output derivative calculation part of the estimation process, only knowledge of the output signal is demanded.
- The theoretical formulation that justifies our proposal is based on differential algebra and, in particular, on non-commutative Weyl algebras. The most significant element of the approach is that we *completely abandon the stochastic formulation of the problem* and rely on computation formulae, *rather than asymptotic estimations*, for the state reconstruction part of the control problem.
- The method presented here has been utilized, through extensions, in various areas of automatic control and signal processing. Among these: fault diagnosis and fault-tolerant control in uncertain systems (Fliess *et al.*, 2004); control of nonlinear systems (Fliess and Sira-Ramírez, 2004; Fliess *et al.*, 2008); and noisy analog signal compression (Fliess *et al.*, 2003). An interesting extension of the output feedback control problem, regarded as a robust control problem through algebraic non-modeled dynamics and disturbances, is due to Fliess and Join (2008), known as intelligent PID control.

References

Bazanella, A. De Silva, A. and Kokotovic, P. (1997) Lyapunov design of excitation control for synchronous machines. *Proceedings of the 36th IEEE CDC-Conference on Decision and Control*, pp. 211–216, San Diego, CA.

Diop, S. and Fliess, M. (1991) Nonlinear observability, identifiability and persistent trajectories. *Proceedings of the 36th IEEE Conference on Decision and Control*, pp. 714–719, Brighton, UK.

Diop, S. Grizzle, J. and Chaplais, F. (2000) On numerical differentiation algorithms for nonlinear estimation. *Proceedings of the 39th IEEE Conference on Decision and Control*, pp. 1133–1138, Sydney, Australia.

Diop, S. Grizzle, J. Moraal, P. and Stefanopoulou, A. (1994) Interpolation and numerical differentiation for observer design. *Proceedings of the American Control Conference*, pp. 1329–1333, Baltimore, MD.

Fliess, M. and Join, C. (2008) Intelligent PID controllers. *16th Mediterranean Conference on Control and Automation, 2008*, pp. 326–331, Ajaccio Corsica, France

Fliess, M. and Sira-Ramírez, H. (2004) Control via state estimations of flat systems. *IFAC Nolcos Conference*, Stuttgart, Germany.

Fliess, M. Join, C. and Sira-Ramírez, H. (2004) Robust residual generation for linear fault diagnosis: An algebraic setting with examples. *International Journal of Control* **77**(14), 1223–1242.

Fliess, M. Join, C. and Sira-Ramírez, H. (2008) Non-linear estimation is easy. *International Journal of Modelling, Identification and Control* **4**(1), 12–27.

Fliess, M. Mboup, M. Mounier, H. and Sira-Ramírez, H. (2003) *Algebraic Methods in Flatness, Signal Processing and State Estimation*. Editiorial Lagaress, Mexico, chapter 1.

Levant, A. (1998) Robust exact differentiation via sliding mode technique. *Automatica* **34**(3), 379–384.

Sira-Ramírez, H. and Agrawal, S. (2004) *Differentially Flat Systems*. Marcel Dekker, New York.

Sira-Ramírez, H. and Fliess, M. (2004) On the output feedback control of a synchronous generator. *Proceedings of the 43rd IEEE Conference on Decision and Control*, Bahamas.

Vukić, Z. Kuljača, L. Dlonlagić, D. and Tešnjak, S. (2003) *Nonlinear Control Systems*. Control Engineering. Marcel Dekker, New York.

7

Miscellaneous Applications

7.1 Introduction

This chapter explores some topics which are associated with the techniques presented in the previous chapters. We propose variants of the linear parameter identification and of the time derivative estimation. Such variations of the approaches and their possible applications are explored, as usual, through physically oriented examples.[1] In Section 7.1.1, we consider the problem of controlling, using passivity-based techniques, a DC motor with an unknown torque input. The same idea of this control strategy is again taken up in Section 7.1.4, for a DC–DC power converter.

The algebraic approach for parameter estimation in linear systems enables a crucial procedure in which the system expressions are multiplied by a suitable power of the time variable. This cancels the effects of the initial conditions. The subsequent use of nested integrations generates the identified parameters as the quotient of two unstable time functions, in the scalar case, or the product of the inverse of a "growing matrix" times a corresponding "growing vector," in the vector case. Although the computation of the unknown parameter can be performed quickly enough, there may be concerns with the unstable nature of the intervening factors. To avoid this effect, in Section 7.2 an alternative methodology based on exponential time functions (instead of time polynomials) is proposed. Section 7.3.2 takes the problem of frequency modulation using the algebraic derivative method. The truncated Taylor series approach to generate the algebraic time derivatives estimation is modified in Section 7.3.3, using some ideas of the parameter estimation procedure.

Dealing with noise when calculating time derivatives can be a tough challenge. A high-gain observer and the algebraic method are put together to develop an alternative algebraic denoising scheme, as shown in Section 7.4.

---

[1] Notice that there are some other problems which are not treated in this book such, as identification in time-delay systems (Belkoura *et al.*, 2009), switched systems (Fliess *et al.*, 2008a), and infinite-dimensional systems (Rudolph, 2006).

Algebraic Identification and Estimation Methods in Feedback Control Systems, First Edition.
Hebertt Sira-Ramírez, Carlos García-Rodríguez, Alberto Luviano-Juárez, John Alexander Cortés-Romero.
© 2014 John Wiley & Sons, Ltd. Published 2014 by John Wiley & Sons, Ltd.

7.1.1 The Separately Excited DC Motor

In this section, we present an algebraic-based fast feedforward adaptation approach for the trajectory-tracking task in a separately excited DC motor with unknown load torque perturbations under available noisy measurements of the state variables. Two cases are presented.

The first case deals with a piecewise constant unknown torque perturbation. The controller is a simple certainty-equivalence, linear, passivity-based controller with feedforward pre-compensation depending explicitly on the unknown torque. A fast adaptation is achieved by online algebraic estimation of the current unknown constant parameter.

The second case deals with an unknown time-varying load perturbation. The versatility of the algebraic approach is demonstrated by now using a flatness Identification and Estimation in Feedback Control Systems based feedback controller with a combination of GPI control and precise continuous online adaptation of the feedforward pre-compensation signal which depends explicitly on the unknown mechanical power and its first-order time derivative.

Consider the following standard model of a separately excited DC motor:

$$L_a \frac{d}{dt} i_a = -R_a i_a - K_m i_f \omega + v_a$$

$$L_f \frac{d}{dt} i_f = -R_f i_f + v_f$$

$$J_m \frac{d}{dt} \omega = K_m i_f i_a - B_m \omega - \tau_L \qquad (7.1)$$

where i_a is the armature circuit current, i_f is the field circuit current, and ω is the angular velocity. The system-independent control inputs are v_a and v_f. The torque τ_L is assumed to adopt unknown piecewise constant values of, possibly, changing signs.

It is primarily desired to track a given reference angular velocity $\omega^\star(t)$. If a suitable reference $i_f^\star$ for i_f is also prescribed, then all nominal trajectories of the system variables, including the control inputs, are easily determined.

We propose the following certainty-equivalence linear controller based on exact tracking error dynamics passive output feedback (ETEDPOF), see Figure 7.1:

$$v_a = v_a^\star(t) - \gamma_1(i_a - i_a^\star(t)) \qquad (7.2)$$

$$v_f = v_f^\star(t) - \gamma_2(i_f - i_f^\star(t)) \qquad (7.3)$$

where $v_a^\star(t)$ and $v_f^\star(t)$ are, respectively, the nominal values of the armature circuit input voltage and the nominal value of the field circuit input voltage.

The design parameters γ_1 and γ_2 are strictly positive constants to be determined later. The nominal values of the input voltages are found thanks to the flatness property of the system:

$$v_a^\star(t) = L_a \frac{(Jm\ddot{F}^\star + B_m \dot{F}^\star)}{K_m G^\star} - \frac{(J_m \dot{F}^\star + B_m F^\star + \tau_L)\dot{G}^\star}{K_m [G^\star]^2} \qquad (7.4)$$

$$+R_a \left[\frac{J_m \dot{F}^\star + B_m F^\star + \tau_L}{K_m G^\star} \right] + K_m G^\star F^\star$$

$$v_f^\star(t) = L_f \dot{G}^\star + R_f G^\star \qquad (7.5)$$

where $F^\star = \omega^\star$ and $G^\star = i_f^\star$ are the nominal reference values of the flat outputs represented by the angular velocity ω and the field current i_f.

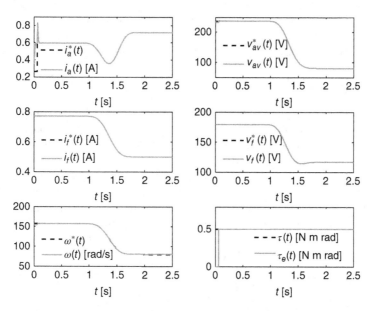

Figure 7.1 Closed-loop response of separately excited DC motor with ETEDPOF controller using online identification of piecewise constant torque load

Note that since $v_a^\star$ depends on the torque load τ_L, we need to perform, thanks to the simple form of the controller, fast adaptation only on the feedforward part of the controller.

Clearly, we need to perform trajectory planning on the flat outputs according to the overall system objectives. In order to emphasize that the feedforward part of the controller needs adaptation, we place a "hat" over τ_L to imply online identification of such an unknown piecewise constant perturbation. We have:

$$v_a^\star(t) = L_a \frac{(J_m \ddot{F}^\star + B_m \dot{F}^\star)}{K_m G^\star} - \frac{(J_m \dot{F}^\star + B_m F^\star + \hat{\tau}_L)\dot{G}^\star}{K_m [G^\star]^2} \tag{7.6}$$

$$+ R_a \frac{J_m \dot{F}^\star + B_m F^\star + \hat{\tau}_L}{K_m G^\star} + K_m G^\star F^\star \tag{7.7}$$

We implement a fast online algebraic estimation procedure for τ_L, yielding an accurate estimate $\hat{\tau}_L$ within the interval of time $[0, \epsilon)$. Once $\hat{\tau}_L$ is found to converge to a constant value, we will replace it in the certainty-equivalence controller.

We compute an estimate $\hat{\tau}_L$ of the uncertain parameter τ_L as follows. We set:

$$\hat{\tau}_L = \begin{cases} \text{arbitrary for } t \in [0, \epsilon) \\[2mm] \dfrac{2}{t^2} \left[-J_m t \omega + z \right] \text{ for } t \in [\epsilon, +\infty) \end{cases} \tag{7.8}$$

where

$$\frac{d}{dt} z = [J_m - B_m \sigma] \omega(\sigma) + K_m \sigma i_a(\sigma) i_f(\sigma) \tag{7.9}$$

with $z(0) = 0$.

We used the following parameter values taken from a real motor:

$$R_a = 3.5 \ \Omega, \quad L_a = 0.0432 \ \text{H}, \quad R_f = 233 \ \Omega, \quad L_f = 25.5 \ \text{F}$$

$$K_m = 1.9469 \ \frac{\text{V s}}{\text{A}}, \quad B_m = 0.0025 \ \text{N m s}, \quad J_m = 0.0017 \ \text{N m s}^2$$

It is desired to bring the angular velocity of the motor from an equilibrium value of 157.08 rad/s to the lower equilibrium value of 80 rad/s in 1 s ($t_{initial} = 1$ s, $t_{final} = 2$ s). The field current was weakened from the initial equilibrium value of 0.77 A to the final value 0.50 A in 1.2 s, starting the maneuver at $t = 0.8$ and ending it at $t = 2$.

The unknown load torque was taken to be 0.5 N m. The parameter ϵ representing the time interval required for an accurate estimate of τ_L was just 0.05 s. Prior to this instant, the arbitrary value bestowed on the load torque parameter, and its value in the controller, was set to zero.

In order to make the scheme robust with respect to high-frequency noises in the measurements of the required state variable signals, we used an invariant filtering approach. We used first-order integration filtering as follows:

$$\hat{\tau}_L = \begin{cases} \text{arbitrary for } t \in [0, \epsilon) \\ \dfrac{6}{t^3} [z_0] \ \text{for } t \in [\epsilon, +\infty) \end{cases} \tag{7.10}$$

where

$$\frac{d}{dt} z_0 = -J_m t \omega_{meas}(t) + z_1$$

$$\frac{d}{dt} z_1 = [J_m - B_m \sigma] \omega_{meas}(\sigma) + K_m \sigma i_{a \ meas}(\sigma) i_{f \ meas}(\sigma) \tag{7.11}$$

with $z_0(0) = 0$, $z_1(0) = 0$, and

$$i_{a \ meas} = i_a + \xi(t), \quad i_{f \ meas} = i_f + \eta(t), \quad \omega_{meas} = \omega + \chi(t) \tag{7.12}$$

where $\xi(t)$, $\eta(t)$, and $\chi(t)$ are zero-mean random processes affecting the state measurements. The noisy measurements were also used in the controller.

We let the noises be zero-mean computer-generated random processes constituted by a sequence of uniformly distributed random variables in the closed interval $[-0.5, 0.5]$. This random process is denoted "$rect(t) - 0.5$."

We set

$$\xi(t) = 0.08(rect(t) - 0.5)$$

$$\eta(t) = -0.05(rect(t) - 0.5)$$

$$\chi(t) = -10(rect(t) - 0.5)$$

We have also let the unknown load torque τ_L change its values and sign. A simple monitoring mechanism, based on the assumption of a constant value of the actual torque, triggers a reinitialization of the online estimator when the estimated torque starts drifting from the prevailing current constant value.

The reinitializations, at time t_i, are accomplished as follows:

$$\hat{\tau}_L(i) = \begin{cases} \text{arbitrary for } t \in [t_i, t_i + \epsilon) \\ \frac{6}{(t-t_i)^3}\left[z_0\right] \text{ for } t \in [t_i + \epsilon, t_{i+1}) \end{cases} \tag{7.13}$$

where

$$\frac{d}{dt}z_0 = -J_m(t - t_i)\omega_{meas}(t) + z_1$$

$$\frac{d}{dt}z_1 = [J_m - B_m(\sigma - t_i)]\omega_{meas}(\sigma) + K_m \sigma i_{a\ meas}(\sigma)i_{f\ meas}(\sigma) \tag{7.14}$$

The following procedure yields an automatic reinitialization on the basis of the drift of the current value of the estimated value of L.

With $\delta > 0$, set for $t > t_i + \epsilon$:

$$\text{if} \quad |\hat{\tau}_L(t) - \hat{\tau}_L(t_i + \epsilon)| < \delta \quad \text{continue}$$

$$\text{else}$$

$$\text{re-initialize}$$

Figure 7.2 depicts the controlled responses of the separately excited DC motor under severe load changes to unexpected constant values. The fast estimation of the load torque allows a robustness feature of the feedback control scheme to keep the angular velocity around the desired trajectory.

7.1.2 Justification of the ETEDPOF Controller

In order to justify the previously proposed linear controller, consider the following model of the separately excited DC motor corresponding to the nominal desired behavior:

$$L_a \frac{d}{dt}i_a^\star(t) = -R_a i_a^\star(t) - K_m i_f^\star(t)\omega^\star(t) + v_a^\star(t)$$

$$L_f \frac{d}{dt}i_f^\star(t) = -R_f i_f^\star(t) + v_f^\star(t)$$

$$J_m \frac{d}{dt}\omega^\star(t) = K_m i_a^\star(t)i_f^\star(t) - B_m \omega^\star(t) - \tau_L \tag{7.15}$$

Thanks to the fast estimation of the torque, we may assume that τ_L is perfectly known. If such is the case, the exact tracking error dynamics may be written as:

$$L_a \frac{d}{dt}e_{i_a} = -R_a e_{i_a} - K_m i_f e_\omega - K_m e_{i_f}\omega^\star(t) + e_{v_a}$$

$$L_f \frac{d}{dt}e_{i_f} = -R_f e_{i_f} + e_{v_f}$$

$$J_m \frac{d}{dt}e_\omega = K_m i_f e_{i_a} + K_m e_{i_f}i_a^\star(t) - B_m e_\omega \tag{7.16}$$

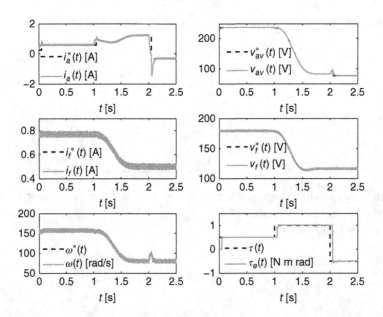

Figure 7.2 Closed-loop response of separately excited DC motor with ETEDPOF controller using online identification of piecewise constant torque load in the presence of high-frequency noise in the measurements

where $e_{i_a} = i_a - i_a^{\star}(t)$, $e_{i_f} = i_f - i_f^{\star}(t)$, and $e_{\omega} = \omega - \omega^{\star}(t)$. The control input errors are denoted by $e_{v_f} = v_f - v_f^{\star}(t)$ and $e_{v_a} = v_a - v_a^{\star}(t)$.

Decomposing the term $K_m e_{i_f} \omega^{\star}(t)$ into the sum

$$\frac{1}{2}K_m e_{i_f} \omega^{\star}(t) + \frac{1}{2}K_m e_{i_f} \omega^{\star}(t)$$

adding to and subtracting from the second equation the quantity $\frac{1}{2}K_m i_a^{\star} e_{\omega}$, and adding to and subtracting from the last equation the quantity $\frac{1}{2}K_m i_a^{\star} e_{i_f}$, we can write the exact tracking error system in the form

$$\mathcal{A}\dot{e} = -\mathcal{R}e + \mathcal{J}(t, i_f)e + \mathcal{B}e_v \tag{7.17}$$

where $\mathcal{A}$ is symmetric and positive definite, $\mathcal{R}$ is a symmetric matrix. The matrix $\mathcal{J}(t, i_f)$ is skew-symmetric and $\mathcal{B}$ is constant.

Consider the following Lyapunov function candidate:

$$V(e) = \frac{1}{2}e^T \mathcal{A}e \tag{7.18}$$

Along the solutions of the exact tracking error dynamics, we have

$$\dot{V}(e) = -e^T \mathcal{R}e + e^T \mathcal{B}e_v \tag{7.19}$$

Taking the control error to be $e_v = -\Gamma \mathcal{B}^T e$ leads to the following expression for $\dot{V}(e)$, $\Gamma = diag[\gamma_1, \gamma_2]$:

$$\dot{V} = -e^T[\mathcal{R} + \mathcal{B}\Gamma \mathcal{B}^T]e \tag{7.20}$$

Thus, $\dot{V}(e) < 0$, provided γ_1 and γ_2 are chosen to satisfy

$$R_a + \gamma_1 > 0$$

$$(R_a + \gamma_1)(R_f + \gamma_2) \geq \frac{1}{4}K_m^2\omega^\star(t), \quad \forall\, t$$

$$(R_a + \gamma_1)(R_f + \gamma_2)B_m \geq \frac{1}{4}K_m^2[B_m\omega^\star(t) + (R_a + \gamma_1)i_a^\star(t)], \quad \forall\, t$$

In the simulation example we have set $\gamma_1 = 10$, $\gamma_2 = 60$ to satisfy all the required inequalities for the maximum values of $\omega^\star(t)$ and $i_f^\star(t)$ in the planning horizon.

7.1.3 A Sensorless Scheme Based on Fast Adaptive Observation

Since the controller does not require measurements of ω, we would also like to free the parameter identification from such a measurement. We propose the following reduced-order adaptive observer for such an objective.

From the armature current equation, take the expression

$$K_m i_f \omega = v_a - R_a i_a - L_a \frac{di_a}{dt} \tag{7.21}$$

as an indirect measurement of the angular velocity ω. We thus propose, thanks to our possibility of online fast estimation of τ_L, the following adaptive observer for ω:

$$J_m \frac{d\hat{\omega}}{dt} = K_m i_f i_a - B_m\hat{\omega} - \hat{\tau}_L + \lambda(K_m i_f \omega - K_m i_f \hat{\omega}) \tag{7.22}$$

The estimation error evolves, after the fast and accurate convergence of $\hat{\tau}_L$ to τ_L, as

$$J_m \frac{de_{o\,\omega}}{dt} = -[B_m + \lambda K_m i_f(t)]e_{o\,\omega} \tag{7.23}$$

where $e_{o\,\omega} = \omega - \hat{\omega}$. Clearly, for $\lambda \gg 0$ and strictly positive planning of $i_f(t)$, the observation error $e_{o\,\omega}$ asymptotically and exponentially converges to zero.

We thus have the following reduced-order observer after replacing the expression for $K_m i_f \omega$:

$$J_m \frac{d\hat{\omega}}{dt} = K_m i_f(t)i_a(t) - B_m\hat{\omega} - \hat{\tau}_L$$
$$+\lambda\left[v_a(t) - R_a i_a(t) - L_a \frac{di_a}{dt} - K_m i_f(t)\hat{\omega}\right] \tag{7.24}$$

Setting $\zeta = J_m\hat{\omega} + \lambda L_a i_a$, we have

$$\hat{\omega} = \frac{1}{J_m}[\zeta - \lambda L_a i_a]$$

$$\frac{d\zeta}{dt} = K_m i_f(t)i_a(t) - \left[\frac{B_m}{J_m} + \lambda\frac{K_m}{J_m}i_f(t)\right]\zeta - \hat{\tau}_L$$

$$+\lambda\left\{v_a - R_a i_a(t) + \frac{L_a}{J_m}i_a(t)\left[B_m + \lambda K_m i_f(t)\right]\right\} \tag{7.25}$$

7.1.3.1 Simulation Results

Simulations were performed for the sensorless feedback control scheme of the separately excited DC motor with a fast adaptive reduced-order angular velocity observer for the online algebraic identification of the piecewise constant unknown load torques (Figures 7.3 and 7.4).

We set $\lambda = 1$ in the proposed reduced-order observer. Since the time-varying gain of the observer is determined by the function

$$\frac{B_m}{J_m} + \lambda \frac{K_m}{J_m} i_f(t) \tag{7.26}$$

which is rather large, the convergence of the observed angular velocity $\hat{\omega}(t)$ to its actual value $\hat{\omega}$ is rather fast, as can be seen.

7.1.3.2 Time-Varying Load Perturbations

In order to deal with this challenging case, we resort to the following more general model of the perturbed separately excited DC motor:

$$L_a \frac{d}{dt} i_a = -R_a i_a - K_m i_f \omega + v_a$$

$$L_f \frac{d}{dt} i_f = -R_f i_f + v_f$$

$$J_m \frac{d}{dt} \omega = K_m i_f i_a - B_m \omega - \frac{P_m(t)}{w} \tag{7.27}$$

where $P_m(t)$ is the time-varying unknown mechanical power demanded by the load.

The load mechanical power is defined as

$$P_m(t) = \tau_L(t)\omega \tag{7.28}$$

where $\tau_L(t)$ is the unknown time-varying load torque. Thus, the estimate of the mechanical power leads to an estimate of the load torque.

In this instance, and to illustrate the versatility of the algebraic approach, we propose the following flatness-based linear time-varying GPI controller including feedforward compensation:

$$v_a^\star(t) = L_a \left(\frac{J_m v_{a,\,GPI} + B_m \dot{F}^\star + \frac{\dot{P}_m(t)}{F^\star} - \frac{P_m(t)\dot{F}^\star}{[F^\star]^2}}{K_m G^\star} \right.$$

$$\left. - \frac{\left(J_m \dot{F}^\star + B_m F^\star + \frac{P_m(t)}{F^\star} \right) \dot{G}^\star}{K_m [G^\star]^2} + R_a \left(\frac{J_m \dot{F}^\star + B_m F^\star + \frac{P_m(t)}{F^\star}}{K_m G^\star} \right) \right)$$

$$+ K_m G^\star F^\star \tag{7.29}$$

The controller for the field current does not depend on $P_m(t)$, and is simply given by

$$v_f^\star(t) = L_f v_{f,\,GPI} + R_f G^\star \tag{7.30}$$

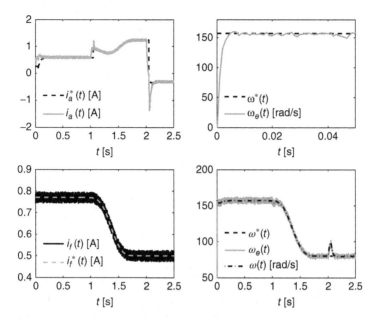

Figure 7.3 Closed-loop response of separately excited DC motor with an adaptive sensorless scheme for trajectory tracking and time-varying piecewise constant loads

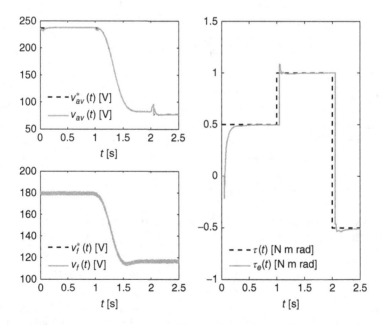

Figure 7.4 Closed-loop response of separately excited DC motor with an adaptive sensorless scheme for trajectory tracking and time-varying piecewise constant loads

The auxiliary control inputs $v_{a\ GPI}$ and $v_{f\ GPI}$ are given, with an abuse of notation combining expressions in the time domain with expressions in the frequency domain, by

$$v_{f,\ GPI} = \dot{G}^{\star} - \left[\frac{k_1 s + k_0}{s + k_2} \right] (G - G^{\star}(t)) \tag{7.31}$$

$$v_{a,\ GPI} = \ddot{F}^{\star}(t) - \left[\frac{\gamma_2 s^2 + \gamma_1 s + \gamma_0}{s(s + \gamma_3)} \right] (F - F^{\star}(t)) \tag{7.32}$$

The design parameters k_2, k_1, k_0 are chosen so that the characteristic polynomial

$$p_f(s) = s^3 + k_2 s^2 + k_1 s + k_0 \tag{7.33}$$

has all its roots in the left half of the complex plane. Similarly, the design constants $\gamma_3, \dots, \gamma_0$ are chosen so that

$$p_\omega(s) = s^4 + \gamma_3 s^3 + \gamma_2 s^2 + \gamma_1 s + \gamma_0 \tag{7.34}$$

has all its roots in the stable portion of the complex plane.

The value of $P_m(t)$, being unknown, requires a reliable online estimation for continuous adaptation of the feedforward part of the exactly local linearizing controller.

Consider the mechanical equation of the machine:

$$J_m \frac{d}{dt} \omega = K_m i_f i_a - B_m \omega - \frac{P_m(t)}{w} \tag{7.35}$$

Clearly,

$$P_m(t) = -\frac{1}{2} J_m \frac{d}{dt}(\omega^2) + K_m i_f i_a \omega - B_m \omega^2 \tag{7.36}$$

Equation (7.36) is a power balance equation for the mechanical part, with the input torque being $K_m i_a i_f$. The negative signs indicate power consumed by both the acquired kinetic energy of the rotating motor inertia and the overcoming of the viscous friction. $P_m(t)$ is the mechanical power delivered to the load, or demanded by the load. In principle, $P_m(t)$ may also be negative.

The estimation of the mechanical power $P_m(t)$ of the load may be carried out online when all state variables are measured. We need to take only the time derivative of the signal $\omega^2(t)$:

$$P_m(t) = -\frac{1}{2} J_m \frac{d}{dt}(\omega^2) + K_m i_f i_a \omega - B_m \omega^2 \tag{7.37}$$

The feedforward part of the feedback controller, v_a, also depends on the time derivative of the unknown mechanical power, namely $\dot{P}_m(t)$. We also online compute the required signal as

$$\dot{P}_m(t) = -\frac{1}{2} J_m \frac{d^2}{dt^2}(\omega^2) + K_m \frac{d}{dt}[i_f i_a \omega] - B_m \frac{d}{dt}(\omega^2) \tag{7.38}$$

The time derivative calculation of the involved signals, say $\omega^2(t)$, is carried out using the algebraic derivative method on the transform of a truncated Taylor series approximation for the signal to be derived. Let the subindex e denote an "estimated" quantity. We then have the following algebraic-based algorithm for derivative computation after the reinitialization time t_i.

Let

$$\dot{z}_1 = z_2 - 450(t - t_i)(\omega_{meas}^2)$$
$$\dot{z}_2 = z_3 + 2400(t - t_i)^3(\omega_{meas}^2)$$
$$\dot{z}_3 = z_4 - 5400(t - t_i)^2(\omega_{meas}^2)$$
$$\dot{z}_4 = z_5 + 4320(t - t_i)(\omega_{meas}^2)$$
$$\dot{z}_5 = -720(\omega_{meas}^2) \tag{7.39}$$

where $\omega_{meas} = \omega + \chi(t)$ stands for the noisy measurement of the angular velocity ω.

The reinitialization process is absolutely necessary due to the fact that the time derivatives of the signal are only approximately computed on the basis of a Taylor polynomial (truncated Taylor series). As time evolves, the adopted approximation deteriorates.

The first time derivative is computed, after the reinitialization instant t_i, as

$$\left[\frac{d\omega^2}{dt}\right]_e (t) = \begin{cases} \left[\frac{d}{dt}(\omega^2)\right]_e (t_i^-) + \left[\frac{d^2}{dt^2}(\omega^2)\right]_e (t_i^-)(t - t_i) + \frac{1}{2}\left[\frac{d^3}{dt^3}(\omega^2)\right]_e (t_i^-)(t - t_i)^2 \\ \qquad\qquad\qquad\qquad\qquad\qquad\qquad \text{for } t \in [t_i, t_i + \epsilon) \\[2ex] \dfrac{30(t - t_i)^5 \omega_{meas}^2 + z_1}{(t - t_i)^6} \qquad\qquad\qquad \text{for } t \in [t_i + \epsilon, t_{i+1}] \end{cases}$$

The second and third time derivatives are computed, after the reinitialization instant t_i, as

$$\left[\frac{d^2\omega^2}{dt^2}\right]_e (t) = \begin{cases} \left[\frac{d^2}{dt^2}(\omega^2)\right]_e (t_i^-) + \left[\frac{d^3}{dt^3}(\omega^2)\right]_e (t_i^-)(t - t_i) \quad \text{for } t \in [t_i, t_i + \epsilon) \\[2ex] \dfrac{-300(t - t_i)^4 \omega_{meas}^2 + 24(t - t_i)^5 \left[\frac{d}{dt}(\omega^2)\right]_e (t) + z_2}{(t - t_i)^6} \\ \qquad\qquad\qquad\qquad\qquad\qquad \text{for } t \in [t_i + \epsilon, t_{i+1}] \end{cases}$$

$$\left[\frac{d^3\omega^2}{dt^3}\right]_e (t) = \begin{cases} \left[\frac{d^3}{dt^3}(\omega^2)\right]_e (t_i^-) \qquad\qquad\qquad\qquad\qquad \text{for } t \in [t_i, t_i + \epsilon) \\[2ex] \dfrac{1200(t - t_i)^3 \omega_{meas}^2 - 180(t - t_i)^4 \left[\frac{d}{dt}(\omega^2)\right]_e (t)}{(t - t_i)^6} \\[3ex] \quad + \dfrac{16(t - t_i)^5 \left[\frac{d^2}{dt^2}(\omega^2)\right]_e (t) + z_3}{(t - t_i)^6} \\[2ex] \qquad\qquad\qquad\qquad\qquad\qquad \text{for } t \in [t_i + \epsilon, t_{i+1}] \end{cases}$$

We proceed in a similar fashion to compute the required derivative of the signal $i_{a}i_{f}\omega$.

7.1.3.3 Simulation Results

Figure 7.5 depicts the performance of the trajectory-tracking controller for the separately excited DC motor with continuous adaptation of the feedforward compensation term. The online estimation of the time-varying mechanical power $P_m(t)$, which is allowed to change sign along the evolution of the system, is carried out under significant noise affecting the measurement of the required state variables. Such noisy measurements were also replaced in the proposed linear time-varying feedback controller, as needed.

7.1.4   Control of the Boost Converter

We deal with the following average boost converter model:

$$L\frac{d}{dt}i = -uv + E$$

$$C\frac{d}{dt}v = ui - \frac{\tilde{P}(t)}{v}$$

where u is the average control input, i is the average inductor current. The variable v represents the average output voltage and $\tilde{P}(t)$ is the average load power $= v^2/R(t)$.

Introduce the following time and state scaling transformation:

$$x_1 = \frac{i}{E}\sqrt{\frac{L}{C}}, \quad y = \frac{v}{E}, \quad \tau = t/\sqrt{LC}$$

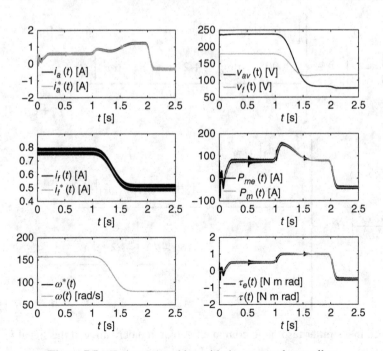

Figure 7.5   Trajectory tracking with the proposed controller

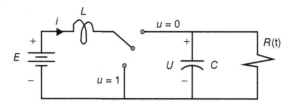

Figure 7.6   Boost converter schematics

Consider, then, the average normalized model (Figure 7.6) of the boost converter ("$\dot{\ } = \frac{d}{d\tau}$"):

$$\dot{x}_1 = -uy + 1$$

$$\dot{y} = ux_1 - \frac{P(\tau)}{y}$$

where

u : average control input
x_1 : average normalized inductor current
y : average normalized output voltage
$P(\tau)$: average normalized load power.

7.1.4.1   Problem Formulation

Given a desired constant equilibrium value for the average capacitor voltage $y = V_d$, find a feedback control law, u, that keeps the voltage at the desired value in spite of the unknown variations of the strictly positive load.

7.1.4.2   Assumptions

- The load parameter is assumed to be of a resistive, but unknown, time-varying nature. The load power $P(\tau)$ is therefore of the form

$$P(\tau) = \frac{y^2}{Q(\tau)}, \quad Q(\tau) > 0, \ \forall t$$

  where $Q(\tau)$ is the time-varying normalized load.
- Both the output voltage and the input current are measurable. Hence, the total stored energy is also measurable.

In order to derive a *certainty-equivalence* feedback controller, we temporarily assume that the load is known. Under such circumstances we derive an ETEDPOF controller for the following system model:

$$\dot{x}_1 = -uy + 1$$

$$\dot{y} = ux_1 - \frac{y}{Q(\tau)}$$

A nominal behavior of the system is given by

$$\dot{x}_1^\star = -u^\star(\tau)y^\star(\tau) + 1$$

$$\dot{y}^\star = u^\star x_1^\star(\tau) - \frac{y^\star(\tau)}{Q(\tau)}$$

Defining $e_1 = x_1 - x_1^\star(\tau)$, $e_2 = y - y^\star(\tau)$, and $e_u = u - u^\star(\tau)$, we have

$$\dot{e} = \mathcal{J}(u)e - \mathcal{R}(\tau)e + \mathcal{B}^\star(\tau)e_u$$

with

$$\mathcal{J}(u) = \begin{bmatrix} 0 & -1 \\ 1 & 0 \end{bmatrix}, \quad \mathcal{R}(\tau) = \begin{bmatrix} 0 & 0 \\ 0 & \frac{1}{Q(\tau)} \end{bmatrix}$$

and

$$\mathcal{B}^\star(\tau) = \begin{bmatrix} -y^\star(\tau) \\ x_1^\star(\tau) \end{bmatrix}$$

An ETEDPOF controller is then simply given by $e_u = -\gamma[\mathcal{B}^\star(\tau)]^T e$, with $\gamma > 0$. That is,

$$u = u^\star(\tau) - \gamma[-y^\star(\tau)(x_1 - x_1^\star(\tau)) + x_1^\star(\tau)(y - y^\star(\tau))]$$

$$= u^\star(\tau) - \gamma[-y^\star(\tau)x_1 + x_1^\star(\tau)y]$$

We take, evidently, $y^\star(\tau) = V_d$. The choice of $x_1^\star(\tau)$ is not so obvious, since $Q(\tau)$ is unknown.

We propose an online estimate for the load power $P(\tau)$. The power balance of the system readily leads to

$$\frac{dE(\tau)}{d\tau} = x_1(\tau) - P(t)$$

where $E(\tau)$ is the average normalized stored energy, defined by

$$E(\tau) = \frac{1}{2}\left[x_1(\tau)^2 + y(\tau)^2\right]$$

Hence, an estimate $P_e(\tau)$ of the load power $P(\tau)$ is obtained directly from the above expression as

$$P_e(\tau) = x_1(\tau) - \frac{d}{d\tau}E(\tau)$$

$$= x_1(\tau) - \frac{1}{2}\frac{d}{d\tau}\left[x_1(\tau)^2 + x_2(\tau)^2\right]$$

where it is required to obtain an online estimate of the time derivative of the measured energy. This is done via algebraic means.

7.1.4.3 Online Time-Derivative Calculation

$$\dot{E}_e(\tau) = \begin{cases} \frac{1}{2}E_e^{(3)}(\tau_i^-)(\tau - \tau_i)^2 + \ddot{E}_e(\tau_i^-)(\tau - \tau_i) + \dot{E}_e(\tau_i^-), \\ \qquad\qquad\qquad\qquad\qquad\qquad \tau \in [\tau_i, \tau_i + \epsilon) \\ \frac{n_1(\tau)}{d(\tau)}, \qquad\qquad\qquad\qquad \tau > \tau_i + \epsilon \end{cases}$$

where

$$n_1(\tau) = 30(\tau - \tau_i)^5 E(\tau) + z_1, \quad d(\tau) = (\tau - \tau_i)^6$$

$$\ddot{E}_e(\tau) = \begin{cases} E_e^{(3)}(\tau_i^-)(\tau - \tau_i) + \ddot{E}_e(\tau_i^-), & \tau \in [t_i, t_i + \epsilon) \\ \dfrac{n_2(\tau)}{d(\tau)}, & \tau > t_i + \epsilon \end{cases}$$

where

$$n_2(\tau) = 300(\tau - \tau_i)^4 E(\tau) + 24(\tau - \tau_i)^5 \dot{E}_e(\tau) + z_2,$$

$$\ddot{E}_e(\tau) = \begin{cases} E_e^{(3)}(\tau_i^-) & \tau \in [\tau_i, \tau_i + \epsilon) \\ \dfrac{n_3(\tau)}{d(\tau)}, & \tau > \tau_i + \epsilon \end{cases}$$

where

$$n_3(\tau) = 1200(\tau - \tau_i)^3 E(\tau) - 180(\tau - \tau_i)^4 \dot{E}_e(\tau)$$
$$+ 18(\tau - \tau_i)^5 \ddot{E}_e(\tau) + z_3$$

and z_3, z_2, z_1 are given by

$$\dot{z}_1 = z_2 - 450(\tau - \tau_i)^4 E(\tau)$$
$$\dot{z}_2 = z_3 + 2400(\tau - \tau_i)^3 E(\tau)$$
$$\dot{z}_3 = z_4 - 5400(\tau - \tau_i)^2 E(\tau)$$
$$\dot{z}_4 = z_5 + 4320(\tau - \tau_i)E(\tau)$$
$$\dot{z}_5 = -720E(\tau)$$

τ_i is a calculation-resetting instant decided upon by either an integral square-error criterion, comparing E and E_e, or taken to be periodical.

In order to complete the controller specification, we proceed to compute a nominal current $x_1^{\star}(\tau)$, given that the average normalized output voltage is constant, $y = V_d$, and that the load power is estimated as $P_e(\tau) = x_1(\tau) - \dot{E}_e(\tau)$.

We consider the following iterative functional operation of Chaplygin type:

$$x_{1,k+1}^{\star}(\tau) = P(\tau) + \frac{1}{2}\frac{d}{d\tau}\left[\left[x_{1,k}^{\star}(\tau)\right]^2 + V_d^2\right]$$

If we set

$$x_{1,0}^{\star}(\tau) = \text{constant}$$

then we obtain

$$x_{1,1}^{\star}(\tau) = P(\tau)$$

Continuing with the iterations, we find

$$x_{1,2}^{\star}(\tau) = P(\tau) + P(\tau)\dot{P}(\tau)$$
$$x_{1,3}^{\star}(\tau) = P(\tau) + [P(\tau) + P(\tau)\dot{P}(\tau)] \times [\dot{P}(\tau) + (\dot{P}(\tau))^2 + P(\tau)\ddot{P}(\tau)]$$
$$x_{1,4}^{\star}(\tau) = \text{etc.}$$

Surprisingly, it turns out that the first iteration is a sufficiently close approximation since further iterations do not modify it substantially.

Given that our knowledge of $P(\tau)$ is through its estimate $P_e(\tau)$, it is reasonable to propose

$$x_1^\star(\tau) = P_e(\tau)$$

for the ETEDPOF controller.

7.1.4.4 Simulation Results

We set a typical converter with the following parameters:

$$L = 10 \text{ mH}, \quad C = 10 \text{ μF}, \quad E = 30 \text{ V}$$

It is desired to keep the output voltage at the normalized constant value $V_d = 2$, which corresponds to $v_d = 60$ V. The normalized resistor $Q(t)$ exhibits a growing variation from an initial value Q_0 to a constant value $Q_0 + \Delta_Q$ and, then, it decreases exponentially to a new constant value $Q_0 + 0.7\Delta_Q$. That is,

$$Q(t) = Q_0 + \Delta_Q(1 - \exp(-\tau)), \quad \tau \in [0, 30)$$

$$Q(t) = Q_0 + \Delta_Q - 0.3\Delta_Q(1 - \exp(-(\tau - 30))), \quad \tau \in [30, +\infty)$$

where $Q_0 = 1$, $\Delta_Q = 4$.

We have chosen to reset the derivative estimator every 10 normalized units of time, that is, approximately every 3.162 ms. See Figure 7.7.

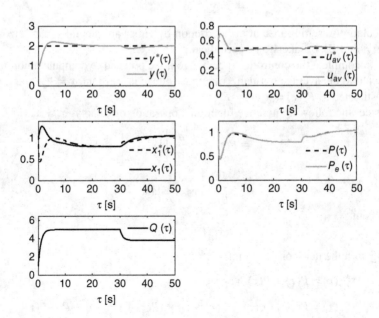

Figure 7.7 Average normalized responses of boost converter with unknown time-varying load

7.1.4.5 Sigma–Delta Modulation Implementation

The average control signal $u_{av}(t)$ is fed into a Σ–Δ modulator block (Figure 7.8) whose output obeys the following discontinuous differential equation:

$$u = \frac{1}{2}(1 + sign(e)), \quad \dot{e} = u_{av}(t) - u$$

The sigma–delta modulator exhibits a sliding regime on $e = 0$ and reproduces at the output a switched signal whose average behavior coincides with the input signal ($u_{eq} = u_{av}$).

Figures 7.9 and 7.10 depict the response of the switched implementation of the proposed control scheme in the normalized and non-normalized instances. Notice that the system response is equal to that with the continuous-time average controller.

An interesting analysis of the algebraic load estimation is presented in the work of Gensior *et al.* (2008), where a comparison with other disturbance approaches in an experimental framework is shown.

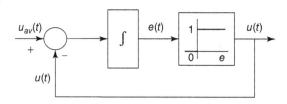

Figure 7.8 Sigma–Delta modulator

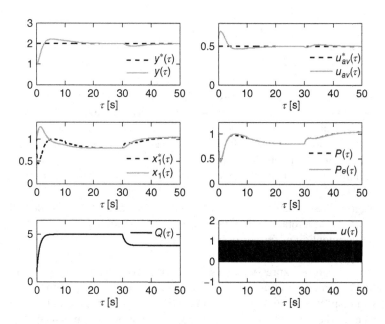

Figure 7.9 Normalized responses of switched boost converter with unknown time-varying load

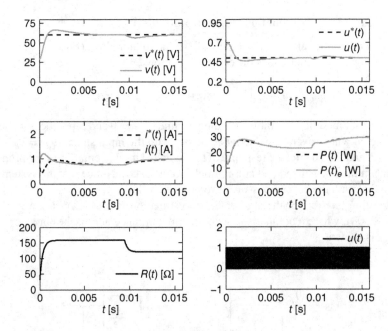

Figure 7.10   Switched responses of non-normalized boost converter with unknown time-varying resistive load

7.2   Alternative Elimination of Initial Conditions

This section is inspired by the algebraic methodology, as used in the context of the time domain. However, it presents a slight change in time functions used for the required convolution operations that yield the set of unknown parameters. To begin, we will give a brief summary of the time-domain algebraic methodology.

- First, if we have an observable input–output system representation affected by an unknown perturbation input, proceed to approximate it by means of a Taylor polynomial of, say, mth order. As in the case of *classical perturbation* inputs (constants, ramps, parabolas, etc.), differentiate the input–output model $(m + 1)$ times with respect to the time variable t in order to annihilate the effects of this perturbation.
- Second, multiply the model by a suitable power of $(-t)^{m+n}$, where $m + n$ is the order of the derived model.
- Third, integrate (using the formula of integration by parts, where needed) the resulting unperturbed system a total of $(n + m + 1)$ times. It is possible to apply additional integrations with the intention of further improving the signal-to-noise ratio. This last part of the procedure is known as invariant filtering.
- Finally, from the above expression, a regressor form, free from derivatives and independent of the initial conditions, is obtained. We can construct, therefore, a system of linear equations (linear with respect to the unknown parameters) which allows us to obtain the online estimated values of the parameters. This time-varying equation system is solved by the method of determinants and implemented with the help of linear time-varying differential equations.

Until now, we have eliminated the influence of the inherent initial conditions over the algebraic estimator, multiplying the differential equations by a power of time equal to the order of the system (i.e., $t^2\ddot{y} = t^2u$). However, we must mention that there exist other functions which allow us to make the same process of elimination. Shinbrot (1957) proposed carrying out similar convolutions in the system model equations, using special functions of time. This procedure is known as the *equation of motion method*. Eykhoff (1974) made a summary of Shinbrot's methods and the differential approximation method proposed by Bellman *et al.* (1964) for parameter estimation in linear systems. Some of these methods are asymptotic in nature, and others assume that the variables of the regressor can be measured without additive noise. Moreover, in these algorithms there is not a counterpart algebraic treatment in the frequency domain as may be performed in the algebraic method. In this section, we will obtain algebraic estimators independent of the initial conditions using other functions of time, which have the characteristic of being equal to zero at $t = 0$, but with the advantage of being bounded functions.

7.2.1   A Bounded Exponential Function

Consider the exponential function

$$\phi(t) = 1 - e^{-\frac{t}{\lambda}} \tag{7.40}$$

It is easy to see that $\phi(t) = 0$ for $t = 0$, and $\forall t \in [0, \infty), \exists \lambda \in \mathfrak{R}^+ | 0 \le \phi(t) \le 1$. Since this function is zero at the beginning, the initial conditions of the system are eliminated. In addition, as $\phi(t)$ is a bounded function, the internal terms of the estimator, after a short time, do not increase as much as when they are multiplied by t. Then, the behavior of these internal elements will depend, mainly, on the input, the output, and their iterated integrals. Another advantage is the possibility to calculate other regressors by assigning different values to λ. Thus, we can obtain a compatible system of equations in this manner. In the original algebraic methodology, this equation system was built up by successive integrations of the regressor. Now, we will explore this variation of the algebraic method with a simple example. Let a linear time-invariant dynamic system of first order be

$$\dot{y} + ay = u \tag{7.41}$$

We propose an estimator for parameter a using the function ϕ. First, multiply both sides of equation (7.41) by (7.40):

$$\left(1 - e^{-\frac{t}{\lambda}}\right)(\dot{y} + ay) = \left(1 - e^{-\frac{t}{\lambda}}\right)u \tag{7.42}$$

Next, integrate this expression by parts:

$$\left(1 - e^{-\frac{T_1}{\lambda}}\right)y(T_1)\bigg]_0^t - \frac{1}{\lambda}\int_0^t e^{-\frac{T_1}{\lambda}}y(T_1)dT_1$$

$$+a\int_0^t \left(1 - e^{-\frac{T_1}{\lambda}}\right)y(T_1)dT_1 = \int_0^t \left(1 - e^{-\frac{T_1}{\lambda}}\right)u(T_1)dT_1$$

The initial conditions of the system disappear when the integration interval is evaluated. So, the estimator of the parameter a is

$$a_{e,\phi} = \frac{n_\phi(t)}{d_\phi(t)} \tag{7.43}$$

where

$$n_\phi(t) = -\left(1 - e^{-\frac{t}{\lambda}}\right)y(t) + \frac{1}{\lambda}\int_0^t e^{-\frac{T_1}{\lambda}}y(T_1)dT_1 + \int_0^t \left(1 - e^{-\frac{T_1}{\lambda}}\right)u(T_1)dT_1$$

$$d_\phi(t) = \int_0^t \left(1 - e^{-\frac{T_1}{\lambda}}\right)y(T_1)dT_1$$

A typical algebraic estimator for this system would be

$$a_{e,t} = \frac{-ty + \int_0^t y(T_1)dT_1 + \int_0^t T_1 u(T_1)dT_1}{\int_0^t T_1 y(T_1)dT_1}$$

There exists a great similarity between these identifier structures. Figure 7.11 presents the behavior of both estimators with $\lambda = 5$ and $\hat{y}(0) = 0.1$.

7.2.2   Correspondence in the Frequency Domain

The Laplace transform of the product of an exponential function and another function of time is

$$\mathcal{L}\{e^{-at}f(t)\} = F(s + a)$$

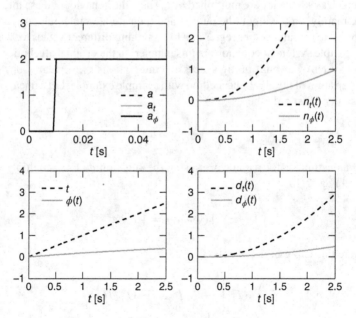

Figure 7.11 Comparison between the typical algebraic estimator and the estimator obtained using function $\phi(t)$

Hence, the product of function ϕ and another function of time is

$$\mathcal{L}\{\phi(t)f(t)\} = \mathcal{L}\left\{f(t) - e^{-\frac{t}{\tau}}f(t)\right\} \tag{7.44}$$

$$= F(s) - F\left(s + \frac{1}{\tau}\right)$$

Applying transformation (7.44) to the system (7.41):

$$\left([sY(s) - Y(0)] - [\sigma Y(\sigma) - Y(0)]_{\sigma=s+\frac{1}{\tau}}\right) + a\left(Y(s) - [Y(\sigma)]_{\sigma=s+\frac{1}{\tau}}\right)$$
$$= \left(U(s) - [U(\sigma)]_{\sigma=s+\frac{1}{\tau}}\right)$$

Expanding the expressions, we notice that the initial conditions disappear:

$$sY(s) - sY\left(s + \frac{1}{\tau}\right) - \frac{1}{\tau}Y\left(s + \frac{1}{\tau}\right) + a\left(Y(s) - Y\left(s + \frac{1}{\tau}\right)\right)$$
$$= U(s) - U\left(s + \frac{1}{\tau}\right)$$

Multiplying by s^{-1}, we obtain

$$Y(s) - Y\left(s + \frac{1}{\tau}\right) - \frac{1}{\tau}s^{-1}Y\left(s + \frac{1}{\tau}\right) + as^{-1}\left(Y(s) - Y\left(s + \frac{1}{\tau}\right)\right)$$
$$= s^{-1}\left(U(s) - U\left(s + \frac{1}{\tau}\right)\right)$$

This equation is expressed in the time domain as

$$y(t)\left(1 - e^{-\frac{t}{\tau}}\right) - \frac{1}{\tau}\int_0^t e^{-\frac{T_1}{\tau}}y(T_1)dT_1 + a\int_0^t y(T_1)\left(1 - e^{-\frac{T_1}{\tau}}y(T_1)\right)dT_1$$
$$= \int_0^t u(T_1)\left(1 - e^{-\frac{T_1}{\tau}}y(T_1)\right)dT_1$$

Finally, the estimator (7.43) can be obtained by returning the last expression to the time domain.

7.2.3 A System of Second Order

Let a mass–spring–damper system with unknown parameter m, c, and k be

$$m\ddot{x} + c\dot{x} + kx = u \tag{7.45}$$

Since (7.45) is a second-order system, we must apply

$$\phi(t)^2(m\ddot{x} + c\dot{x} + kx) = \phi(t)^2 u$$

The transformation to a frequency-domain-type expression is

$$\mathcal{L}\{\phi^2(t)f(t)\} = \mathcal{L}\left\{f(t) - 2e^{-\frac{t}{\tau}}f(t) + e^{-\frac{2t}{\tau}}f(t)\right\} \tag{7.46}$$

$$= F(s) - 2F\left(s + \frac{1}{\tau}\right) + F\left(s + \frac{2}{\tau}\right)$$

Applying transformations (7.46) to the system (7.45):

$$m\left(\begin{array}{l}\left[s^2X(s) - sX(0) - \dot{X}(0)\right] - 2\left[\sigma_1^2X(\sigma_1) - \sigma_1X(0) - \dot{X}(0)\right]_{\sigma_1=s+\frac{1}{\tau}}\\ +\left[\sigma_2^2X(\sigma_2) - \sigma_2X(0) - \dot{X}(0)\right]_{\sigma_2=s+\frac{2}{\tau}}\end{array}\right)$$

$$+c\left(\begin{array}{l}\left[sX(s) - X(0)\right] - 2\left[\sigma X(\sigma) - X(0)\right]_{\sigma=s+\frac{1}{\tau}}\\ +\left[\sigma_2X(\sigma_2) - X(0)\right]_{\sigma_2=s+\frac{2}{\tau}}\end{array}\right)$$

$$+k\left(X(s) - 2\left[X(\sigma_1)\right]_{\sigma_1=s+\frac{1}{\tau}} + \left[X(\sigma_2)\right]_{\sigma_2=s+\frac{2}{\tau}}\right)$$

$$= \left(U(s) - 2\left[U(\sigma_1)\right]_{\sigma_1=s+\frac{1}{\tau}} + \left[U(\sigma_2)\right]_{\sigma_2=s+\frac{2}{\tau}}\right)$$

The initial conditions disappear by expanding the expression

$$m\left[\begin{array}{l}s^2X(s) - 2\left(s+\frac{1}{\tau}\right)^2X\left(s+\frac{1}{\tau}\right)\\ +\left(s+\frac{2}{\tau}\right)^2X\left(s+\frac{2}{\tau}\right)\end{array}\right] + c\left[\begin{array}{l}sX(s) - 2\left(s+\frac{1}{\tau}\right)X\left(s+\frac{1}{\tau}\right)\\ +\left(s+\frac{2}{\tau}\right)X\left(s+\frac{2}{\tau}\right)\end{array}\right]$$

$$+k\left(X(s) - 2X\left(s+\frac{1}{\tau}\right) + X\left(s+\frac{2}{\tau}\right)\right) = \left(U(s) - 2U\left(s+\frac{1}{\tau}\right) + U\left(s+\frac{2}{\tau}\right)\right)$$

Multiplying by s^{-2}, we obtain

$$m\left[X(s) - 2\left(1+s^{-1}\frac{2}{\tau} + s^{-2}\left(\frac{1}{\tau}\right)^2\right)X\left(s+\frac{1}{\tau}\right)\right.$$
$$\left.+\left(1+s^{-1}\frac{4}{\tau} + s^{-2}\left(\frac{2}{\tau}\right)^2\right)X\left(s+\frac{2}{\tau}\right)\right]$$

$$+c\left[s^{-1}X(s) - 2\left(s^{-1} + s^{-2}\frac{1}{\tau}\right)X\left(s+\frac{1}{\tau}\right) + \left(s^{-1} + s^{-2}\frac{2}{\tau}\right)X\left(s+\frac{2}{\tau}\right)\right]$$

$$+s^{-2}k\left[X(s) - 2X\left(s+\frac{1}{\tau}\right) + X\left(s+\frac{2}{\tau}\right)\right]$$

$$= s^{-2}\left[U(s) - 2U\left(s+\frac{1}{\tau}\right) + U\left(s+\frac{2}{\tau}\right)\right]$$

Now, we can propose an equation system to calculate the unknown parameters from the expression returned to the time domain:

$$\begin{bmatrix} p_{11} & p_{12} & p_{13}\\ p_{21} & p_{22} & p_{23}\\ p_{31} & p_{32} & p_{33}\end{bmatrix}\begin{bmatrix} m_e\\ c_e\\ k_e\end{bmatrix} = \begin{bmatrix} q_1\\ q_2\\ q_3\end{bmatrix}$$

$$p_{11} = x(t)\left(1 - e^{-\frac{t}{\tau}}\right) - 2\left[\frac{1}{2}e^{-\frac{t}{\tau}}x(t) + \frac{2}{\tau}\int_0^t e^{-\frac{T_1}{\tau}}x(T_1)dT_1\right.$$
$$\left.+\left(\frac{1}{\tau}\right)^2\int_0^t\int_0^{T_1}e^{-\frac{T_2}{\tau}}x(T_2)dT_2dT_1\right]$$

$$+ e^{-\frac{2t}{\tau}} x(t) + \frac{4}{\tau} \int_0^t e^{-\frac{2T_1}{\tau}} x(T_1) dT_1$$

$$+ \left(\frac{2}{\tau}\right)^2 \int_0^t \int_0^{T_1} e^{-\frac{2T_2}{\tau}} x(T_2) dT_2 dT_1$$

$$p_{12} = \int_0^t x(T_1) \left(1 - e^{-\frac{t}{\tau}}\right) dT_1 - 2 \left[\frac{1}{2} \int_0^t e^{-\frac{T_1}{\tau}} x(T_1) dT_1 \right.$$

$$\left. + \frac{1}{\tau} \int_0^t \int_0^{T_1} e^{-\frac{T_2}{\tau}} x(T_2) dT_2 dT_1\right] + \int_0^t e^{-\frac{2T_1}{\tau}} x(T_1) dT_1$$

$$+ \frac{2}{\tau} \int_0^t \int_0^{T_1} e^{-\frac{2T_2}{\tau}} x(T_2) dT_2 dT_1$$

$$p_{13} = \int_0^t \int_0^{T_1} x(T_2) \left(1 - 2e^{-\frac{T_2}{\tau}} + e^{-\frac{2T_2}{\tau}}\right) dT_2 dT_1$$

$$q_1 = \int_0^t \int_0^{T_1} u(T_2) \left(1 - 2e^{-\frac{T_2}{\tau}} + e^{-\frac{2T_2}{\tau}}\right) dT_2 dT_1$$

$$p_{ij} = \int_0^t p_{(i-1)j}(T_1) dT_1 \qquad i = 2,3 \quad j = 1,2,3$$

$$q_i = \int_0^t q_{(i-1)}(T_1) dT_1 \qquad i = 2,3$$

With $\lambda = 5$, $y(0) = 0.1$, $\dot{y}(0) = -0.1$, the results are shown in Figure 7.12.

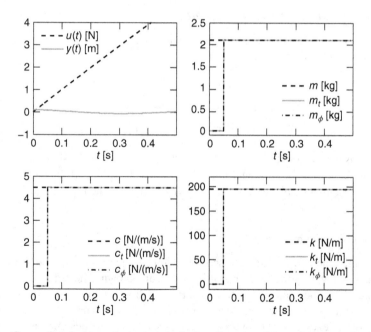

Figure 7.12 Comparison between the typical algebraic estimator and the estimator obtained using function $\phi(t)$ in a system of second order

7.3 Other Functions of Time for Parameter Estimation

We can propose different estimators of parameters from different functions of time if these functions are zero at $t = 0$. For example, a rational function and a sinusoidal function satisfy this requirement and are also bounded:

$$\Upsilon(t) = 1 - \frac{1}{\lambda t + 1}$$

$$\rho(t) = \sin(\lambda t)$$

Multiplying the system (7.41) by these functions and integrating by parts, yields the following estimator of parameter a:

$$a_{e,\Upsilon} = \frac{-\frac{bt}{bt+1}y(t) + \int_0^t \frac{b}{(bT_1+1)^2}y(T_1)dT_1 + \int_0^t \frac{bT_1}{bT_1+1}u(T_1)dT_1}{\int_0^t \frac{bT_1}{bT_1+1}y(T_1)dT_1}$$

$$a_{e,\rho} = \frac{-\sin(\lambda t)y(t) + \int_0^t \lambda\cos(\lambda T_1)y(T_1)dT_1 + \int_0^t \sin(\lambda T_1)u(T_1)dT_1}{\int_0^t \sin(\lambda T_1)y(T_1)dT_1}$$

With $\lambda = 5$ and $\dot{y}(0) = 0.1$, the simulation results are shown in Figure 7.13.

7.3.1 A Mechanical System Example

Consider the rotational mechanical system shown in Figure 7.14. The mathematical model of such a system, using Newton's laws, is given by

$$J_1\ddot{y}_1 + c_1\dot{y}_1 + k_1(y_1 - y_2) = \tau$$
$$J_2\ddot{y}_2 + c_2\dot{y}_2 + k_1(y_2 - y_1) + k_2(y_2 - y_3) = 0 \tag{7.47}$$
$$J_3\ddot{y}_3 + c_3\dot{y}_3 + k_2(y_3 - y_2) = 0$$

where J_i stands for the moment of inertia of the ith disk, c_i denotes a viscous friction coefficient exhibited by the corresponding bearings, k_i is a linear torsional stiffness coefficient, and θ_i represents the angular displacement of the ith disk while τ is the input torque.

7.3.1.1 Problem Formulation

It is assumed that neither one of the viscous friction coefficients c_i, nor the torsional stiffness coefficients k_i, as well as the moments of inertia J_i, are known. The only measured quantities are the angular displacement of the first disk θ_1, which we denote from now on simply by θ and the control input τ, which we denote from now on by u. The unknown inertia value J_1 will similarly be denoted simply by J. It is even possible that the designer does not know how many disks constitute the entire coupled mechanical system. The simulations and experiments are carried out, however, on a rotational system constituted of three inertia disks coupled by

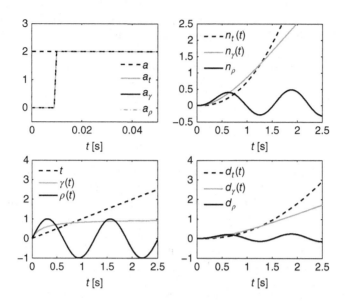

Figure 7.13 Comparison between the typical algebraic estimator and estimators obtained using functions $\Upsilon(t)$ and $\rho(t)$

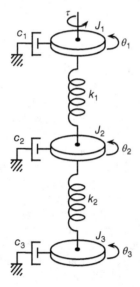

Figure 7.14 Mechanical rotational system of three disks

means of rotational springs and exhibiting some unknown friction effects due to the motions of the bearings. The formulation of the problem is as follows.

It is desired to accurately regulate the output angular position of the first disk, $y = \theta$, toward a desired given angular reference trajectory $y^*(t) = \theta^*(t)$, in spite of all the dynamic uncertainty surrounding the knowledge of the system description.

7.3.1.2   Design of a Perturbation Rejection Controller

We define a *Taylor polynomial* as the finite truncation of a Taylor series expansion around a given instant of time. We consider all signals and functions generating such signals to be sufficiently differentiable and restricted to the class of time functions which pointwise coincide with their convergent Taylor series. So, a Taylor polynomial renders a locally valid approximation to the given signal at some given instant of time.

Consider the following *simplified* model of the mechanical system describing only the angular displacement of the first inertial disk:

$$\ddot{y}(t) = \frac{1}{J}u(t) + \xi(t) \tag{7.48}$$

where $\xi(t)$ is an unknown perturbation which involves the dynamic responses and torques arising from the omitted coupled subsystem terms.

For a perturbation rejection control law design, we consider the term $\xi(t)$ to be locally approximated by a second-degree Taylor polynomial of the form

$$\alpha_2 \frac{t^2}{2!} + \alpha_1 \frac{t}{1!} + \alpha_0$$

with unknown constant coefficients representing a locally valid truncation, centered around $t = 0$, of the unknown perturbation input $\xi(t)$. In other words, the perturbation input signal $\xi(t)$ to the second-order system is represented here by an element of a *family* $\mathcal{F}_2(t)$ of second-degree time polynomials which will *automatically* "tailor itself" around the actual online perturbation input signal at any arbitrarily considered instant of time (under the assumption, of course, that such a perturbation is sufficiently smooth, and bounded, with reasonable high-frequency content). For this, note that a fixed-order, m, family of time polynomials, say $\mathcal{F}_m(t)$, is *invariant* under arbitrary time translations. In other words, the time translation of $p(t) \in \mathcal{F}_m(t)$ by an arbitrary amount $\tau > 0$ produces a polynomial $p(t - \tau)$ which also belongs to the same family $\mathcal{F}_m(t)$. This *self-updating* property of a perturbation function modeled by an element of a fixed-order family of polynomials seems to have passed unnoticed so far, and lies at the heart of our simple approach to the robust control of perturbed systems.

As an example, notice, for instance, that at time t_i, the perturbation term represented by a Taylor polynomial centered around t_i, and locally valid for any later time t, satisfies the following equality:

$$\alpha_2 \frac{(t - t_i)^2}{2!} + \alpha_1 \frac{t - t_i}{1!} + \alpha_0 = \alpha_2 \frac{t^2}{2} + \hat{\alpha}_1 \frac{t}{1!} + \hat{\alpha}_0$$

for some equally unknown constants $\hat{\alpha}_1$ and $\hat{\alpha}_0$. No doubt, then, the previous equality indicates that this new Taylor polynomial is, indeed, included in the *same family* of second-degree polynomials considered to be centered around the instant of time $t_i = 0$.

Consider the unperturbed nominal second-order system $\ddot{y}^*(t) = \frac{1}{J}u^*(t)$, and define the following perturbed tracking-error system:

$$J\ddot{e}_y = e_u + \xi(t)$$

where $e_y = y(t) - y^*(t)$ and $e_u = u(t) - u^*(t)$. The nominal input signal $u^*(t)$ is clearly obtained from the nominal inverse relation $u^*(t) = J\ddot{y}^*(t)$ for a given differentiable reference trajectory $y^*(t)$. A GPI trajectory-tracking controller for this perturbed tracking-error dynamics would be prescribed as follows.

Initially, propose an output feedback controller based on an *integral reconstructor* of the error velocity:

$$u = J(\ddot{y}^*(t) - \kappa_1(\hat{\dot{y}} - \dot{y}^*(t)) - \kappa_0 e_y)$$

for some suitable coefficients $\kappa_1, \kappa_0 > 0$. The integral reconstructor of the angular velocity error, $\hat{\dot{y}} - \dot{y}^*(t)$, is obtained by direct integration of the perturbed tracking-error system equations while duly neglecting the integrated perturbation term (which is now to be considered, in this particular case, as a third-degree polynomial). Define the integral reconstructor of the tracking-error velocity as

$$\hat{\dot{e}}_y = \hat{\dot{y}}(t) - \dot{y}^*(t) = \frac{1}{J}\left(\int^{(1)} u - u^*(t)\right) = \frac{1}{J}\left(\int^{(1)} e_u(t)\right) \qquad (7.49)$$

The perturbation ξ affects the use of this structural estimate in the proposed controller, since $\dot{y} = \frac{1}{J}\int u + \alpha_2\frac{t^3}{3!} + \alpha_1\frac{t^2}{2!} + \alpha_0 t$. Therefore, in order to handle the effects of the local third-order degree polynomial estimation error, as well as those arising from the original second-order degree polynomial input, we propose a a modified GPI control law of the form

$$u = J\left(\ddot{y}^* - k_5\left(\hat{\dot{y}} - [y^*]^{(1)}\right) - k_4 e_y\right.$$

$$\left. -k_3\int^{(1)} e_y - k_2\int^{(2)} e_y - k_1\int^{(3)} e_y - k_0\int^{(4)} e_y\right) \qquad (7.50)$$

which includes up to four integrations of the tracking error e_y in order to compensate for the highest, third-order, polynomial perturbation input appearing in the closed-loop system by effect of the structural angular velocity error estimation in the proposed integral reconstructor.

Substituting equation (7.49) into equation (7.50) and rearranging the terms, one obtains

$$\left[e_u + k_5\left(\int e_u\right)\right] = -J\left[k_4 e_y + k_3\left(\int^{(1)} e_y\right)\right.$$

$$\left. +k_2\left(\int^{(2)} e_y\right) + k_1\left(\int^{(3)} e_y\right) + k_0\left(\int^{(4)} e_y\right)\right]$$

Finally, multiplying both sides by s^4 and solving, we find that the transfer function of the trajectory-tracking feedback controller can be written

$$\frac{e_u(s)}{e_y(s)} = -J\left[\frac{k_4 s^4 + k_3 s^3 + k_2 s^2 + k_1 s + k_0}{s^3(s + k_5)}\right]$$

This rational proper feedback-compensating network determines the following closed-loop characteristic polynomial: $s^6 + k_5 s^5 + k_4 s^4 + k_3 s^3 + k_2 s^2 + k_1 s + k_0$. It is hence possible to determine the necessary controller coefficients by equating the closed-loop characteristic polynomial to those of a desired Hurwitz polynomial in the complex variable s, such as $(s^2 + 2\zeta\omega_n s + \omega_n)^3$. Taking into account that ζ and ω_n are both positive quantities, the expressions for the controller coefficients k_i are readily obtained as

$$k_5 = 6\zeta\omega_n$$
$$k_4 = 3\omega_n^2 + 12\zeta^2\omega^2$$
$$k_3 = 12\zeta\omega_n^3 + 8\zeta^3\omega_n^3$$
$$k_2 = 3\omega_n^4 + 12\zeta^2\omega_n^4$$
$$k_1 = 6\zeta\omega_n^5$$
$$k_0 = \omega_n^6$$

Now, let's return to the rotational system (7.48), where we assumed the perturbation was approximated by an element of a second-degree family of polynomials. We thus differentiate the model three times with respect to time. The new structure of the system may be considered to be

$$Jy^{(5)}(t) = u^{(3)}(t) \tag{7.51}$$

The next step is to multiply equation (7.51) by $\phi(t)^5$ and integrate five times with respect to time:

$$J \int_0^{(5)} \left(1 - e^{-\frac{t}{\lambda}}\right)^5 y^{(5)} = \int_0^{(5)} \left(1 - e^{-\frac{t}{\lambda}}\right)^5 u^{(3)}$$

The iterated integration by parts leads to an expression from which it is possible to synthesize a linear time-varying filter. Finally, the formula for obtaining the estimated value J_e is given by

$$J_e(\phi^5 y + z_1) = (\phi^5 v_1 + \chi_1) \tag{7.52}$$

where

$$\dot{z}_1 = z_2 - \frac{25}{\lambda}\phi^4 e^{\frac{-t}{\lambda}} y$$

$$\dot{z}_2 = z_3 + \frac{1}{\lambda^2}\left(200\phi^3 e^{\frac{-2t}{\lambda}} - 60\phi^4 e^{\frac{-t}{\lambda}}\right) y$$

$$\dot{z}_3 = z_4 + \frac{600}{\lambda^3}\left(-\phi^2 e^{\frac{-3t}{\lambda}} - \phi^3 e^{\frac{-2t}{\lambda}} - \frac{1}{12}\phi^4 e^{\frac{-t}{\lambda}}\right) y$$

$$\dot{z}_4 = z_5 + \frac{25}{\lambda^4}\left(24\phi e^{\frac{-4t}{\lambda}} - 72\phi^2 e^{\frac{-3t}{\lambda}} + 28\phi^3 e^{\frac{-2t}{\lambda}} - \phi^4 e^{\frac{-t}{\lambda}}\right) y$$

$$\dot{z}_5 = \frac{5}{\lambda^5}\left(-24 e^{\frac{-5t}{\lambda}} + 240\phi e^{\frac{-4t}{\lambda}} - 300\phi^2 e^{\frac{-3t}{\lambda}} + 100\phi^3 e^{\frac{-2t}{\lambda}} - \phi^4 e^{\frac{-t}{\lambda}}\right) y$$

$$\dot{\chi}_1 = \chi_2 - \frac{25}{\lambda}\phi^4 e^{\frac{-t}{\lambda}} v_1$$

$$\dot{\chi}_2 = \chi_3 + \frac{1}{\lambda^2}\left(200\phi^3 e^{\frac{-2t}{\lambda}} - 60\phi^4 e^{\frac{-t}{\lambda}}\right)v_1$$

$$\dot{\chi}_3 = \chi_4 + \frac{50}{\lambda^3}\left(-12\phi^2 e^{\frac{-3t}{\lambda}} + 12\phi^3 e^{\frac{-2t}{\lambda}} - \phi^4 e^{\frac{-t}{\lambda}}\right)v_1$$

$$\dot{\chi}_4 = \chi_5 + \frac{25}{\lambda^4}\left(24\phi e^{\frac{-4t}{\lambda}} - 72\phi^2 e^{\frac{-3t}{\lambda}} + 28\phi^3 e^{\frac{-2t}{\lambda}} - \phi^4 e^{\frac{-t}{\lambda}}\right)v_1$$

$$\dot{\chi}_5 = \frac{5}{\lambda^5}\left(-24e^{\frac{-5t}{\lambda}} + 240\phi e^{\frac{-4t}{\lambda}} - 300\phi^2 e^{\frac{-3t}{\lambda}} + 100\phi^3 e^{\frac{-2t}{\lambda}} - \phi^4 e^{\frac{-t}{\lambda}}\right)v_1$$

$$\dot{v}_1 = v_2$$

$$\dot{v}_2 = u$$

7.3.1.3 Numerical Simulations

The parameters used in the digital computer simulations belong to an experimental plant. These parameters were identified using a classic outline method. The parameters of the first disk are

$$J_1 = 0.0187 \text{ kg m}^2$$

$$K_1 = 2.6 \text{ N m/rad}$$

$$C_1 = 0.3619 \text{ N m s/rad}$$

and the parameters of the cascade attached system are

$$K_2 = 2.5 \text{ N m/rad}$$

$$J_2 = 0.01031 \text{ kg m}^2$$

$$J_3 = 0.0019 \text{ kg m}^2$$

$$C_{2,3} = 0.3619 \text{ N m s/rad}$$

The selected controller gains were $\zeta = 2.5$ and $w_n = 35$. The results of parameter identification are shown in Figures 7.15 and 7.16. The parameter identification was obtained in 10 ms with a 2% error, approximately.

7.3.1.4 Experimental Results

The experimental response of the rotational 1-DOF system is shown in Figure 7.17. The arbitrary initial value adopted for J_e was set to $J_e = 0.016$ kg m^2. The algebraic estimation of the inertia parameter was obtained after 10 ms, with an error similar to that of the digital simulation. This parameter value was used adaptively in the proposed certainty-equivalence control law, resulting in excellent tracking results. In Figure 7.18, we show the features of the identification process and the response of the system to the controller actions in the rotational system where only the first disk position is regulated with the effects of the rest of the cascaded system acting as an unmodeled perturbation. The curves depict a satisfactory estimation of the inertia parameter for the first disk, which is the only parameter needed in our control scheme.

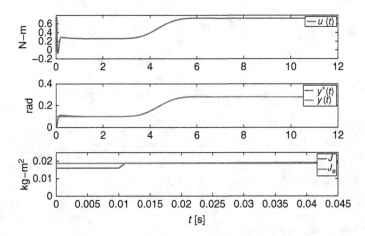

Figure 7.15 Simulation result of the tracking trajectory and inertia identification in the rotational system consisting of one disk

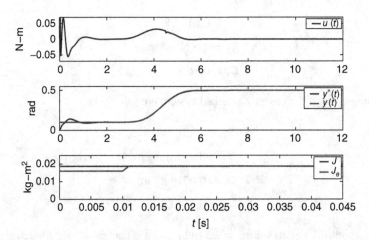

Figure 7.16 Simulation result of the tracking trajectory and inertia identification in the rotational system consisting of three disks

7.3.2 A Derivative Approach to Demodulation

Frequency-modulated signals are often obtained directly using voltage-controlled oscillators (VCOs). The frequency of the standard harmonic oscillator is added to a message signal, denoted by $u(t)$, as follows:

$$\ddot{y} = -[\omega^2 + u(t)]y \tag{7.53}$$

The initial conditions for the generating system are typically given by $y(0) = 0$ and $\dot{y}(0) = 1$.

 The demodulation process involves a so-called *phase-locked loop* approach, which entitles input and feedback signal multiplication, low-pass forward filtering, and integral feedback. Here we give a simple solution to the frequency demodulation problem using differentiations of the measured signal $y(t)$.

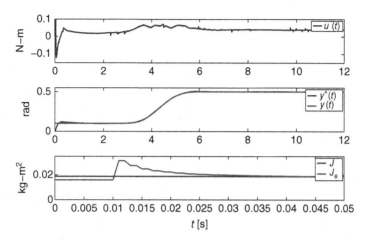

Figure 7.17 Experimental result of the tracking trajectory and inertia identification in the rotational system of one disk

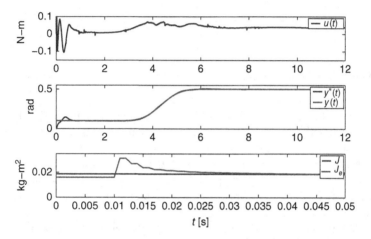

Figure 7.18 Experimental result of the tracking trajectory and inertia identification in the rotational system of three disks

Clearly, in the absence of transmission noises, the message can be found from the model of the carrier signal $y(t)$. We have

$$u(t) = -\frac{\ddot{y}(t) + \omega^2 y(t)}{y(t)} \tag{7.54}$$

Since the signal $y(t)$ is at our disposal, we can attempt an estimation scheme based on the algebraic operations indicated in the formulae for $u(t)$ above. We propose

$$u_e(t) = \begin{cases} \text{arbitrary} & \forall\, y \approx 0 \\[2mm] -\left[\dfrac{\ddot{y} + \omega^2 y(t)}{y(t)}\right] & \forall\, y \neq 0 \end{cases} \tag{7.55}$$

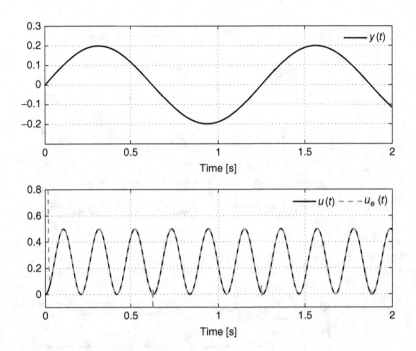

Figure 7.19   Algebraic frequency demodulation using time derivatives calculation

Simulations of the algebraic frequency-demodulation process are depicted in Figure 7.19, where the transmitted noiseless signal $y(t)$ is depicted along with the message signal $u(t)$ and its estimated waveform $u_e(t)$. In this instance, we set the transmission frequency $\omega = 5$ rad/s and the message signal $u(t) = A\sin^2(3\omega t)$, with $A = 0.5$. Resettings of the calculations were performed every 0.2 s, with a waiting period after each resetting of $\epsilon = 0.01$ s.

Singularities are obtained at the instants when $y(t)$ crosses the value 0. These are depicted as spikes in the figure.

The simulations for the proposed algorithm collapse when the signal $y(t)$ contains some additive measurement noise. An invariant filtering approach is then highly recommended.

7.3.3   Time Derivatives via Parameter Identification

Consider a truncated Taylor series expansion model constituting the polynomial approximation of a given signal $y(t)$ which, as before, is assumed to be sufficiently smooth:

$$\tilde{y}(t) = \sum_{k=0}^{N-1} \frac{1}{k!} y^{(k)}(t_i)(t - t_i)^k, \quad t \geq t_i \tag{7.56}$$

Clearly, the set of constant parameters

$$\{y^{(1)}(t_i), y^{(2)}(t_i), \ldots, y^{(N-2)}(t_i), y^{(N-1)}(t_i)\}$$

is linearly identifiable at each arbitrary instant t_i.

Recall that a linear time-invariant homogeneous system representing the truncated Taylor series, with initial conditions at time t_i, is given by

$$\frac{d^N \tilde{y}}{dt^N} = 0$$

Taking transforms in this expression, we obtain

$$s^N \tilde{y}(s) = s^{N-1} y(t_i) + s^{N-2} \dot{y}(t_i) + \cdots + s y^{(N-2)}(t_i) + y^{(N-1)}(t_i)$$

Taking successively $N - 1, N - 2, \ldots, 1$ and 0 derivatives with respect to s, we are led to

$$\frac{d^{N-1}[s^N \tilde{y}]}{ds^{N-1}} = (N - 1)! y(t_i)$$

$$\frac{d^{N-2}[s^N \tilde{y}]}{ds^{N-2}} = (N - 1)! s y(t_i) + (N - 2)! \dot{y}(t_i)$$

$$\frac{d^{N-3}[s^N \tilde{y}]}{ds^{N-3}} = \frac{1}{2!}(N - 1)! s^2 y(t_i) + (N - 2)! s \dot{y}(t_i) + (N - 3)! \ddot{y}(t_i)$$

$$\frac{d^{N-4}[s^N \tilde{y}]}{ds^{N-4}} = \frac{1}{3!}(N - 1)! s^3 y(t_i) + \frac{1}{2!}(N - 2)! s^2 \dot{y}(t_i) + (N - 3)! s \ddot{y}(t_i)$$
$$+ (N - 4)! y^{(3)}(t_i)$$

$$\vdots$$

$$\frac{d^{N-j}[s^N \tilde{y}]}{ds^{N-j}} = \frac{(N - 1)!}{(j - 1)!} s^{j-1} y(t_i) + \frac{(N - 2)!}{(j - 2)!} s^{j-2} \dot{y}(t_i) + \cdots$$
$$\ldots + (N - j)! y^{(j-1)}(t_i)$$

$$\vdots$$

$$\frac{d[s^N \tilde{y}]}{ds} = (N - 1) s^{N-2} y(t_i) + (N - 2) s^{N-3} \dot{y}(t_i) + \cdots + y^{(N-2)}(t_i)$$

$$s^N \tilde{y} = s^{N-1} y(t_i) + s^{N-2} \dot{y}(t_i) + \cdots + s y^{(N-2)}(t_i) + y^{(N-1)}(t_i)$$

By considering $y(t_i)$ itself as an unknown, the previous set of equations is then a set of N equations in N unknowns.

In order to avoid the presence of positive power factors of s in the above expressions, implying iterated time derivatives in the time domain, we need to multiply each of the above equations by s^{-N}, but at least one symbol of the form s^{-1} is to be set aside to acknowledge the constant nature of the unknowns. Thus we must, at least, multiply every equation by s^{-N-1}. A low-pass filtering effect, of invariant nature, may be accomplished by further multiplying out every expression by an additional factor of the form s^{-r}, where r is the number of desired iterated integrals needed to attenuate high-frequency noises and contribute to bettering the signal-to-noise ratio in the final expression for each unknown coefficient of the truncated Taylor series expansion. We then need to use the factor $s^{-(N+r+1)}$, with $r > 0$ in every equation.

We rewrite the set of equations, still in the frequency domain, as

$$P(s) \left[\frac{1}{s} c \right] = q(s) \tag{7.57}$$

where

$$P(s) =$$

$$
\begin{bmatrix}
(N-1)!s^{-N-r} & 0 & 0 & \cdots & 0 & 0 \\[2mm]
(N-1)!\dfrac{s^{-N-r+1}}{1!} & (N-2)!s^{-N-r} & 0 & \cdots & 0 & 0 \\[2mm]
(N-1)!\dfrac{s^{-N-r+2}}{2!} & (N-2)!\dfrac{s^{-N-r+1}}{1!} & (N-3)!s^{-N-r} & \cdots & 0 & 0 \\[2mm]
\vdots & \vdots & \vdots & \cdots & \vdots & \vdots \\[2mm]
\dfrac{(N-1)!}{(j-1)!}s^{\tilde{N}_1+j} & \dfrac{(N-2)!}{(j-2)!}s^{\tilde{N}_2+j} & \dfrac{(N-3)!}{(j-3)!}s^{\tilde{N}_3+j} & \cdots & \vdots & \vdots \\[2mm]
\vdots & \vdots & \vdots & \cdots & \vdots & \vdots \\[2mm]
(N-1)s^{-r-2} & (N-2)s^{-r-3} & (N-3)s^{-r-4} & \cdots & s^{\tilde{N}_0} & 0 \\[2mm]
s^{-r-1} & s^{-r-2} & s^{-r-3} & \cdots & s^{\tilde{N}_{-1}} & s^{\tilde{N}_0}
\end{bmatrix}
$$

$$(7.58)$$

with $\tilde{N}_i = -N - r - i$

$$q(s) = s^{-N-r}\left[\frac{d^{N-1}[s^N\tilde{y}]}{ds^{N-1}}, \frac{d^{N-2}[s^N\tilde{y}]}{ds^{N-2}}, \frac{d^{N-3}[s^N\tilde{y}]}{ds^{N-3}}, \cdots, \frac{d[s^N\tilde{y}]}{ds}, s^N\tilde{y}\right]^T$$

and

$$\frac{1}{s}c = \left[\left[\frac{y(t_i)}{s}\right], \left[\frac{\dot{y}(t_i)}{s}\right], \left[\frac{\ddot{y}(t_i)}{s}\right], \cdots \left[\frac{y^{(N-2)}(t_i)}{s}\right], \left[\frac{y^{(N-1)}(t_i)}{s}\right]\right]^T$$

The vector of constant coefficients of the truncated Taylor series expansion is then obtained, in the transformed domain, as

$$\left[\frac{1}{s}c\right] = P^{-1}(s)q(s)$$

However, one should not cancel the factors s^{-N-r} appearing in the numerator and denominator for each unknown constant $\frac{y^{(j)}(t_i)}{s}$. Instead of giving a general formula, as usual, we prefer a clarifying example.

7.3.4 Example

Consider a fifth-order Taylor series expansion of a smooth signal. Intuitively, such an expansion will accurately provide us with at least two time derivatives of the signal. We want to compute the coefficients at regularly spaced intervals of time t_i, with some provisions for a low-pass filtering represented by two iterated integrations. In other words, we have $N = 6$, $r = 2$. We obtain the following expression in the frequency domain:

$$\left[\frac{y(t_i)}{s}\right] = \frac{1}{120}\left[720y(s) + 1800s\frac{dy}{ds} + 1200s^2\frac{d^2y}{ds^2} + 300s^3\frac{d^3y}{ds^3} + 30s^4\frac{d^4y}{ds^4} + s^5\frac{d^5y}{ds^5}\right]$$

$$\left[\frac{\dot{y}(t_i)}{s}\right] = -\frac{1}{24}\left[360sy(s) + 1320s^2\frac{dy}{ds} + 1020s^3\frac{d^2y}{ds^2} + 276s^4\frac{d^3y}{ds^3}\right.$$
$$\left. +29s^5\frac{d^4y}{ds^4} + s^6\frac{d^5y}{ds^5}\right]$$

$$\left[\frac{\ddot{y}(t_i)}{s}\right] = \frac{1}{12}\left[240s^2y(s) + 1020s^3\frac{dy}{ds} + 876s^4\frac{d^2y}{ds^2} + 254s^5\frac{d^3y}{ds^3}\right.$$
$$\left. +28s^6\frac{d^4y}{ds^4} + s^7\frac{d^5y}{ds^5}\right]$$

$$\left[\frac{y^{(3)}(t_i)}{s}\right] = -\frac{1}{12}\left[180s^3y(s) + 828s^4\frac{dy}{ds} + 762s^5\frac{d^2y}{ds^2} + 234s^6\frac{d^3y}{ds^3}\right.$$
$$\left. +27s^7\frac{d^4y}{ds^4} + s^8\frac{d^5y}{ds^5}\right]$$

$$\left[\frac{y^{(4)}(t_i)}{s}\right] = \frac{1}{24}\left[144s^4y(s) + 696s^5\frac{dy}{ds} + 672s^6\frac{d^2y}{ds^2} + 216s^7\frac{d^3y}{ds^3}\right.$$
$$\left. +26s^8\frac{d^4y}{ds^4} + s^9\frac{d^5y}{ds^5}\right]$$

$$\left[\frac{y^{(5)}(t_i)}{s}\right] = -\frac{1}{120}\left[120s^5y(s) + 600s^6\frac{dy}{ds} + 600s^7\frac{d^2y}{ds^2} + 200s^8\frac{d^3y}{ds^3}\right.$$
$$\left. +25s^9\frac{d^4y}{ds^4} + s^{10}\frac{d^5y}{ds^5}\right]$$

We multiply out each equation to eliminate positive powers of s and further adjoin a second-order integration as a low-pass filter. Thus, the first expression is multiplied by s^{-7}, the second by s^{-8}, and so on. We obtain

$$s^{-7}\left[\frac{y(t_i)}{s}\right] = \frac{1}{120}\left[720s^{-7}y(s) + 1800s^{-6}\frac{dy}{ds} + 1200s^{-5}\frac{d^2y}{ds^2} + 300s^{-4}\frac{d^3y}{ds^3}\right.$$
$$\left. +30s^{-3}\frac{d^4y}{ds^4} + s^{-2}\frac{d^5y}{ds^5}\right]$$

$$s^{-8}\left[\frac{\dot{y}(t_i)}{s}\right] = -\frac{1}{24}\left[360s^{-7}y(s) + 1320s^{-6}\frac{dy}{ds} + 1020s^{-5}\frac{d^2y}{ds^2} + 276s^{-4}\frac{d^3y}{ds^3}\right.$$
$$\left. +29s^{-3}\frac{d^4y}{ds^4} + s^{-2}\frac{d^5y}{ds^5}\right]$$

$$s^{-9}\left[\frac{\ddot{y}(t_i)}{s}\right] = \frac{1}{12}\left[240s^{-7}y(s) + 1020s^{-6}\frac{dy}{ds} + 876s^{-5}\frac{d^2y}{ds^2} + 254s^{-4}\frac{d^3y}{ds^3}\right.$$
$$\left. +28s^{-3}\frac{d^4y}{ds^4} + s^{-2}\frac{d^5y}{ds^5}\right]$$

$$s^{-10}\left[\frac{y^{(3)}(t_i)}{s}\right] = -\frac{1}{12}\left[180s^{-7}y(s) + 828s^{-6}\frac{dy}{ds} + 762s^{-5}\frac{d^2y}{ds^2} + 234s^{-4}\frac{d^3y}{ds^3}\right.$$
$$\left. +27s^{-3}\frac{d^4y}{ds^4} + s^{-2}\frac{d^5y}{ds^5}\right]$$

$$s^{-11}\left[\frac{y^{(4)}(t_i)}{s}\right] = \frac{1}{24}\left[144s^{-7}y(s) + 696s^{-6}\frac{dy}{ds} + 672s^{-5}\frac{d^2y}{ds^2} + 216s^{-4}\frac{d^3y}{ds^3}\right.$$
$$\left. +26s^{-3}\frac{d^4y}{ds^4} + s^{-2}\frac{d^5y}{ds^5}\right]$$

$$s^{-12}\left[\frac{y^{(5)}(t_i)}{s}\right] = -\frac{1}{120}\left[120s^{-7}y(s) + 600s^{-6}\frac{dy}{ds} + 600s^{-5}\frac{d^2y}{ds^2} + 200s^{-4}\frac{d^3y}{ds^3}\right.$$
$$\left. +25s^{-3}\frac{d^4y}{ds^4} + s^{-2}\frac{d^5y}{ds^5}\right]$$

In the time domain, we have

$$y(t_i) = \frac{42}{(t-t_i)^7}\left[720\left(\int_{t_i}^{(7)} y\right) - 1800\left(\int_{t_i}^{(6)}(t-t_i)y\right) + 1200\left(\int^{(5)}(t-t_i)^2 y\right)\right.$$
$$\left. -300\left(\int_{t_i}^{(4)}(t-t_i)^3 y\right) + 30\left(\int_{t_i}^{(3)}(t-t_i)^4 y\right) - \left(\int_{t_i}^{(2)}(t-t_i)^5 y\right)\right]$$

$$\dot{y}(t_i) = -\frac{1680}{(t-t_i)^8}\left[360\left(\int_{t_i}^{(7)} y\right) - 1320\left(\int_{t_i}^{(6)}(t-t_i)y\right) + 1020\left(\int_{t_i}^{(5)}(t-t_i)^2 y\right)\right.$$
$$\left. -276\left(\int_{t_i}^{(4)}(t-t_i)^3 y\right) + 29\left(\int_{t_i}^{(3)}(t-t_i)^4 y\right) - \left(\int_{t_i}^{(2)}(t-t_i)^5 y\right)\right]$$

$$\ddot{y}(t_i) = \frac{30240}{(t-t_i)^9}\left[240\left(\int_{t_i}^{(7)} y\right) - 1020\left(\int_{t_i}^{(6)}(t-t_i)y\right) + 876\left(\int_{t_i}^{(5)}(t-t_i)^2 y\right)\right.$$
$$\left. -254\left(\int_{t_i}^{(4)}(t-t_i)^3 y\right) + 28\left(\int_{t_i}^{(3)}(t-t_i)^4 y\right) - \left(\int_{t_i}^{(2)}(t-t_i)^5 y\right)\right]$$

$$y^{(3)}(t_i) = -\frac{302400}{(t-t_i)^{10}}\left[180\left(\int_{t_i}^{(7)}\right) - 828\left(\int_{t_i}^{(6)}(t-t_i)y\right) + 762\left(\int_{t_i}^{(5)}(t-t_i)^2 y\right)\right.$$
$$\left. -234\left(\int_{t_i}^{(4)}(t-t_i)^3 y\right) + 27\left(\int_{t_i}^{(3)}(t-t_i)^4 y\right) - \left(\int_{t_i}^{(2)}(t-t_i)^5 y\right)\right]$$

$$y^{(4)}(t_i) = \frac{1663200}{(t-t_i)^{11}}\left[144\left(\int_{t_i}^{(7)} y\right) - 696\left(\int_{t_i}^{(6)}(t-t_i)y\right) + 672\left(\int_{t_i}^{(5)}(t-t_i)^2 y\right)\right.$$
$$\left. -216\left(\int_{t_i}^{(4)}(t-t_i)^3 y\right) + 26\left(\int_{t_i}^{(3)}(t-t_i)^4 y\right) - \left(\int_{t_i}^{(2)}(t-t_i)^5 y\right)\right]$$

$$y^{(5)}(t_i) = -\frac{3991680}{(t-t_i)^{12}} \left[120 \left(\int_{t_i}^{\cdot} {}^{(7)} y \right) - 600 \left(\int_{t_i}^{\cdot} {}^{(6)} (t-t_i)y \right) + 600 \left(\int_{t_i}^{\cdot} {}^{(5)} (t-t_i)^2 y \right) \right.$$

$$\left. -200 \left(\int_{t_i}^{\cdot} {}^{(4)} (t-t_i)^3 y \right) + 25 \left(\int_{t_i}^{\cdot} {}^{(3)} (t-t_i)^4 y \right) - \left(\int_{t_i}^{\cdot} {}^{(2)} (t-t_i)^5 y \right) \right]$$

With the above estimates we may reconstruct, for all $t \geq t_i$, with t_i being a resetting time, the signal $y(t)$ itself and the first three time derivatives as follows (Figure 7.20):

$$y(t) \approx y(t_i) + (t-t_i)\dot{y}(t_i) + \frac{1}{2}(t-t_i)^2 \ddot{y}(t_i) + \frac{1}{6}(t-t_i)^3 y^{(3)}(t_i)$$

$$+ \frac{1}{24}(t-t_i)^4 y^{(4)}(t_i) + \frac{1}{120}(t-t_i)^5 y^{(5)}(t_i)$$

$$\dot{y}(t) \approx \dot{y}(t_i) + (t-t_i)\ddot{y}(t_i) + \frac{1}{2}(t-t_i)^2 y^{(3)}(t_i) + \frac{1}{6}(t-t_i)^3 y^{(4)}(t_i)$$

$$+ \frac{1}{24}(t-t_i)^4 y^{(5)}(t_i)$$

$$\ddot{y}(t) \approx \ddot{y}(t_i) + (t-t_i)y^{(3)}(t_i) + \frac{1}{2}(t-t_i)^2 y^{(4)}(t_i) + \frac{1}{6}(t-t_i)^3 y^{(5)}(t_i)$$

$$y^{(3)}(t) \approx y^{(3)}(t_i) + (t-t_i)y^{(4)}(t_i) + \frac{1}{2}(t-t_i)^2 y^{(5)}(t_i)$$

Below, we show simulations using these formulae for the computation of the time derivatives of the same signal in the previous examples. Namely, $y(t) = \exp(\sin(3t))$ when the measured

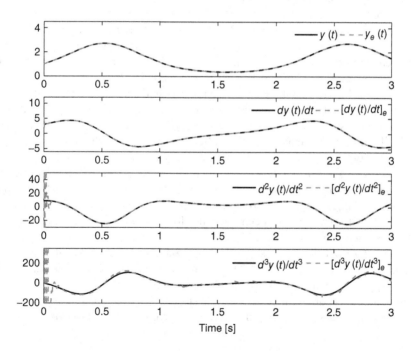

Figure 7.20 Time derivatives calculation

signal undergoes significant measurement noise perturbation given by the function $\xi(t) = 10^{-3}[rect(t) - 0.5]$, where $rect(t)$ is a random process consisting of piecewise constant random variables uniformly distributed in the interval $[0, 1]$ of the real line. With the above approximation formulae, including an extra second-order integration acting as an invariant filtering process, we can reliably compute up to, say, a third-order time derivative of the given signal $y(t)$.

7.4   An Algebraic Denoising Scheme

High-gain observers (HGOs) constitute an effective signal derivative estimation technique in a variety of application fields, such as robotics (Lee and Khalil, 1997), motion control (Tan and Zhao, 2004), position/force control, motor load estimation, induction motor control (Cortes-Romero et al., 2009; Solsona and Valla, 2003), disturbance and fault estimation (Martínez-Guerra and Diop, 2004), among many others. This class of observers offers a certain degree of design freedom which can be utilized to achieve properties such as robustness against external perturbations and other uncertainties. In addition, when the observer is designed, the number of observed time derivatives can be chosen in connection with the observer dimension.

Two problems arising in connection with these HGOs are: the initial "peaking" phenomenon and the undesired noise amplification. In relation to the peaking phenomenon, several approaches have been proposed to solve it. See, for instance, Chitour (2002), where the high-gain differentiators are used in combination with a traditional filtering scheme to estimate the higher-order time derivatives of a signal.

On the contrary, high-gain observers have been implemented to estimate output time derivatives and unknown plant inputs, such as faults and plant perturbations. In applications where environmental disturbances and measurement noises play an important rôle, the tuning of the observer gain becomes an important and difficult issue. The undesired amplification tends to grow in the calculation of each consecutive higher-order time derivative. A possible solution to this problem uses the well-known extended Kalman filter, under the assumption that the noise is a Gaussian process with known statistics (see Alessandri, 2000; Lewis 1986). In Busawon and Kabore (2001), an alternative is proposed to reduce the noise effects. A class of "proportional integral observers" is considered, which consists of the use of an integral gain in the observer injection structure to attenuate the effects of noise in the reconstruction error dynamics. This filtering scheme may achieve an effective reduction of the noise effects. Thus, the problem relies on applying adequate filtering techniques to the signal and its time derivatives. If classical filtering schemes are selected, then undesired signal time delays often arise. In contrast, numerical algorithms are slow and also known to be computationally complex.

Algebraic techniques have been proposed to solve the problem of parameter estimation, for linear systems with constant parameters, and also for linear systems with time-varying parameters (see Fliess and Sira-Ramírez, 2003; Tian et al., 2008). There also exists an algebraic approach to carry out the state estimation task in a class of algebraically observable non-linear systems (Fliess et al., 2008b). These approaches are based on the so-called algebraic derivative approach. In this technique, the problem of state estimation is closely related to the problem of numerical differentiation, which is an active area of research in numerous science and engineering areas such as control theory and signal processing (see Mboup et al., 2007).

Some numerical differentiation algorithms such as linear differentiation, backward finite differences, and the Savitzky–Golay differentiation have been analyzed in Braci and Diop (2003). Taylor expansion approaches have been implemented to obtain numerical differentiation from a noisy signal, see Mboup et al. (2007, 2009). However, this method becomes ill-conditioned for higher-order truncation of the Taylor series expansion.

Even if the time derivative estimations obtained from a high-gain observer are very noisy, their average signals are "good" in the sense that they tend to fit the actual underlying signal derivative. This fact can be exploited by means of algebraic manipulations similar to those applied in the linear parameter identification problem. The aim of our approach is to provide a filtering scheme for a noisy measured or estimated derivative signal. In particular, we take the estimated expressions from a HGO (Cortes-Romero et al., 2009) and apply an algebraic filtering scheme. This approach is inspired by the algebraic state identification technique presented for continuous-time systems by Fliess and Sira-Ramírez (2003).

7.4.0.1 Problem Formulation

Consider a smooth signal $x(t)$, affected by an additive, zero-mean, high-frequency noise, $\mu(t)$:

$$y(t) = x(t) + \mu(t) \tag{7.59}$$

Carrying out any signal processing, or implementing a control law through this noisy measurement is undesirable because of the noise effects downstream. There exist low-pass filtering alternatives, but they need a certain knowledge of the noise characteristics while producing a significant delay in the filtered signal.

Given a noisy measurement $y(t)$ of the smooth signal $x(t)$ as in (7.59), with $\mu(t)$ being an additive, zero-mean, high-frequency measurement noise, devise an online filtering process for the measured signal $y(t)$ such that it recovers the signal $x(t)$ with no delay.

7.4.0.2 An Observer-Based Filtering Algorithm: An Algebraic Approach

The proposed algorithm is carried out in two steps. The first step consists of devising an asymptotic estimation scheme for the time derivatives of $y(t)$ via a traditional high-gain observer as if y were uncorrupted by noise. Since the additive signal $\mu(t)$ is a zero-mean noise, we know that the observer-based estimated derivatives are also crucially affected by the noise, but their average values are asymptotically coincident with the actual derivatives of $x(t)$. The second step consists of considering the algebraic expression relating the jth time derivative of $y(t)$ and its estimated value from the observer equations. This expression receives an algebraic treatment to extract the required filtered signal. One generally proceeds by multiplying the appropriate expression by the jth positive power of the time variable, $(-t)^j$, then the resulting product is integrated by parts j times, thus obtaining an expression which is free from the initial conditions and any time derivatives of $y(t)$. From the obtained expression, the signal $y(t)$ can be expressed as the quotient of two time functions. This filtering can be carried out quite fast and with no involved delays, as are common in many of the traditional low-pass filtering schemes.

7.4.0.3 Obtaining an Average Derivative

Consider a Taylor series expansion of the noise-free signal $x(t)$, which is assumed to be infinitely differentiable in a neighborhood of t_0:

$$x(t) = x(t_0) + \frac{(t - t_0)}{1!} \dot{x}(t_0) + \frac{(t - t_0)^2}{2!} \ddot{x}(t_0) + \cdots \qquad (7.60)$$

For an $(n - 1)$th-order approximation of $x(t)$, it is possible to consider the following nth-order model:

$$\frac{d^n x(t)}{dt^n} = 0 \qquad (7.61)$$

Rewriting (7.61) in state variable form, we obtain

$$\begin{aligned} \dot{x}_1 &= x_2 \\ \dot{x}_2 &= x_3 \\ &\vdots \quad \vdots \\ \dot{x}_n &= 0 \end{aligned} \qquad (7.62)$$

$$x(t) = x_1$$

A Luenberger observer can be proposed in order to estimate the unknown states of (7.62), as follows:

$$\begin{aligned} \dot{\hat{x}}_1 &= \hat{x}_2 + \lambda_{n-1}\varepsilon \\ \dot{\hat{x}}_2 &= x_3 + \lambda_{n-2}\varepsilon \\ &\vdots \quad \vdots \\ \dot{\hat{x}}_n &= \lambda_0 \varepsilon \end{aligned} \qquad (7.63)$$

where the output estimation error is defined by $\varepsilon = y - \hat{x}_1$. The asymptotic convergence of this estimation error depends on the parameters λ_i, which can be selected according to a prespecified Hurwitz polynomial governing the resulting output estimation error dynamics.

7.4.0.4 Obtaining an Online Smooth Signal

As is known, under noisy output measurement conditions, the output time derivative estimates obtained from a HGO are significantly distorted. Nevertheless, their realizations are close *on average* to their actual values, provided a zero-mean assumption has been established on the nature of the noise. In this subsection, an algebraic filtering process is proposed using the estimated derivatives supplied by an observer of high-gain type.

Abusing the notation, a jth-order algebraic filtering is obtained from

$$\int^{(j)} t^j y^{(j)} = \int^{(j)} t^j \hat{x}_{j+1} \quad \text{for} \quad 1 < j < n \qquad (7.64)$$

where $\hat{x}_{j+1}$ is the jth estimated derivative of $y(t)$ supplied by (7.63). Integrating by parts $\int_0^{(j)} t^j y^{(j)}$, it is possible to find a filtered version for $y(t)$.

In order to illustrate, we consider the particular case when $j = 3$. Integration by parts of the corresponding expression allows us to check that

$$\int^{(3)} (-t)^3 y^{(3)} = 6 \int^{(3)} y - 18 \int^{(2)} ty + 9 \int^{(1)} t^2 y - t^3 y$$

and

$$y = \frac{\int^{(3)} t^3 y^{(3)} + 6\int^{(3)} y - 18\int^{(2)} ty + 9\int^{(1)} t^2 y}{t^3}$$

Given that the third time derivative estimate of y is $\hat{x}_4$, a filtered version of y is given by

$$y_{Filt_3} = \frac{\int^{(3)} t^3 \hat{x}_4 + 6\int^{(3)} y - 18\int^{(2)} ty + 9\int^{(1)} t^2 y}{t^3}$$

Remark 7.4.1 *The algebraic filtering expressions are in terms of the time variable t and some finite power terms which may produce high numerical values as time elapses. To avoid having numerical computing problems, instead of multiplying by t^j, the expression can be multiplied by an admissible bounded time function (i.e., a function which allows us to apply the algebraic manipulations to generate a valid filtering expression free from initial conditions) so that the numerator and denominator signals remain bounded. For instance, multiplying by $(1 - e^{-t})^j$ instead of $(-t)^j$.*

7.4.1   Example

Let $y(t)$ be a signal of the form (7.59). It is desired to filter the noisy output signal by means of the proposed combination of a Luenberger observer and the algebraic filter approach. Proceeding with the methodology, an approximation to the signal $x(t)$ can be modeled as

$$\frac{d^7 x}{dt^7} = 0 \tag{7.65}$$

7.4.1.1   Observer Design

An associated state-space representation of (7.65) can readily be conformed as

$$\dot{x}_1 = x_2$$
$$\dot{x}_2 = x_3$$
$$\dot{x}_3 = x_4$$
$$\dot{x}_4 = x_5$$
$$\dot{x}_5 = x_6$$
$$\dot{x}_6 = x_7$$
$$\dot{x}_7 = 0$$
$$y = x_1 + \mu$$

A Luenberger observer for the polynomial model of $x(t)$ is then synthesized as follows:

$$\dot{\hat{x}}_1 = \hat{x}_2 + \lambda_6 \varepsilon(t)$$
$$\dot{\hat{x}}_2 = \hat{x}_3 + \lambda_5 \varepsilon(t)$$

$$\hat{\dot{x}}_3 = \hat{x}_4 + \lambda_4 \varepsilon(t)$$

$$\hat{\dot{x}}_4 = \hat{x}_5 + \lambda_3 \varepsilon(t)$$

$$\hat{\dot{x}}_5 = \hat{x}_6 + \lambda_2 \varepsilon(t)$$

$$\hat{\dot{x}}_6 = \hat{x}_7 + \lambda_1 \varepsilon(t)$$

$$\hat{\dot{x}}_7 = \lambda_0 \varepsilon(t)$$

$$\varepsilon(t) = y - x_1$$

where the corresponding output estimation error dynamics is clearly given by the linear dynamics

$$\varepsilon^{(7)} + \lambda_6 \varepsilon^{(6)} + \lambda_5 \varepsilon^{(5)} + \cdots + \lambda_1 \dot{\varepsilon} + \lambda_0 = 0$$

whose corresponding characteristic polynomial is

$$p(s) = s^{(7)} + \lambda_6 s^6 + \lambda_5 s^5 + \lambda_4 s^4 + \lambda_3 s^3 + \lambda_2 s^2 + \lambda_1 s + \lambda_0$$

The gain parameters of the observer are chosen such that the characteristic polynomial of the output observation error matches the known polynomial $(s^2 + 2\zeta \omega_n s + \omega_n^2)^3 (s + p)$, with ζ, ω_n, p being positive real values.

7.4.1.2   Filter Design

In this subsection, two different algebraic filters are presented. Each algebraic filter arises from a different time derivative estimate expression found from the observer. Namely, a first-order time derivative estimate-based filter and a fourth-order time derivative estimate-based filter will be examined.

First-order time derivative-based filter. From the estimation of the first-order time derivative $\hat{x}_2$, we have

$$y_{Filt_1} = \frac{\int t \hat{x}_2 + \int y}{t} \tag{7.66}$$

Fourth-order time derivative-based filter. From the estimation of the fourth-order time derivative $\hat{x}_5$, we have

$$y_{Filt_4} = \frac{n_4(t)}{d_4(t)} \tag{7.67}$$

with $n_4(t) = \int^{(4)} t^5 \hat{x}_5 - 24 \int^{(4)} y + 96 \int^{(3)} ty - 72 \int^{(2)} t^2 y + 16 \int t^3 y$ and $d_4(t) = t^4$, respectively.

7.4.2   Numerical Results

Numerical simulations are performed in order to show the performance of the proposed approach. A comparison is also carried out with the performances of two classic filtering methods, one of them consisting of an eighth-order Butterworth low-pass filter and the second consisting of an eighth-order Chebyshev low-pass filter; both filtering schemes are provided with a cutoff frequency of 700 s^{-1}. The signal to be taken has the following form:

$$y(t) = \frac{10}{1 + 0.5 \cos(5t^2)} + \mu(t) \qquad (7.68)$$

with $\mu(t)$ being a zero-mean noise.

The observer gain parameters were selected as a function of the desired characteristic polynomial, with the desired parameters $\omega_n = 200$, $p = 200$, and $\zeta = 1$. Two different noise levels were taken for the simulations, one consisting of a 20-dB amplitude and the other 10 dB. Figure 7.21 depicts the analyzed signal. Figures 7.22 and 7.23 show the comparison between

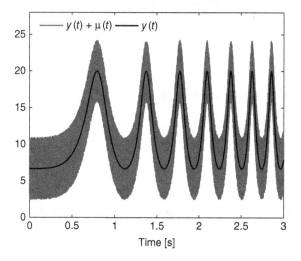

Figure 7.21 Noisy chirp signal

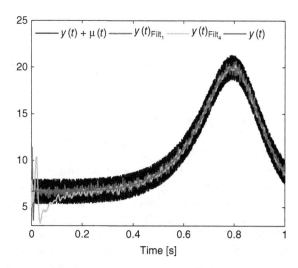

Figure 7.22 Different-order algebraic filtering response: SNR = 20 dB

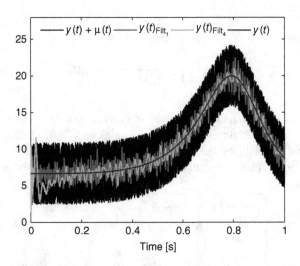

Figure 7.23   Different-order algebraic filtering response: SNR = 10 dB

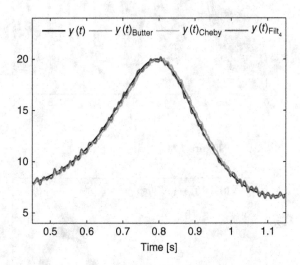

Figure 7.24   Comparison with classical low-pass filters: SNR = 20 dB

the different-order algebraic filters for the 20-dB additive noise and the 10-dB additive noise, respectively. The fourth-order filtering scheme achieves less harmonic components in relation to the first-order filter. The comparison regarding the classic filtering approaches is given in Figures 7.24 and 7.25 for the 20-dB noise and finally, the same comparison for the 10-dB noise is shown in Figures 7.26 and 7.27. To highlight the filtering effects, the images show a zoom of the filtered response. Notice that classic filtering responses exhibit a significant delay effect. The delay effects are absent in the proposed combination of observer plus algebraic filtering approach.

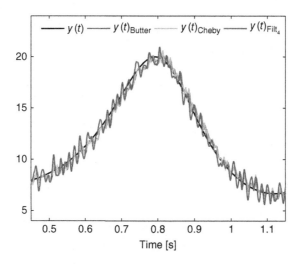

Figure 7.25   Comparison with classical low-pass filters: SNR = 10 dB

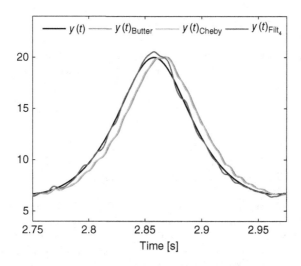

Figure 7.26   Comparison with classical low-pass filters: SNR = 20 dB

7.5   Remarks

In this section, an alternative algebraic filtering technique for noisy signals has been proposed. The filtering scheme is based on a high-gain observer and algebraic manipulations on some of its expressions to obtain estimates of the required signals. The algebraic filtering approach has several advantages over classical filtering methods. First, there is no delay in the filtered signal, which is a major advantage for practical applications. Besides, the algebraic filter does not need any special feature, as for classical filter design, like cutoff frequency or the need for gain tuning. Thus, once the filter has been built, it can work over a wide range of signal

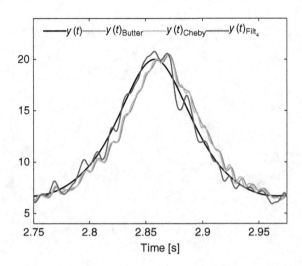

Figure 7.27   Comparison with classical low-pass filters: SNR = 10 dB

frequencies without any redesign. The comparison with some classical filters demonstrates the above conclusions. Moreover, the presented approach can be extended to time derivatives filtering, which is still an open problem. In addition, the proposed scheme is easy to implement, and hence is ideal for engineering applications like signal processing and automatic control.

References

Alessandri, A. (2000) Design of sliding-mode observers and filters for nonlinear dynamic systems. *Proceedings of 39th IEEE Conference on Decision and Control*, vol. 3, pp. 2593–2598, Sydney, Australia.

Belkoura, L., Richard, J. and Fliess, M. (2009) Parameters estimation of systems with delayed and structured entries. *Automatica* **45**, 1117–1125.

Bellman, R., Kalaba, R. and Kotkin, B. (1964) Differential approximation applied to the solution of convolution equations. *Mathematics of Computation* **18**, 487–491.

Braci, M. and Diop, S. (2003) On numerical differentiation algorithms for nonlinear estimation. *Proceedings of 42nd IEEE Conference on Decision and Control*, pp. 1565–1569, Hawaii.

Busawon, K. and Kabore, P. (2001) Disturbance attenuation using proportional integral observers. *International Journal of Control* **74**(6), 618–627.

Chitour, Y. (2002) Time-varying high-gain observers for numerical differentiation. *IEEE Transactions on Automatic Control* **47**(9), 1565–1569.

Cortes-Romero, J., Luviano-Juarez, A. and Sira-Ramirez, H. (2009) Robust GPI controller for trajectory tracking for induction motors. *IEEE International Conference on Mechatronics*, pp. 1–6, Malaga, Spain.

Eykhoff, P. (1974) *System Identification. Parameter and State Estimation*. Wiley-Interscience, New York, chapter 8.

Fliess, M. and Sira-Ramírez, H. (2003) An algebraic framework for linear identification. *ESAIM, Control, Optimization and Calculus of Variations* **9**(1), 151–168.

Fliess, M., Join, C. and Perruquetti, W. (2008a) Real-time estimation for switched linear systems. *Proceedings of 47th IEEE Conference on Decision and Control, 2008*, pp. 941–946, Cancun, Mexico.

Fliess, M., Join, C. and Sira-Ramirez, H. (2008b) Non-linear estimation is easy. *International Journal of Modelling, Identification and Control* **4**(1), 12–27.

Gensior, A., Weber, J., Rudolph, J. and Guldner, H. (2008) Algebraic parameter identification and asymptotic estimation of the load of a boost converter. *IEEE Transactions on Industrial Electronics* **55**(9), 3352–3360.

Lee, K. and Khalil, H. (1997) Adaptive output feedback control of robot manipulators using high-gain observer. *International Journal of Control* **67**(6), 869–886.

Lewis, F. (1986) *Optimal Estimation With An Introduction To Stochastic Control Theory*. John Wiley & Sons, Inc., New York.

Martínez-Guerra, R. and Diop, S. (2004) Diagnosis of nonlinear systems: An algebraic and differential approach. *IEE Control Theory and Applications* **151**, 130–135.

Mboup, M., Join, C. and Fliess, M. (2007) A revised look at numerical differentiation with an application to nonlinear feedback control. *Mediterranean Conference on Control and Automation*, pp. 6–26.

Mboup, M., Join, C. and Fliess, M. (2009) Numerical differentiation with annihilators in noisy environment. *Numerical Algorithms* **50**, 439–467. DOI 10.1007/s11075-008-9236-1.

Rudolph, J. (2006) Identifying parameters in infinite-dimensional systems. *École d'Été d'Automatique, Session 27: Méthodes d'Estimation et d'Identification Rapides en Automatique et Signal*, Grenoble.

Shinbrot, M. (1957) On the analysis of linear and nonlinear systems. *Transactions of the ASME* **79**, 547–552.

Solsona, JA. and Valla, MI. (2003) Disturbance and nonlinear Luenberger observers for estimating mechanical variables in permanent magnet synchronous motors under mechanical parameters uncertainties. *IEEE Transactions on Industrial Electronics* **50**(4), 717–725.

Tan, K. and Zhao, S. (2004) Precision motion control with a high gain disturbance compensator for linear motors. *ISA Transactions* **43**, 399–412.

Tian, Y., Floquet, T. and Perruquetti, W. (2008) Fast state estimation in linear time-varying systems: An algebraic approach. *Proceedings of the 47th IEEE Conference on Decision and Control*, pp. 2539–2544, Cancun, Mexico.

Appendix A

Parameter Identification in Linear Continuous Systems: A Module Approach

A.1 Generalities on Linear Systems Identification

In this section, in a rather tutorial fashion, we provide the interested reader with the rigorous mathematical developments on which the online algebraic parameter estimation technique is based. The fundamental developments are based on the module-theoretic approach to linear systems. This section is taken, with some modification, from the work of Fliess and Sira-Ramírez (2003, 2008).

We deal with linear time-invariant systems in the context of finitely generated free $k[d/dt]$ modules over the ring of linear differential operators:

$$\sum_{\text{finite}} \alpha_i \frac{d^i}{dt^i}, \qquad \alpha_i \in k$$

We distinguish between the *nominal* system Λ^{nom} and the *perturbed* system Λ. We let $\pi = (\pi_1, \ldots, \pi_q)$ be the set of *perturbations*.

The short exact sequence

$$0 \to \text{span}_{k[d/dt]}(\pi) \to \Lambda \to \Lambda^{\text{nom}} \to 0$$

defines the nominal system

$$\Lambda^{\text{nom}} = \Lambda / \text{span}_{k[d/dt]}(\pi)$$

A *linear dynamics* is a linear system Λ equipped with a *control*, that is, a finite set $u = (u_1, \ldots, u_m)$.

1. Control and perturbation inputs <u>do not</u> interact.
2. A nominal system with no control inputs is *torsion*.

Algebraic Identification and Estimation Methods in Feedback Control Systems, First Edition.
Hebertt Sira-Ramírez, Carlos García-Rodríguez, Alberto Luviano-Juárez, John Alexander Cortés-Romero.
© 2014 John Wiley & Sons, Ltd. Published 2014 by John Wiley & Sons, Ltd.

A.1.1 Example

A DC motor subject to load perturbation inputs may be modeled by means of the perturbed linear system

$$\frac{d^2\omega}{dt^2} + \left(\frac{B}{J} + \frac{R}{L}\right)\frac{d\omega}{dt} + \left(\frac{k_e k_m + RB}{LJ}\right)\omega = \left(\frac{k_m E}{LJ}\right)u - \left(\frac{k_m R}{LJ}\tau + \frac{1}{J}\dot\tau\right)$$

Load perturbations are *independent* of the input voltage. If no inputs and no perturbations exist, the system evolves according to

$$\frac{d^2\omega}{dt^2} + \left(\frac{B}{J} + \frac{R}{L}\right)\frac{d\omega}{dt} + \left(\frac{k_e k_m + RB}{LJ}\right)\omega = 0$$

A.2 Some Definitions and Results

We let k denote either the field of real numbers, $\mathbb{R}$, or the field of complex numbers, $\mathbb{C}$. K is a finite algebraic extension of $k(a)$ generated by a finite set $a = (a_1, \dots, a_r)$ of *unknown parameters*.

A linear parameter uncertain system is a finitely generated $K[d/dt]$ module Λ over the ring $K[d/dt]$ of linear differential operators,

$$\sum_{\text{finite}} \alpha_i \frac{d^i}{dt^i}, \qquad \alpha_i \in K$$

A.2.1 Example

In the DC motor case, the unperturbed system is just

$$\frac{d^2\omega}{dt^2} + \left(\frac{B}{J} + \frac{R}{L}\right)\frac{d\omega}{dt} + \left(\frac{k_e k_m + RB}{LJ}\right)\omega = \left(\frac{k_m E}{LJ}\right)u$$

Let $a = (B, J, L, R, k_e, k_m, E)$ and consider $K = \mathbb{R}(a)$, then

$$\frac{d^2}{dt^2} + \left(\frac{B}{J} + \frac{R}{L}\right)\frac{d}{dt} + \left(\frac{k_e k_m + RB}{LJ}\right) \in K[d/dt]$$

$$\left(\frac{k_m E}{LJ}\right) \in K[d/dt]$$

$\Lambda^{\text{nom}} = \text{span}_{K[d/dt]}(\omega, u)$, $\Lambda = \text{span}_{K[d/dt]}(\omega, u, \tau)$ where τ is the load perturbation input.

A.2.2   Example

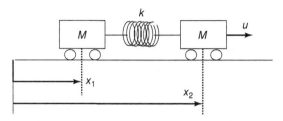

Consider the two Identification and Estimation in Feedback Control Systems mass system joined by a spring:

$$M\ddot{x}_1 = -k(x_1 - x_2)$$

$$M\ddot{x}_2 = k(x_1 - x_2) + u$$

$$y = x_1$$

The input–output relation

$$\left(\frac{M}{k}\right)\frac{d^4 y}{dt^4} + 2\frac{d^2 y}{dt^2} = \left(\frac{1}{M}\right)u$$

$$\left(\frac{M}{k}\right)\frac{d^4}{dt^4} + 2\frac{d^2}{dt^2} \in \mathbb{R}(M,k)[d/dt],$$

$$\left(\frac{1}{M}\right) \in \mathbb{R}(M,k)[d/dt]$$

A.3   Linear Identifiability

A.3.1   Definition

The set of parameters $a = (a_1, a_2, \ldots, a_r)$ is said to be *linearly identifiable* if, and only if,

$$P\begin{pmatrix} a_1 \\ \vdots \\ a_r \end{pmatrix} = Q$$

where

- P and Q are, respectively, $r \times r$ and $r \times 1$ matrices.
- The entries of P and Q belong to $\text{span}_{k[d/dt]}(u, y)$.
- $\det P \neq 0$.

A.3.2 Example (continued)

$$\frac{d^2\omega}{dt^2} + \left(\frac{B}{J} + \frac{R}{L}\right)\frac{d\omega}{dt} + \left(\frac{k_e k_m + RB}{LJ}\right)\omega = \left(\frac{k_m E}{LJ}\right)u$$

The parameters

$$\gamma_1 = \left(\frac{B}{J} + \frac{R}{L}\right), \qquad \gamma_0 = \left(\frac{k_e k_m + RB}{LJ}\right), \qquad \gamma = \left(\frac{k_m E}{LJ}\right)$$

are all linearly identifiable:

$$\begin{bmatrix} \dfrac{d\omega}{dt} & \omega & -u \\[2ex] \dfrac{d^2\omega}{dt^2} & \dfrac{d\omega}{dt} & -\dfrac{du}{dt} \\[2ex] \dfrac{d^3\omega}{dt^3} & \dfrac{d^2\omega}{dt^2} & -\dfrac{d^2u}{dt^2} \end{bmatrix} \begin{bmatrix} \gamma_1 \\[2ex] \gamma_0 \\[2ex] \gamma \end{bmatrix} = \begin{bmatrix} -\dfrac{d^2\omega}{dt} \\[2ex] -\dfrac{d^3\omega}{dt^3} \\[2ex] -\dfrac{d^4\omega}{dt^4} \end{bmatrix}$$

The set of parameters a is *weakly linearly identifiable* if and only if there exists a finite set $a = (a'_1, \ldots, a'_r)$ of related unknown parameters s.t.

- the components of a' (respectively a) are algebraic over $k(a)$ (respectively $k(a')$);
- a' is linearly identifiable.

A.3.3 Example

In the two-mass–spring system, M/k and $1/M$ are linearly identifiable:

$$\begin{bmatrix} y^{(4)} & -u \\ y^{(5)} & -\dot{u} \end{bmatrix} \begin{bmatrix} M/k \\ 1/M \end{bmatrix} = \begin{bmatrix} -2\ddot{y} \\ -2y^{(3)} \end{bmatrix}$$

M and k are, however, weakly linearly identifiable since $M = (1/M)^{-1}$ and $k = (M/k)^{-1}$ $(1/M)^{-1}$.

We need the symmetric K-algebra of the nominal system $\mathrm{sym}(\Lambda^{\mathrm{nom}})$, viewed as a K-vector space. This vector space may be endowed with the canonical structure of a *differential ring* by extending the time derivation as usual:

$$\xi, \eta \in \Lambda^{\mathrm{nom}}, \qquad \frac{d}{dt}(\xi\eta) = \frac{d\xi}{dt}\eta + \xi\frac{d\eta}{dt}$$

The quotient field $Q(\mathrm{sym}(\Lambda^{\mathrm{nom}}))$ is a *differential field*. The differential field generated by u^{nom} and y^{nom} is contained in $Q(\mathrm{sym}(\Lambda^{\mathrm{nom}}))$.

A.3.4   Example (continued)

For the two-mass−spring system:

$$\left(\frac{M}{k}\right) = \frac{2\ddot{u}\ddot{y} - 2uy^{(3)}}{uy^{(5)} - \dot{u}y^{(4)}}$$

$$\left(\frac{1}{M}\right) = \frac{2y^{(5)}\ddot{y} - 2y^{(3)}y^{(4)}}{uy^{(5)} - \dot{u}y^{(4)}}$$

More generally, the set of unknown parameters a is said to be *algebraically identifiable* (respectively rationally identifiable) if and only if any component of a is algebraic over the differential over-field Œ (respectively belongs to Œ):

$$\text{Œ} \subset Q(\text{sym}(\Lambda^{\text{nom}}))$$

A.4   Structured Perturbations

Consider $\xi = K\mathbf{1}(t)$, K constant. Then

$$t\frac{d\xi}{dt} = 0 \quad \forall\, t$$

Similarly, let $\xi = Kt\mathbf{1}(t)$, with K constant. It follows that

$$t\frac{d^2\xi}{dt^2} = 0 \quad \forall\, t$$

Similarly, if $\xi = K(t - L)^{n-1}\mathbf{1}(t)$, one has

$$(t - L)\frac{d^n\xi}{dt^n} = 0 \quad \forall\, t$$

Other frequent perturbations are annihilated by suitable operations involving t and d/dt:

$$\xi = A\sin(\omega t + \phi)\mathbf{1}(t)$$

$$\left[2 - 4\frac{d}{dt}t + \left(\frac{d^2}{dt^2} + \omega^2\right)t^2\right]\xi = 0$$

We consider then the (non-commutative) ring of differential operators $k\left[t, \dfrac{d}{dt}\right]$.
   A subset S of indeterminates represents a *structured perturbation* if the module

$$\text{span}_{k\left[t, \frac{d}{dt}\right]}(S)$$

is a torsion module.
   In essence, then, a structured perturbation is one that satisfies a linear, possibly time-varying, homogeneous differential equation.

A.4.1   Example

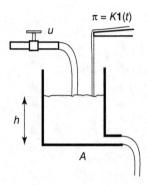

$$\pi = K1(t)$$

The linearized model of the system around an equilibrium point yields the following parameter-uncertain system:

$$A\frac{dh_\delta}{dt} + Bh_\delta = u_\delta + K1(t)$$

where A and B are the unknown parameters and $K1(t)$ represents a structured perturbation.
   Differentiating once the previous expression:

$$A\frac{dh_\delta}{dt} + Bh_\delta = u_\delta + K1(t)$$

and multiplying by t^2 we obtain an equation free from perturbations:

$$At^2\frac{d^2h_\delta}{dt^2} + Bt^2\frac{dh_\delta}{dt} = t^2\dot{u}_\delta$$

Differentiating once more, we find

$$\begin{bmatrix} t^2\frac{d^2h_\delta}{dt^2} & t^2\frac{dh_\delta}{dt} \\ t^2\frac{d^3h_\delta}{dt^3} + 2t\frac{d^2h_\delta}{dt^2} & t^2\frac{d^2h_\delta}{dt^2} + 2t\frac{dh_\delta}{dt} \end{bmatrix}\begin{bmatrix} A \\ B \end{bmatrix} = \begin{bmatrix} t^2\dot{u}_\delta \\ t^2\frac{d^2u_\delta}{dt^2} + 2t\frac{du_\delta}{dt} \end{bmatrix}$$

We can annihilate structured perturbations by appropriate operations involving t, $\frac{d}{dt}$ while preserving linear identifiability.
   It is natural then to consider the *Weyl algebra* of the ring of linear differential operators $k\left[t, \frac{d}{dt}\right]$ with polynomial coefficients:

$$\sum_{\text{finite}} \alpha_i\frac{d^i}{dt^i}, \qquad \alpha_i \in k[t]$$

This is a *non-commutative algebra*, since

$$\frac{d}{dt}t = 1 + t\frac{d}{dt}$$

In other words: the commutator of t and $\frac{d}{dt}$ is,

$$\left[t, \frac{d}{dt}\right] = t\frac{d}{dt} - \frac{d}{dt}t = -1 \neq 0$$

The problem of parameter identification in structurally perturbed linear systems is translated into the context of systems defined now with a $K\left[t, \frac{d}{dt}\right]$ left module, where $K = k(a)$ is a finite extension of the base field k ($\mathbb{R}$ or $\mathbb{C}$).

A linearly identifiable system with structured perturbations satisfies

$$P\begin{pmatrix} a_1 \\ a_2 \\ \vdots \\ a_r \end{pmatrix} = Q + Q'$$

with

$$Q' \in \mathrm{span}_{k\left[t, \frac{d}{dt}\right]} \pi$$

That is, there exists $\Delta \in k\left[t, \frac{d}{dt}\right]$ s.t. $\Delta Q' = 0$:

$$\Delta P\begin{pmatrix} a_1 \\ a_2 \\ \vdots \\ a_r \end{pmatrix} = \Delta Q \qquad \text{(linear identifier)}$$

The influence of a structured perturbation on the linear identifiability can always be annihilated through the action of a differential operator with polynomial coefficients.

A.4.2 Example

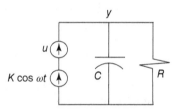

Consider an RC circuit feed with a current source u and an additive sinusoidal perturbation current source, described by

$$C\frac{dy}{dt} + \frac{1}{R}y = u + K\cos\omega t\,\mathbf{1}(t)$$

The sinusoidal perturbation is annihilated by differentiating twice, adding the original equation times ω^2, and getting rid of the impulses by multiplying by, say, t^3. We obtain

$$Ct^3\left[\frac{d^3y}{dt^3} + \omega^2\frac{dy}{dt}\right] + \frac{1}{R}t^3\left(\frac{d^2y}{dt^2} + \omega^2 y\right) = t^3\left(\frac{d^2u}{dt^2} + \omega^2 u\right)$$

$$\left[\begin{array}{cc} t^3\left(\frac{d^3y}{dt^3} + \omega^2\frac{dy}{dt}\right) & t^3\left(\frac{d^2y}{dt^2} + \omega^2 y\right) \\ \frac{d}{dt}\left[t^3\left(\frac{d^3y}{dt^3} + \omega^2\frac{dy}{dt}\right)\right] & \frac{d}{dt}\left[t^3\left(\frac{d^2y}{dt^2} + \omega^2 y\right)\right] \end{array}\right]\left[\begin{array}{c} C \\ \frac{1}{R} \end{array}\right]$$

$$= \left[\begin{array}{c} t^3\left(\frac{d^2u}{dt^2} + \omega^2 u\right) \\ \frac{d}{dt}\left[t^3\left(\frac{d^2u}{dt^2} + \omega^2 u\right)\right] \end{array}\right]$$

To the submodule $\Upsilon \subset \Lambda$ spanned by u and y there corresponds a *linear differential algebraic variety* $\mathcal{T} \subset \mathcal{U}^{m+p}$ where $\mathcal{U}$ is the differential over-field of k.

Any element τ of $\mathcal{T}$ will be called *a trajectory*. A trajectory τ is *persistent* for the linear identifier if and only if

$$\det \Delta P \neq 0$$

The following is the fundamental result of the approach.

The set of persistent trajectories of a linear identifier is open with respect to the differential Zariski topology.

For the linear identifier

$$\Delta P \begin{pmatrix} a_1 \\ a_2 \\ \vdots \\ a_r \end{pmatrix} = \Delta Q$$

a trajectory $(y(t), u(t))$ is said to be *non-persistent* if and only if $\det P \equiv 0$. Otherwise, it is called *persistent*.

A.4.3 Example

$$\dot{y} = ay + bu$$

$$\begin{bmatrix} y & u \\ \dot{y} & \dot{u} \end{bmatrix} \begin{bmatrix} a \\ b \end{bmatrix} = \begin{bmatrix} \dot{y} \\ \ddot{y} \end{bmatrix}$$

The non-persistent trajectories satisfy

$$y\dot{u} - \dot{y}u = 0$$

In order to be able to compute parameters in terms of *integrals* rather than *derivatives*, we must learn to turn products of the form $t^\gamma \frac{d^\alpha}{dt^\alpha}$ into *exact differentiations* of the form $\frac{d^\beta}{dt^\beta} t^\delta$.

The identity

$$t\frac{d}{dt} = \frac{d}{dt}t - 1$$

allows one to write

$$\int t\frac{d}{dt} = t - \int$$

independently of initial conditions. Thus,

$$\int \int t^2 \frac{d^2}{dt^2} = t^2 - 4 \int t + 2 \int \int$$

We can, therefore, identify parameters:

1. independently of structured perturbations,
2. independently of initial conditions, and
3. based only on inputs and outputs.

A.4.4   Example

Consider

$$\ddot{y} = ay + bu$$

with a and b unknown, initial conditions also unknown

$$t^2\ddot{y} = at^2y + bt^2u$$

Integrating by parts, we obtain the linear identifier

$$\begin{bmatrix} \int_0^t \int_0^\sigma \theta^2 y(\theta)d\theta d\sigma & \int_0^t \int_0^\sigma \theta^2 u(\theta)d\theta d\sigma \\ \int_0^t \int_0^\sigma \int_0^\theta \lambda^2 y(\lambda)d\lambda d\theta d\sigma & \int_0^t \int_0^\sigma \int_0^\theta \lambda^2 u(\lambda)d\lambda d\theta d\sigma \end{bmatrix} \begin{bmatrix} a \\ b \end{bmatrix} =$$

$$\begin{bmatrix} t^2 y - 4\int_0^t \sigma y(\sigma)d\sigma + 2\int_0^t \int_0^\sigma y(\theta)d\theta d\sigma \\ \int_0^t \sigma^2 y(\sigma)d\sigma - 4\int_0^t \int_0^\sigma \theta y(\theta)d\theta d\sigma + 2\int_0^t \int_0^\sigma \int_0^\theta y(\lambda)d\lambda d\theta d\sigma \end{bmatrix}$$

$\det P(0) = 0$ but $\det P(\epsilon) \neq 0$ for $\epsilon > 0$.

A.5   The Frequency Domain Alternative

A parallel theory may also be proposed with an operational calculus setting. One now deals with $k(a)[s, \frac{d}{ds}]$ rings of differential polynomials. The setting is better understood through an example.

Consider the perturbed version of the previous example with a and b unknown constants:

$$\ddot{y} = a\dot{y} + bu + K\mathbf{1}(t)$$

In Laplace transform notation, we have

$$s^2 Y(s) - sy_0 - \dot{y}_0 = a(sY(s) - y_0) + bU(s) + \frac{K}{s}$$

Multiplying out by s (equivalent to time differentiation):

$$s^3 Y(s) - s^2 y_0 - s\dot{y}_0 = a(s^2 Y(s) - sy_0) + bsU(s) + K$$

Differentiating three times with respect to s (equivalent to multiplication by $-t^3$):

$$s^3\frac{d^3 Y(s)}{ds^3} + 9s^2\frac{d^2 Y(s)}{ds^2} + 18s\frac{dY(s)}{ds} + 6Y(s)$$

$$= a\left(s^2\frac{d^3 Y(s)}{ds^3} + 6s\frac{d^2 Y(s)}{ds^2}\right) + b\left(s\frac{d^3 U(s)}{ds^3} + 3\frac{d^2 U(s)}{ds^2}\right)$$

Multiplying out by s^{-3} (equivalent to iterated integration three times):

$$\frac{d^3 Y(s)}{ds^3} + 9s^{-1}\frac{d^2 Y(s)}{ds^2} + 18s^{-2}\frac{dY(s)}{ds} + 6s^{-3}Y(s)$$

$$= a\left(s^{-1}\frac{d^3 Y(s)}{ds^3} + 6s^{-2}\frac{d^2 Y(s)}{ds^2}\right) + b\left(s^{-2}\frac{d^3 U(s)}{ds^3} + 3s^{-3}\frac{d^2 U(s)}{ds^2}\right)$$

Reverting to the time domain:

$$-t^3 y + 9 \left(\int t^2 y \right) - 18 \left(\int^{(2)} ty \right) + 6 \left(\int^{(3)} y \right)$$

$$= a \left(- \left(\int t^3 y \right) + 6 \left(\int^{(2)} t^2 y \right) - 6 \left(\int^{(3)} ty \right) \right) + b \left(- \left(\int^{(2)} t^3 u \right) + 3 \left(\int^{(3)} t^2 u \right) \right)$$

Integrating once more, we obtain the same linear identifier as before:

$$\begin{bmatrix} p_{11}(t) & p_{12}(t) \\ (\int p_{11}) & (\int p_{12}) \end{bmatrix} \begin{bmatrix} a \\ b \end{bmatrix} = \begin{bmatrix} q_1(t) \\ (\int q_1) \end{bmatrix}$$

$$p_{11}(t) = \left(\int t^3 y \right) + 6 \left(\int^{(2)} t^2 y \right) - 6 \left(\int^{(3)} ty \right)$$

$$p_{12}(t) = - \left(\int^{(2)} t^3 u \right) + 3 \left(\int^{(3)} t^2 u \right)$$

$$q_1(t) = -t^3 y + 9 \left(\int t^2 y \right) - 18 \left(\int^{(2)} ty \right) + 6 \left(\int^{(3)} y \right)$$

References

Fliess, M. and Sira-Ramírez, H. (2003) An algebraic framework for linear identification. *ESAIM, Control, Optimization and Calculus of Variations* **9**(1), 151–168.

Fliess, M. and Sira-Ramirez, H. (2008) Closed loop parameter identification for continuous-time linear systems via new algebraic techniques in *Identification of Continuous-time Models from Sampled Data*, H. Granier and L. Wang Editors vol. XXVI of Advances in Industrial Control. Springer-Verlag, Berlin, pp. 363–391.

Appendix B

Parameter Identification in Linear Discrete Systems: A Module Approach

In this section, as in the continuous-time instance, we provide the interested reader with rigorous mathematical developments in the online discrete-time algebraic parameter estimation approach. The fundamental developments are based on the module theory over principal ideals. This section is taken from the work of Fliess *et al.* (2008).

B.1 A Short Review of Module Theory over Principal Ideal Rings

B.1.1 Principal Ideal Rings

Let R be a commutative principal ideal ring (see, e.g., Jacobson, 1974). Well-known examples of such rings are:

- the ring $K[\delta]$ of polynomials $\sum_{\text{finite}} c_\alpha \delta^\alpha$, $\alpha \in \mathbf{N}$, in the indeterminate δ, with coefficients in a field K, i.e., $c_\alpha \in K$;
- the ring $K[\delta, \delta^{-1}]$ of Laurent polynomials $\sum_{\text{finite}} c_\alpha \delta^\alpha$, $c_\alpha \in K$, $\alpha \in \mathbf{Z}$, in the indeterminate δ, over the field K, i.e., $c_\alpha \in K$.

B.1.2 Modules

Let M be a finitely generated module (see, e.g., Jacobson, 1974) over R. An element $m \in M$ is said to be *torsion* if, and only if, there exists $r \in R$, $r \neq 0$ such that $rm = 0$. The subset M^{tor} of torsion elements is a submodule of M, called its *torsion submodule*. The module M is said to be *torsion* if, and only if, $M^{\text{tor}} = M$. It is said to be *torsion free* if, and only if, $M^{tor} = \{0\}$.

Algebraic Identification and Estimation Methods in Feedback Control Systems, First Edition.
Hebertt Sira-Ramírez, Carlos García-Rodríguez, Alberto Luviano-Juárez, John Alexander Cortés-Romero.
© 2014 John Wiley & Sons, Ltd. Published 2014 by John Wiley & Sons, Ltd.

A module M is torsion free if, and only if, it is *free*, that is, there exists a *basis* [i.e., a μ-tuple $\{b_1, \ldots, b_\mu\}$ of elements in M] such that:

1. $M = \text{span}_R(b_1, \ldots, b_\mu)$;
2. $b_1, \ldots, b_\mu$ are R-linearly independent.

Any finitely generated R-module M may be written as a direct sum $M = M^{\text{tor}} \oplus F$, where $F = M/M^{\text{tor}}$ is a free module which is uniquely defined up to isomorphism.

B.1.3   A Connection with Vector Spaces

Let $Q(R)$ be the quotient field of R. We want to enrich the module structure of M in order to get a vector space structure over $Q(R)$. This procedure, called *localization* (Jacobson, 1974), which consists of formally multiplying elements of M by elements of $Q(R)$, is written in tensor notation as

$$\hat{M} = Q(R) \otimes_R M$$

where $\hat{M}$ is a $Q(R)$-vector space. Note that the kernel of the R-linear morphism $M \to \hat{M}$ $m \mapsto \hat{m} = 1 \otimes m$ is the torsion submodule M^{tor}. This morphism is thus injective if, and only if, M is free. The *rank* of the module M, written $\text{rk}(M)$, which is a new kind of dimension, is by definition the dimension of the vector space $\hat{M}$, that is, $\text{rk}(M) = \dim_{Q(R)}(\hat{M})$. The module M is thus torsion if, and only if, its rank is 0.

B.2   Systems

From now on, δ denotes the backward time-shift operator, that is, $\delta\phi(t) = \phi(t-1)$. A linear discrete-time system is a finitely generated $K[\delta, \delta^{-1}]$-module Λ.

Example B.2.1 *Compute system Λ corresponding to the $K[\delta, \delta^{-1}]$-linear equations*

$$\sum_{\kappa=1}^{\mu} a_{\iota\kappa}\xi_\kappa = 0, \quad a_{\iota\kappa} \in K[\delta, \delta^{-1}], \quad \iota = 1, \ldots, \nu \tag{B.1}$$

in the unknowns $\xi_1, \ldots, \xi_\mu$. Let E and F be the free modules spanned by $e_1, \ldots, e_\nu$ and $f_1, \ldots, f_\mu$. Define the morphism $\varrho : E \to F$ by $e_\iota \mapsto \sum_{\kappa=1}^{\mu} a_{\iota\kappa}f_\kappa$. Then $\Lambda = F/\varrho(E)$. Write again (B.1) in the form

$$P_\Lambda \begin{pmatrix} \xi_1 \\ \vdots \\ \xi_\mu \end{pmatrix} = 0$$

$P_\Lambda \in K[\delta, \delta^{-1}]^{\nu \times \mu}$ *is called a* presentation matrix *of Λ.*

B.3   Perturbations

Distinguish in Λ a finite set $\pi = (\pi_1, \ldots, \pi_r)$ of *perturbation variables*. The *nominal* system Λ^{nom} is defined by the quotient module $\Lambda/\text{span}_{K[\delta,\delta^{-1}]}(\pi)$. The canonical image of any $\lambda \in \Lambda$ is written $\lambda^{\text{nom}} \in \Lambda^{\text{nom}}$. If $\pi = \emptyset$, that is, if $\Lambda = \Lambda^{\text{nom}}$, system Λ is said to be *unperturbed*. If $\pi \neq \emptyset$, Λ is called a *perturbed* system.

B.4 Dynamics and Input–Output Systems

A *linear discrete-time dynamics* is a system equipped with a finite set $u = (u_1, \ldots, u_m)$ of *control variables*, which satisfy the following properties.

- The control and perturbation variables do not *interact*, that is,

$$\text{span}_{K[\delta,\delta^{-1}]}(u) \cap \text{span}_{K[\delta,\delta^{-1}]}(\pi) = \{0\} \tag{B.2}$$

- $\Lambda^{\text{nom}}/\text{span}_{K[\delta,\delta^{-1}]}(u^{\text{nom}})$ is torsion.

If $u = \emptyset$, the last condition implies that Λ^{nom} is torsion. The control variables are said to be *independent* if, and only if, $u_1, \ldots, u_m$ are $K[\delta, \delta^{-1}]$-linearly independent. Note that equation (B.2) yields the following property: The restriction of the epimorphism $\Lambda \rightarrow \Lambda^{\text{nom}}$ to $\text{span}_{K[\delta,\delta^{-1}]}(u)$ defines an isomorphism $\text{span}_{K[\delta,\delta^{-1}]}(u) \rightarrow \text{span}_{K[\delta,\delta^{-1}]}(u^{\text{nom}})$. This implies that u^{nom} is independent if, and only if, u is independent.

Pick up a finite subset $y = (y_1, \ldots, y_p)$ of *output variables* in the dynamics Λ. Then Λ is called a *linear discrete-time input–output system*. The system is said to be *mono-variable* if, and only if, $m = p = 1$. If not, it is said to be *multi-variable*.

Example B.4.1 *Consider the unperturbed mono-variable input–output system* $\alpha y = \beta u$, $\alpha, \beta \in K[\delta, \delta^{-1}]$. *The quotient module* $\Lambda/\text{span}_{K[\delta,\delta^{-1}]}(u)$, *which corresponds to* $\alpha\zeta = 0$, *is clearly torsion.*

Remark B.4.2 *Our definition comprises non-causal systems such as* $y_{k+1} = y_k + u_{k+2}$, *that is,* $\delta^{-1}y = y + \delta^{-2}u$. *See Fliess (1992b) for a complete module-theoretic treatment of this question, which does not play any rôle in the mathematical definition of identifiability.*

B.5 Transfer Matrices

Write as $K(z)$ the quotient field of $K[\delta, \delta^{-1}]$, where $z = \delta^{-1}$. The $K(z)$-vector space (see Section B.1)

$$\hat{\Lambda} = K(z) \otimes_{K[\delta,\delta^{-1}]} \Lambda$$

is called the *transfer vector space*[1] of system Λ. The $K[\delta, \delta^{-1}]$-linear morphism $\Lambda \rightarrow \hat{\Lambda}$, $\lambda \mapsto \hat{\lambda} = 1 \otimes \lambda$, is called the *(formal) z-transform*, or the *(formal) discrete Laplace transform* (compare with Jury, 1973). Assume that Λ is a dynamics. From the fact that $\Lambda^{\text{nom}}/\text{span}_{K[\delta,\delta^{-1}]}(u^{\text{nom}})$ is torsion, it follows that $\hat{\Lambda}^{\text{nom}} = \widehat{\text{span}_{K[\delta,\delta^{-1}]}(u^{\text{nom}})}$, that is, $\hat{u}^{\text{nom}} = (\hat{u}_1^{\text{nom}}, \ldots, \hat{u}_m^{\text{nom}})$ is spanning $\hat{\Lambda}^{\text{nom}}$. If, moreover, u is independent, $\hat{u}^{\text{nom}}$ is a basis of $\hat{\Lambda}^{\text{nom}}$. Thus

$$\begin{pmatrix} \hat{y}_1^{\text{nom}} \\ \vdots \\ \hat{y}_p^{\text{nom}} \end{pmatrix} = T \begin{pmatrix} \hat{u}_1^{\text{nom}} \\ \vdots \\ \hat{u}_m^{\text{nom}} \end{pmatrix} \tag{B.3}$$

---

[1] This section is a discrete-time adaptation of Fliess (1992b) (see also Fliess, 2002).

where $T \in K(z)^{p \times m}$ is the *(nominal) transfer matrix*, which is uniquely defined if u is independent. For a perturbed system, formula (B.3) becomes

$$\begin{pmatrix} \hat{y}_1 \\ \vdots \\ \hat{y}_p \end{pmatrix} = T \begin{pmatrix} \hat{u}_1 \\ \vdots \\ \hat{u}_m \end{pmatrix} + P \begin{pmatrix} \hat{\pi}_1 \\ \vdots \\ \hat{\pi}_r \end{pmatrix}$$

where $P \in K(z)^{p \times r}$.

B.6 Identifiability

Definition B.6.1 (Field extensions) *Let L/K be a* field extension *(Jacobson, 1974), that is, two fields K, L such that $K \subseteq L$. This extension is said to be* algebraic *if, and only if, any element $\xi \in L$ is algebraic over K, that is, satisfies an algebraic equation: there exists a polynomial $P \in K[x]$ such that $P(\xi) = 0$. An algebraic extension L/K is said to be* finite *if, and only if, it is generated by a finite set. An extension L/K, which is not algebraic, is called* transcendental.

Definition B.6.2 (Symmetric algebra) *It is possible to enrich the R-module structure of M by introducing formally commutative products of its elements. It results in an R-algebra, called the* symmetric algebra *(Jacobson, 1974) of M, and written $Sym(M)$. This algebra, which is integral (i.e., has no zero divisor $\neq 0$) possesses a quotient field.*

Definition B.6.3 (Difference algebra) *A* difference ring *R (Cohn, 1965) is a commutative ring equipped with a* transforming operator, *that is, a monomorphism $\delta : R \to R$. Thus, for $a, b \in R$, $\delta(a + b) = \delta a + \delta b$, $\delta(ab) = (\delta a)(\delta b)$. Moreover, $\delta a = 0$ is equivalent to $a = 0$. A* constant *$c \in R$ is an element such that $\delta c = c$. The set of constants is a difference subring of R. A difference ring which is a field is called a* difference field. *If δ is an isomorphism, that is, if δ^{-1} is defined, the difference ring or field is said to be* inversive.

B.7 An Algebraic Setting for Identifiability

B.7.1 Unknown Parameters

From now on the field K is a finite algebraic extension of the field $k(\Theta)$ generated over a given ground field k by a finite set $\Theta = (\theta_1, \dots, \theta_r)$ of *unknown parameters*.

B.7.2 Various Types of Identifiability

Write as **Sym**(Λ^{nom}) the symmetric K-algebra generated by Λ^{nom} and $Q(\mathbf{Sym}(\Lambda^{\mathrm{nom}}))$ its quotient field. Any element of **Sym**(Λ^{nom}) may be written as a finite sum $c \sum \lambda_1^{\mathrm{nom}} \dots \lambda_\iota^{\mathrm{nom}}, c \in K$, $\lambda_1^{\mathrm{nom}}, \dots, \lambda_\iota^{\mathrm{nom}} \in \Lambda^{\mathrm{nom}}, \iota \geq 1$. Set:

- $\delta^\nu(c) = c$, $\nu \in Z$ (this condition implies that the unknown parameters are assumed to be constant);
- $\delta^\nu(\lambda_1^{\mathrm{nom}} \dots \lambda_\iota^{\mathrm{nom}}) = (\delta^\nu \lambda_1^{\mathrm{nom}}) \dots (\delta^\nu \lambda_\iota^{\mathrm{nom}})$.

Assume that, for $\vartheta \in \mathbf{Sym}(\Lambda^{\text{nom}})$, $\delta\vartheta = 0$. Multiplying both sides by δ^{-1} yields $\vartheta = 0$. Thus, δ is a monomorphism of $Sym(\Lambda^{\text{nom}})$. We have proved the following result:

Proposition B.7.1 $Sym(\Lambda^{\text{nom}})$ *(respectively* $Q(\mathrm{Sym}(\Lambda^{\text{nom}}))$*) may be endowed with a structure of inversive difference ring (respectively field), where any element of K is a constant.*

Let $\mathbf{\mathcal{Œ}} \subseteq Q(\mathbf{Sym}(\Lambda^{\text{nom}}))$ be the differential over-field of k generated by u^{nom} and y^{nom}. The set Θ of unknown parameters is said to be *algebraically identifiable* (respectively *rationally identifiable*) if, and only if, any component of Θ is algebraic over (respectively belongs to) $\mathbf{\mathcal{Œ}}$. It is said to be *linearly identifiable* if, and only if,

$$P \begin{pmatrix} \theta_1 \\ \vdots \\ \theta_r \end{pmatrix} = Q \tag{B.4}$$

where

- P and Q are respectively $r \times r$ and $r \times 1$ matrices;
- the entries of P and Q belong to $\mathrm{span}_{k[\delta,\delta^{-1}]}(u^{\text{nom}}, y^{\text{nom}})$;
- $\det(P) \neq 0$.

The next result is obvious.

Proposition B.7.2 *Linear (respectively rational) identifiability implies rational (respectively algebraic) identifiability. The converse does not hold.*

The set Θ of unknown parameters is said to be *weakly linearly identifiable* if, and only if, there exists a finite set $\Theta' = (\theta'_1, \ldots, \theta'_{q'})$ of *related unknown parameters* such that

- the components of Θ' (respectively Θ) are algebraic over $k(\Theta)$ (respectively $k(\Theta')$);
- Θ' is linearly identifiable.

The next result is clear:

Proposition B.7.3 *Weak linear identifiability implies algebraic identifiability.*

Example B.7.4 *Consider the two nominal systems* $(\delta^{-1} - a)y^{\text{nom}} = u^{\text{nom}}$ *and* $(\delta^{-1} - b^2)y^{\text{nom}} = u^{\text{nom}}$, $m = p = 1$, *where a and b are the unknown parameters. It is clear that a is linearly identifiable, but not b which is algebraically identifiable. Note that b^2 is linearly identifiable and therefore that b is weakly linearly identifiable.*

Remark B.7.5 *Most examples encountered in practice, if they are not linearly identifiable, seem to be weakly linearly identifiable.*

B.8   Linear Identifiability of Transfer Functions

The nominal transfer function

$$\frac{b_0 + b_1 z^{-1} + \cdots + b_\mu z^{-\mu}}{a_0 + a_1 z^{-1} + \cdots + a_{\sigma-1} z^{-\sigma+1} + z^{-\sigma}} \tag{B.5}$$

where $a_0, a_1, \ldots, a_{\sigma-1}, b_0, b_1, \ldots, b_\mu$ are constant unknown parameters is said to be *linearly identifiable* if, and only if, those coefficients in the nominal SISO system

$$(a_0 + a_1 \delta + \cdots + \delta^\sigma) y^{\text{nom}} = (b_0 + b_1 \delta + \cdots + b_\mu \delta^\mu) u^{\text{nom}} \tag{B.6}$$

are linearly identifiable. Set

$$P_{SISO} = \begin{pmatrix} y^{\text{nom}} & \cdots & \delta^{\sigma-1} y^{\text{nom}} & -u^{\text{nom}} & \cdots & -\delta^\mu u^{\text{nom}} \\ \vdots & \vdots & \vdots & \vdots & \vdots & \vdots \\ \delta^{\sigma+\mu} y^{\text{nom}} & \cdots & \delta^{2\sigma+\mu-1} y^{\text{nom}} & -\delta^{\sigma+\mu} u^{\text{nom}} & \cdots & -\delta^{\sigma+2\mu} u^{\text{nom}} \end{pmatrix}$$

and

$$Q_{SISO} = \begin{pmatrix} -\delta^\sigma y^{\text{nom}} \\ \vdots \\ -\delta^{2\sigma+\mu} y^{\text{nom}} \end{pmatrix}$$

The equality

$$P_{SISO} \begin{pmatrix} a_0 \\ a_1 \\ \vdots \\ a_{\sigma-1} \\ b_0 \\ \vdots \\ b_\mu \end{pmatrix} = Q_{SISO}$$

yields the linear identifiability of $a_0, a_1, \ldots, a_{\sigma-1}, b_0, b_1, \ldots, b_\mu$ in (B.6). Let us rephrase this result as follows:

Proposition B.8.1 *The nominal transfer function (B.5) is linearly identifiable.*

B.8.1   An Example of a Transfer Matrix

It would be easy to extend Proposition B.1.12 to the MIMO nominal system

$$D(\delta) \begin{pmatrix} y_1^{\text{nom}} \\ \vdots \\ y_p^{\text{nom}} \end{pmatrix} = N(\delta) \begin{pmatrix} u_1^{\text{nom}} \\ \vdots \\ u_m^{\text{nom}} \end{pmatrix}$$

where

- the polynomial $p \times p$ and $p \times m$ matrices D and N are left co-prime;
- $\det(D) \neq 0$;
- the degrees of D and Q are bounded by Ω_D and Ω_N;
- there exists a nonzero entry of D of degree Ω_D (assume that the coefficients of δ^{Ω_D} are 1). All the other coefficients are unknown.

The linear identifiability of those unknown coefficients may be interpreted as a linear identifiability of the transfer matrix $T(z^{-1}) = D^{-1}N$. We prefer to sketch a more concrete approach by considering the 2×2 transfer matrix which often occurs in practice:

$$\begin{pmatrix} \dfrac{z^2}{a_{11}z^2+b_{11}z+c_{11}} & \dfrac{z^2}{a_{12}z^2+b_{12}z+c_{12}} \\[2mm] \dfrac{z^2}{a_{21}z^2+b_{21}z+c_{21}} & \dfrac{z^2}{a_{22}z^2+b_{22}z+c_{22}} \end{pmatrix} \tag{B.7}$$

with all its coefficients being unknown. Associate with matrix (B.7) the nominal input–output system

$$(a_{11} + b_{11}\delta + c_{11}\delta^2)(a_{12} + b_{12}\delta + c_{12}\delta^2)y_1^{\mathrm{nom}} =$$
$$(a_{12} + b_{12}\delta + c_{12}\delta^2)u_1^{\mathrm{nom}} + (a_{11} + b_{11}\delta + c_{11}\delta^2)u_2^{\mathrm{nom}}$$
$$(a_{21} + b_{21}\delta + c_{21}\delta^2)(a_{22} + b_{22}\delta + c_{22}\delta^2)y_1^{\mathrm{nom}} =$$
$$(a_{22} + b_{22}\delta + c_{22}\delta^2)u_1^{\mathrm{nom}} + (a_{21} + b_{21}\delta + c_{21}\delta^2)u_2^{\mathrm{nom}} \tag{B.8}$$

The same reasoning as in Proposition B.1.12 shows that $a_{\kappa,1}, b_{\kappa,1}, c_{\kappa,1}, a_{\kappa,2}, b_{\kappa,2}, c_{\kappa,2}, a_{\kappa,1}a_{\kappa,2},$ $a_{\kappa,1}b_{\kappa,2}, a_{\kappa,1}c_{\kappa,2}, b_{\kappa,1}a_{\kappa,2}, b_{\kappa,1}b_{\kappa,2}, b_{\kappa,1}c_{\kappa,2}, c_{\kappa,1}a_{\kappa,2}, c_{\kappa,1}b_{\kappa,2}, c_{\kappa,1}c_{\kappa,2}, \kappa = 1, 2$ are linearly identifiable *via* system (B.8). It yields the following result:

Proposition B.8.2 *The parameters $a_{\kappa,\iota}, b_{\kappa,\iota}, c_{\kappa,\iota}, \kappa, \iota = 1, 2$ of the nominal transfer matrix (B.7) are weakly linearly and rationally identifiable via system (B.8).*

Remark B.8.3 *The fact that we compute the unknown coefficients of matrix (B.7) and their products may be interpreted as a possibility to check the modeling (B.7).*

B.9   Linear Identification of Perturbed Systems

B.9.1   Linear Identifiability and Perturbations

By pulling back equation (B.4) from Λ^{nom} to Λ we obtain the following perturbation-dependent characterization of linear identifiability:

Proposition B.9.1 *The unknown parameters Θ of equation (B.4) satisfy*

$$P^{\mathrm{pert}} \begin{pmatrix} \theta_1 \\ \vdots \\ \theta_r \end{pmatrix} = Q^{\mathrm{pert}} + \Pi \tag{B.9}$$

where

- *P^{pert} is a $r \times r$ matrix, Q^{pert} and Π are $r \times 1$ matrices;*
- *the entries of P^{pert} and Q^{pert} are obtained by replacing u_i^{nom} and y_j^{nom} in P and Q by u_i and y_j, $i = 1, \ldots, m, j = 1, \ldots, p$;*
- *the entries of Π belong to $\mathrm{span}_{K[\delta, \delta^{-1}]}(\pi)$.*

B.9.2   Annihilating Some Peculiar Perturbations

Let us consider perturbations which are proportional to classic functions $\mathbf{Z} \to \mathbb{R}$. Define the unit-step Heaviside function $\mathbf{1}_k$ by

$$\mathbf{1}_k = \begin{cases} 0 & \text{if } k < 0 \\ 1 & \text{if } k \geq 0 \end{cases}$$

Thus, $(1 - \delta)\mathbf{1} = \mathbf{1}_k - \mathbf{1}_{k-1}$ is equal to 1 if $k = 0$ and to 0 if $k \neq 0$. This implies that the function $k(1 - \delta)\mathbf{1} = k(\mathbf{1}_k - \mathbf{1}_{k-1})$ is identically zero. We will say that $k(1 - \delta)$ is an *annihilator* (see, e.g., Jacobson, 1974; McConnell and Robson, 1987) of $\mathbf{1}$. Here is a short list of other functions and their annihilators (compare with Jury, 1973):

- $\mathbf{1}_{\alpha k}$, $\alpha \in \mathbf{N}$, $\alpha \neq 0$, is again annihilated by $k(1 - \delta)$.
- $\delta^\beta \mathbf{1} = \mathbf{1}_{k-\beta}$, $\beta \in \mathbf{Z}$, is annihilated by $(k - \beta)(1 - \delta)$.
- $k^\gamma H(t)$, $\gamma \in \mathbf{N}$, is annihilated by $k(1 - \delta)^\gamma$.
- The function

$$E_{\omega_k} = \begin{cases} 0 & \text{if } k < 0 \\ \omega^k & \text{if } k \geq 0 \end{cases}$$

  where $\omega \in \mathbb{R}$, is annihilated by $k(1 - \omega\delta)$.
- Set complex functions (i.e., $\mathbf{Z} \to \mathbb{C}$). The function

$$\text{SIN}(\alpha k + \beta) = \begin{cases} 0 & \text{if } k < 0 \\ \sin(\alpha k + \beta) & \text{if } k \geq 0 \end{cases}$$

  where $\alpha, \beta \in \mathbb{C}$, $\alpha \neq 0$, is annihilated by $k(1 - 2\delta \cos \alpha + \delta^2)$.
- A function which is equal to 0 at $k \neq k_0, k_1$, $k_0 \neq k_1$, is annihilated by $(k - k_0)(k - k_1)$.

B.9.3   Introducing Non-commutative Algebra

The previous subsection shows the necessity of abandoning the commutative ring $K[\delta, \delta^{-1}]$ and of working with the non-commutative ring $K[k, \delta, \delta^{-1}]$ of skew Laurent polynomials (i.e., of finite sums $\sum c_\alpha \delta^\alpha$, $c_\alpha \in K[k]$, $\alpha \in \mathbf{Z}$). The non-commutativity may be seen from the inequality $k\delta \neq \delta k = (k - 1)\delta$. Note (McConnell and Robson, 1987) that $K[k, \delta, \delta^{-1}]$ is a left principal ideal ring. It is therefore a *left Ore ring*, that is, it admits a left field of quotients (McConnell and Robson, 1987). It makes possible a straightforward extension of standard definitions and properties of commutative algebra, like torsion elements and modules.

The Ore property us of $K[k, \delta, \delta^{-1}]$ permits us to extend the ring of scalars in order to obtain the following left $K[k, \delta, \delta^{-1}]$-module from the $K[\delta, \delta^{-1}]$-module Λ in the same manner as in Section B.1:

$$\Lambda_{K[k,\delta,\delta^{-1}]} = K[t, \delta, \delta^{-1}] \otimes_{K[\delta,\delta^{-1}]} \Lambda$$

For the purpose of dealing with peculiar perturbations, we will from now on define a system L as a finitely generated left $K[k, \delta, \delta^{-1}]$-module in the following manner:

$$\mathbf{L} = \Lambda_{K[k,\delta,\delta^{-1}]}/M$$

where M is a finitely generated module spanned by elements of

$$\mathrm{span}_{K[k,\delta,\delta^{-1}]}(\pi).$$

Call again a perturbation the q-tuple $\pi = (\pi_1, \ldots, \pi_q) \subset \mathbf{L}$, which is the canonical image of $\pi = (\pi_1, \ldots, \pi_q)$. The nominal system is $\mathbf{L}^{\mathrm{nom}} = \mathbf{L}/\mathrm{span}_{K[k,\delta,\delta^{-1}]}(\pi)$. The canonical image of any element $\ell \in \mathbf{L}$ in $\mathbf{L}^{\mathrm{nom}}$ is written ℓ^{nom}. The next properties are straightforward:

1. $\mathbf{L}^{\mathrm{nom}} \simeq K[k, \delta, \delta^{-1}] \otimes_{K[\delta,\delta^{-1}]} \Lambda^{\mathrm{nom}}$.
2. The canonical mapping $\Lambda^{\mathrm{nom}} \to \mathbf{L}^{\mathrm{nom}}$, $\lambda^{\mathrm{nom}} \mapsto \ell^{nom} = 1 \otimes \lambda^{\mathrm{nom}}$, is injective. With a slight abuse of notation, Λ^{nom} will therefore be considered as a subset of $\mathbf{L}^{\mathrm{nom}}$.

B.9.4   The Identifier

Following Section B.1.8, we say that the perturbations π are *structured* if, and only if, the module $\mathrm{span}_{\mathbb{R}[k,\delta,\delta^{-1}]}(\pi)$ is torsion. This property will be assumed in the sequel. Equation (B.10) in the next proposition is called an *identifier* of the unknown parameters.

Proposition B.9.2 *There exists* $\Delta \in \mathbb{R}[k, \delta, \delta^{-1}]$ *such that equation (B.9) becomes*

$$\Delta P^{\mathrm{pert}} \begin{pmatrix} \theta_1 \\ \vdots \\ \theta_r \end{pmatrix} = \Delta Q^{\mathrm{pert}} \tag{B.10}$$

Proof. The existence of Δ is an immediate consequence of the torsion property of $\mathrm{span}_{\mathbb{R}[k,\delta,\delta^{-1}]}(\pi)$ and of the assumption that the parameters are constant.

Remark B.9.3 *From the examples of basic structured perturbations in Section B.1.8, we see that in equation (B.10) the difference operator* Δ *will often be of the form* $\Delta = \varpi \Delta'$, $\varpi \in \mathbb{R}[k]$, $\Delta' \in \mathbb{R}[\delta]$. *Then we will rewrite the identifier (B.10) as*

$$\Delta' P^{\mathrm{pert}} \begin{pmatrix} \theta_1 \\ \vdots \\ \theta_r \end{pmatrix} = \Delta' Q^{\mathrm{pert}} \tag{B.11}$$

Note that any entry of the matrices $\Delta' P^{\mathrm{pert}}$, $\Delta' Q^{\mathrm{pert}}$ *may be written*

$$\mathrm{finite} \sum \delta^\alpha x$$

where $x \in \mathrm{span}_{\mathbb{R}}(u, y)$, $\alpha \in \mathbf{Z}$. *By multiplying both sides by some positive power of* δ, *it can be assumed that* $\alpha \geq 0$.

B.10   Persistent Trajectories

B.10.1   Trajectories

Let $\mathcal{T}$ be a left $K[k, \delta, \delta^{-1}]$-module. A $\mathcal{T}$-*trajectory*[2] of system L is a $K[t, \delta, \delta^{-1}]$-module morphism $\tau : \mathbf{L} \to \mathcal{T}$. Here, $\mathcal{T}$ will be exclusively the left $K[k, \delta, \delta^{-1}]$-module of two-sided sequences $\{x_k | x_k \in K, \quad k \in \mathbf{Z}\}$.

---

[2] Compare with Fliess (1992a); Fliess and Marquez (2001).

B.10.2   Persistent Trajectories of an Identifier

A trajectory τ is said to be *persistent for the identifier* (B.10) if, and only if,

$$\det(\tau(\Delta P^{\text{pert}})) \not\equiv 0 \tag{B.12}$$

In other words, persistent trajectories are those trajectories which do not satisfy a given nonlinear difference equation.

Write $\det(\tau(\Delta P^{\text{pert}}))$ as a time function f_k. The trajectory τ is said to *lack persistency* at time k_0 if, and only if, $f_{k_0} = 0$.

Example B.10.1 *Consider the SISO perturbed system $\delta^{-1} y = au + \xi \mathbf{1}_k$, or $y_{k+1} = au_k + \xi \mathbf{1}_k$, where a is an unknown parameter and $\xi \mathbf{1}_k$ a constant perturbation of unknown intensity. An identifier for $k \geq 1$ is*

$$(u_k - u_{k-1})a = y_{k+1} - y_k$$

Trajectories corresponding to a constant control variable (i.e., $u_0 = u_{\pm 1} = u_{\pm 2} = \dots$) are not persistent. A trajectory lacks persistency at time k_0 if, and only if, $u_{k_0} = u_{k_0 - 1}$.

References

Cohn, R. (1965) *Difference Algebra*. John Wiley & Sons, Inc., New York.

Fliess, M. (1992a) A remark on Willems' trajectory characterization of linear controllability. *System Control Letters* **19**, 43–45.

Fliess, M. (1992b) Reversible linear and nonlinear discrete-time dynamics. *IEEE Transactions on Automatic Control* **37**, 1144–1153.

Fliess, M. (2002) On the structure of linear recurrent error-control codes. *ESAIM, Control, Optimization and Calculus of Variations* **8**, 703–713.

Fliess, M. and Marquez, R. (2001) Une approche intrinsèque de la commande prédictive linéaire discrète. *APII - Journal Européen des Systèmes Automatisés* **35**, 127–147.

Fliess, M., Fuchsummer, S., Schöberl, M., Schlacher, K. and Sira-Ramirez, H. (2008) An introduction to algebraic discrete-time linear parametric identification with a concrete application. *Journal Européen des Systèmes Automatisés* **42**(2&3), 210–232.

Jacobson, N. (1974) *Basic Algebra*, vols I & II. W. H. Freeman & Company, San Francisco.

Jury, E. (1973) *Theory and Application of the z-Transform Method*. Krieger Publishing Co., Malabar, FL.

McConnell, J. and Robson, J. (1987) *Noncommutative Noetherian Rings*. Pure and Applied Mathematics. Wiley-Interscience, New York.

Appendix C

Simultaneous State and Parameter Estimation: An Algebraic Approach

This section presents some definitions in order to establish a basic algebraic framework for nonlinear state estimation. This framework is taken from the work of Fliess *et al.* (2008). We recommend any interested reader to check this paper and the references cited therein in order to have a detailed description of this differential algebraic setting. The algebraic method for nonlinear estimation is an extension of algebraic techniques for linear closed-loop parametric estimation (Fliess and Sira-Ramírez, 2003, 2008). The technique applies to linear closed-loop fault diagnosis (Fliess *et al.*, 2004), and to linear state reconstructors (Fliess and Sira-Ramírez, 2004).

C.1 Rings, Fields, and Extensions

Set R an *ordinary differential ring*, that is, a commutative ring equipped with a single *derivation*

$$\frac{d}{dt} : R \to R$$

such that

$$\forall x, y \in R, \ \frac{d}{dt}(x + y) = \frac{dx}{dt} + \frac{dy}{dt}, \ \frac{d}{dt}(xy) = \frac{dx}{dt}y + x\frac{dy}{dt}$$

where $d^\nu x / dt^\nu = x^{(\nu)}$ for $\nu \geq 0$. The elements of R with derivation equal to zero are called *constants*. The set of all constant elements of R is the *subring of constants*, which is a subfield if R is a field.

A field is an algebraic structure where the operations of addition and multiplication can be carried out; these operations satisfy the associativity, commutativity, and distributivity axioms. Additionally, there exist the additive and multiplicative identity elements, the additive and multiplicative inverse elements, which permit the operations of subtraction and division, except

Algebraic Identification and Estimation Methods in Feedback Control Systems, First Edition.
Hebertt Sira-Ramírez, Carlos García-Rodríguez, Alberto Luviano-Juárez, John Alexander Cortés-Romero.
© 2014 John Wiley & Sons, Ltd. Published 2014 by John Wiley & Sons, Ltd.

division by zero. A field is a commutative quotient ring, that is, a commutative ring in which every nonzero element is invertible with respect to the product.

A *differential ring (respectively field) extension*, denoted R_2/R_1, is given by two differential rings (respectively fields) R_1, R_2 such that $R_1 \subseteq R_2$ and the derivation of R_1 is the restriction to R_1 of the derivation of R_2.

C.1.1 Notation

- Let S be a subset of R_2. The expression $R_1\{S\}$ (respectively $R_1\langle S\rangle$) denotes the differential subring (respectively subfield) of R_2 generated by R_1 and S.
- Let k be a differential field and $X = \{x_\iota | \iota \in I\}$ a set of differential indeterminates, that is, a set of indeterminates and their derivatives of any order. The expression $k\{X\}$ denotes the differential ring of differential polynomials, that is, of polynomials belonging to $k[x_i^{(\nu_\iota)} | \iota \in I;$ $\nu_\iota \geq 0]$. Any differential polynomial is of the form $\sum_{\text{finite}} c \prod_{\text{finite}} (x_i^{(\mu_\iota)})^{\alpha_{\mu_\iota}}$, $c \in k$.

A *differential ideal* $\mathfrak{I}$ of R is an ideal which is also a differential subring. It is said to be *prime* if, and only if, $\mathfrak{I}$ is prime in the usual sense.

All fields are assumed to be of characteristic zero. Assume also that the differential field extension K/k is *finitely generated*, that is, there exists a finite subset $S \subset K$ such that $K = k\langle S\rangle$. An element a of K is said to be *differentially algebraic* over k if, and only if, it satisfies an algebraic differential equation with coefficients in k. It is said to be *differentially transcendental* over k if, and only if, it is not differentially algebraic. The extension K/k is said to be *differentially algebraic* if, and only if, any element of K is differentially algebraic over k. An extension which is not differentially algebraic is said to be *differentially transcendental*.

A set $\{\xi_\iota | \iota \in I\}$ of elements in K is said to be *differentially algebraically independent* over k if, and only if, the set $\{\xi_\iota^{(\nu)} | \iota \in I, \nu > 0\}$ of derivatives of any order is algebraically independent over k. If a set is not differentially algebraically independent over k, it is *differentially algebraically dependent* over k. An independent set which is maximal with respect to inclusion is called a *differential transcendence basis*. The cardinalities, that is, the numbers of elements, of two such bases are equal. This cardinality is the *differential transcendence degree* of the extension K/k; it is written diff tr deg (K/k). The extension K/k is differentially algebraic if, and only if, its transcendence degree is finite; this degree is 0.

C.2   Nonlinear Systems

Let k be a given differential ground field. A *nonlinear input–output system* is a finitely generated differential extension K/k. Set $K = k\langle S, \mathbf{W}, \pi\rangle$, where:

- S is a finite set of system variables, which contains the set of control variables $\mathbf{u} = (u_1, \dots, u_m)$ and the set of output variables $\mathbf{y} = (y_1, \dots, y_p)$;
- $\mathbf{W} = \{\mathbf{w}_1, \dots, \mathbf{w}_q\}$ denotes the *fault variables* (which are not a topic of this book);
- $\pi = (\pi_1), \dots, \pi_r$ denotes the *perturbation*, or *disturbance*, variables.

These satisfy the following conditions.

- There is no interaction between the control, fault, and perturbation variables, that is, the differential extensions $k\langle \mathbf{u}\rangle/k$, $k\langle \mathbf{W}\rangle/k$, and $k\langle \pi\rangle/k$ are *linearly disjoint*.
- The control and fault variables are assumed to be independent, that is, $\mathbf{u}$ is a differential transcendence basis of $k\langle \mathbf{u}\rangle/k$ and $\mathbf{W}$ is a differential transcendence basis of $k\langle \mathbf{W}\rangle/k$.

- The extension $K/k\langle \mathbf{u}, \mathbf{W}, \pi \rangle$ is differentially algebraic.
- Assume that the differential ideal $(\pi) \subset k\{S, \pi, \mathbf{W}\}$ generated by π is prime. Write the quotient differential ring

$$k\{S^{\text{nom}}, \mathbf{W}^{\text{nom}}\} = k\{S, \pi, \mathbf{W}\}/(\pi)$$

where the *nominal* system and fault variables S^{nom}, $\mathbf{W}^{\text{nom}}$ are canonical images of S, $\mathbf{W}$. Ignoring the perturbation variables in the original system yields the *nominal system* K^{nom}/k, where $K^{\text{nom}} = k\langle S^{\text{nom}}, \mathbf{W}^{\text{nom}} \rangle$ is the quotient field of $k\langle S^{\text{nom}}, \mathbf{W}^{\text{nom}} \rangle$, which is an *integral domain* (i.e., without zero divisors). The extension

$$K^{\text{nom}}/k\langle \mathbf{u}^{\text{nom}}, \mathbf{W}^{\text{nom}} \rangle$$

is differentially algebraic.

- Assume as above that the differential ideal $(\mathbf{W}^{\text{nom}}) \subset k\{S^{\text{nom}}, \mathbf{W}^{\text{nom}}\}$ generated by $\mathbf{W}^{\text{nom}}$ is prime. Write

$$k\{S^{\text{pure}}\} = k\{S^{\text{nom}}, \mathbf{W}^{\text{nom}}\}/\mathbf{W}^{\text{nom}}$$

where the *pure* system variables S^{pure} are the canonical images of S^{nom}. Ignoring the fault variables in the nominal system yields the *pure system* K^{pure}/k, where $K^{\text{pure}} = k\langle S^{\text{pure}} \rangle$ is the quotient field of $k\{S^{\text{pure}}\}$. The extension

$$K^{\text{pure}}/k\langle \mathbf{u}^{\text{pure}} \rangle$$

is differential algebraic.

C.2.1   Remark

Take into account that differential algebra deals with algebraic differential equations, which only contain polynomial functions of the variables and their derivatives up to some finite order. Sometimes the system of interest contains transcendent functions, in those cases there exists the possibility of recovering algebraic differential equations using another transcendental function (see, e.g., Fliess *et al.*, 1995).

C.3   Differential Flatness

Considering the definition of a nonlinear system as given in Section C.2, the system K/k is said to be differentially flat if, and only if, the pure system K^{pure}/k is differentially flat (see Fliess *et al.*, 1995). The algebraic closure $\overline{K}^{\text{pure}}$ of K^{pure} is equal to the algebraic closure of a purely differentially transcendental extension of k, which means that there exists a finite subset

$$z^{\text{pure}} = \{z_1^{\text{pure}}, \ldots, z_m^{\text{pure}}\}$$

of $\overline{K}^{\text{pure}}$, where the number m of its elements is equal to the number of independent control variables, such that

- z^{pure} is differentially algebraically independent over k,
- z^{pure} is algebraic over K^{pure},
- any pure system variable is algebraic over $k\langle z^{\text{pure}} \rangle$.

The set z^{pure} is a *(pure) flat*, or *linearizing, output*.

C.4   Observability and Identifiability

Diop and Fliess (1991a,b) and also Diop (2002), established that a nonlinear input–output system is observable if, and only if, any system variable, a state variable for example, is a *differential function* of $\mathbf{u}$ and $\mathbf{y}$, that is, a function of control and output variables and their derivatives up to some finite order. This definition can be generalized to parametric identifiability and fault isolability; more generally, an unknown quantity may be determined if, and only if, it is expressible as a differential function of the control and output variables. According to the nonlinear system definition of Section C.2, these concepts are presented as follows.

C.5   Observability

A system K/k is said to be *observable* if, and only if, the extension

$$K^{\text{pure}}/k\langle \mathbf{u}^{\text{pure}}, \mathbf{y}^{\text{pure}}\rangle$$

is algebraic.

C.6   Identifiable Parameters

Set $k = k_0\langle\Theta\rangle$, where k_0 is a differential field and $\Theta = \{\theta_1, \ldots, \theta_r\}$ a finite set of unknown parameters, which might not be constant. According to Diop and Fliess (1991a,b), a parameter $\theta_i, i = 1, \ldots, r$ is said to be

- *rationally identifiable* if, and only if, it is equal to a differential rational function over k_0 of the variables $\mathbf{u}$ and $\mathbf{y}$, that is, to a rational function of $\mathbf{u}$ and $\mathbf{y}$ and their derivatives up to some finite order, with coefficients in k_0;
- *algebraically identifiable* if, and only if, it is algebraic over k_0, that is, it satisfies an algebraic equation with coefficients in k_0.

C.7   Determinable Variables

A variable $\Upsilon \in K$ is said to be

- *rationally determinable* if, and only if, Υ^{pure} belongs to $k\langle \mathbf{u}^{\text{pure}}, \mathbf{y}^{\text{pure}}\rangle$;
- *algebraically determinable* if, and only if, Υ^{pure} is algebraic over $k\langle \mathbf{u}^{\text{pure}}, \mathbf{y}^{\text{pure}}\rangle$.

A system variable χ^{pure} is then said to be

- *rationally observable* if, and only if, χ^{pure} belongs to $k\langle \mathbf{u}^{\text{pure}}, \mathbf{y}^{\text{pure}}\rangle$;
- *algebraically observable* if, and only if, χ^{pure} is algebraic over $k\langle \mathbf{u}^{\text{pure}}, \mathbf{y}^{\text{pure}}\rangle$.

C.7.1   Remark

In the case of algebraic determinability, the corresponding algebraic equation might possess several roots which are not easily discriminated (see Li *et al.*, 2006 for a concrete example).

C.8   Numerical Differentiation

In the algebraic approach, differentiators are not of asymptotic nature, and do not require any statistical knowledge of the corrupting noises. There are, of course, situations where the noise can be very strong and maybe with low frequency components, so that the algebraic technique may be insufficient.

C.8.1   Polynomial Time Signals

A real-valued polynomial function of degree N described by

$$x_N(t) = \sum_{v=0}^{N} x^{(v)}(0) \frac{t^v}{v!} \in \mathbb{R}[t] \quad \text{for} \quad t \geq 0$$

can be rewritten, using classical operational calculus, as

$$x_N(s) = \sum_{v=0}^{N} \frac{x^{(v)}(0)}{s^{v+1}} \tag{C.1}$$

Now, both sides are multiplied by $(d^\alpha/ds^\alpha)s^{N+1}$ with $\alpha = 0, 1, \ldots, N$, where the operator d/ds, sometimes called the algebraic derivative, corresponds in the time domain to multiplication by $-t$ (see Mikusiński, 1983; Mikusiński and Boehme, 1987; Yosida, 1984). The quantities $x^{(v)}(0)$, $v = 0, 1, \ldots, N$ can be obtained via the triangular system of linear equations

$$\frac{d^\alpha s^{N+1} X_N}{ds^\alpha} = \frac{d^\alpha}{ds^\alpha} \left(\sum_{v=0}^{N} x^{(v)}(0) s^{N-v} \right) \tag{C.2}$$

Following Fliess and Sira-Ramírez (2003, 2008), the terms $x^{(v)}(0)$ are said to be *linearly identifiable*. The time derivatives of the form $s^\mu (d^\iota X_N/ds^\iota)$, $\mu = 1, \ldots, N$, $0 \leq \iota \leq N$ presented in (C.2) are removed by multiplying both sides by $s^{-\overline{N}}$ for $\overline{N} > N$.

C.8.2   Analytic Time Signals

Set a real-valued analytic time function described by the convergent power series

$$x(t) = \sum_{v=0}^{\infty} x^{(v)}(0) \frac{t^v}{v!} \quad \text{for} \quad 0 \leq t < \rho \tag{C.3}$$

The truncated Taylor expansion of (C.3) is

$$x(t) = \sum_{v=0}^{N} x^{(v)}(0) \frac{t^v}{v!} + \mathcal{O}(t^{N+1}) \tag{C.4}$$

Approximating $x(t)$ by a truncated Taylor expansion of order N, $x_N(t) = \sum_{v=0}^{N} x^v(0)(t^v/v!)$, in the interval $(0, \epsilon)$, $0 < \epsilon \leq \rho$, we can apply the operational analog (C.1) of $x(t)$, that is, $X(s) = \sum_{v \geq 0}(x^v(0)/s^{v+1})$, this being an operationally convergent series in the sense of Mikusiński

(1983) and Mikusiński and Boehme (1987). The numerical estimated value of $x^{(\nu)}(0)$, obtained by replacing $X_N(s)$ by $X(s)$ in (C.2), is denoted by

$$[x^{(\nu)}(0)]_{eN}(t) \quad \text{for} \quad 0 \leq \nu \leq N$$

Notice that, for $0 < t < \epsilon$,

$$\lim_{t \downarrow 0} [x^{(\nu)}(0)]_{eN}(t) = \lim_{N \to +\infty} [x^{(\nu)}(0)]_{eN}(t) = x^{(\nu)}(0)$$

The term $\mathcal{O}(t^{N+1})$ in (C.4) becomes negligible if $t \downarrow 0$ or $N \to +\infty$ and $x_N(t)$ becomes $x(t)$. To estimate derivatives up to some finite order of a given smooth function $f : [0, +\infty] \to \mathbb{R}$, it is necessary to take a suitable truncated Taylor expansion around a given time instant t_0 and apply the previous methodology. Resetting and using sliding time windows permits us to estimate derivatives of various orders at any sample time instant.

C.8.3   Noisy Signals

For a precise mathematical foundation on the treatment of noise through iterated time integrals, see the work of, Fliess (2006), which is based on non-standard analysis. Considering that the integral over a finite time interval of a zero-mean, highly fluctuating function is infinitesimal, that is, "very small," we assume that our signals are affected by additive noises with a highly fluctuating, or oscillatory, nature. They may therefore be attenuated by low-pass filters, like iterated time integrals. Those iterated integrations are carried out when both sides of (C.1) are multiplied by $s^{-\overline{N}}$, for $\overline{N} > 0$ large enough.

References

Diop, S. (2002) From the geometry to the algebra of nonlinear observability. In A. Anzaldo-Meneses, F. Monroy-Prez, B. Bonnard, and J.P. Gauthier, (eds), *Contemporary Trends in Nonlinear Geometric Control Theory and its Applications*, pp. 305–345. World Scientific, Singapore.

Diop, S. and Fliess, M. (1991a) Nonlinear observability, identifiability and persistent trajectories. *Proceedings of the 36th IEEE Conference on Decision and Control*, pp. 714–719, Brighton, UK.

Diop, S. and Fliess, M. (1991b) On nonlinear observability. *Proceedings of the European Control Conference*, pp. 152–157, Hermès, Paris.

Fliess, M. (2006) Analyse non standard du bruit. *C. R. Math. Academie des Sciences de Paris* **342**(10), 797–802.

Fliess, M. and Sira-Ramírez, H. (2003) An algebraic framework for linear identification. *ESAIM, Control, Optimization and Calculus of Variations* **9**(1), 151–168.

Fliess, M. and Sira-Ramírez, H. (2004) Reconstructeurs d'etat. *C.R. Academie des Sciences de Paris, Série I* **338**(1), 91–96.

Fliess, M. and Sira-Ramirez, H. (2008) *Identification of Continuous-time Models from Sampled Data*, vol. XXVI of Advances in Industrial Control. Springer-Verlag, Berlin, pp. 363–391.

Fliess, M., Join, C. and Sira-Ramírez, H. (2004) Robust residual generation for linear fault diagnosis: An algebraic setting with examples. *International Journal of Control* **77**(14), 1223–1242.

Fliess, M., Join, C. and Sira-Ramirez, H. (2008) Non-linear estimation is easy. *International Journal of Modelling, Identification and Control* **4**(1), 12–27.

Fliess, M., Lévine, J., Martin, P. and Rouchon, P. (1995) Flatness and defect of non-linear systems: Introductory theory and applications. *International Journal of Control* **61**, 1327–1361.

Li, M., Chiasson, J., Bodson, M. and Tolbert, L. (2006) A differential-algebraic approach to speed estimation in an induction motor. *IEEE Transactions on Automatic Control* **51**, 1172–1177.

Mikusiński, J. (1983) *Operational Calculus*, vol. 1, 2nd edn. PWN&Pergamon, Warsaw.

Mikusiński, J. and Boehme, T. (1987) *Operational Calculus*, vol. 2, 2nd edn. PWN&Pergamon, Warsaw.

Yosida, K. (1984) *Operational Calculus: A Theory of Hyperfunctions* (translated from the Japanese). Springer, New York.

Appendix D

Generalized Proportional Integral Control

D.1 Generalities on GPI Control

GPI control, or control based on *integral reconstructors*, is a relatively recent development in the literature on automatic control (see Fliess *et al.*, 2002). Its main line of development rests within the finite-dimensional linear systems case, with some extensions to linear-delay differential systems and nonlinear systems.

D.1.1 Main Idea

The fundamental departure of GPI control from traditional state feedback controller design lies in the absence of asymptotic observers. Instead, structural reconstruction of the state is proposed. The state of the system is obtained on the basis of inputs and outputs alone, modulo the effect of initial conditions and modulo the effect of classical perturbations: that is, constant, ramp, parabolic disturbances. As a result of these errors and thanks to the superposition principle, *a posteriori* added iterated output, or input, integral error compensation completes the stable feedback design.

As a consequence of this approach, control laws, based on state feedback, are implemented with no need for asymptotic observers nor for digital computations based on output samplings. The control laws are purely analog.

D.1.2 An Illustrative Example

Consider the following linear system:

$$\ddot{y} = u, \quad y(0) = y_0, \quad \dot{y}(0) = \dot{y}_0$$

Algebraic Identification and Estimation Methods in Feedback Control Systems, First Edition.
Hebertt Sira-Ramírez, Carlos García-Rodríguez, Alberto Luviano-Juárez, John Alexander Cortés-Romero.
© 2014 John Wiley & Sons, Ltd. Published 2014 by John Wiley & Sons, Ltd.

A stabilizing *state* feedback control law is given by

$$u = -k_2\dot{y} - k_1 y$$

In practice, $\dot{y}$ must be asymptotically estimated or approximately computed by means of samplings of $y(t)$. Note, however, that from the model

$$\dot{y} = \int_0^t u(\tau)d\tau + \dot{y}(0)$$

Let us see what happens if we <u>insist</u> on using, in the proposed control law, the following *integral reconstruction* of $\dot{y}$:

$$\widehat{\dot{y}} = \int_0^t u(\tau)d\tau$$

That is, we use as the control law

$$u = -k_2\widehat{\dot{y}} - k_1 y = -k_2 \int_0^t u(\tau)d\tau - k_1 y$$

Since the relation between the structural estimate of $\dot{y}$ and its actual value is given by

$$\dot{y} = \widehat{\dot{y}} + \dot{y}_0$$

the closed-loop system would be governed by

$$\ddot{y} = -k_2\widehat{\dot{y}} - k_1 y, \quad = -k_2(\dot{y} - \dot{y}_0) - k_1 y$$

that is,

$$\ddot{y} + k_2\dot{y} + k_1 y = k_2\dot{y}_0$$

As a result:

$$y \to (k_2/k_1)\dot{y}_0 \neq 0$$

Classically, a constant stabilization error is corrected by means of an *integral control action* based on the output error.

We proceed to propose a controller which includes integral compensation:

$$u = -k_2 \int_0^t u(\tau)d\tau - k_1 y - k_0 \int_0^t y(\tau)d\tau$$

or

$$u = -\int_0^t [k_2 u(\tau)d\tau + k_0 y(\tau)]d\tau - k_1 y$$

The closed-loop system is now given by

$$\ddot{y} + k_2\dot{y} + k_1 y + k_0 \int_0^t y(\tau)d\tau = k_2\dot{y}_0$$

which, defining $\xi = \int_0^t y(\tau)d\tau - (k_2/k_0)\dot{y}_0$, may be rewritten as

$$\ddot{y} + k_2\dot{y} + k_1 y = -k_0\xi$$

$$\dot{\xi} = y$$

$$\xi(0) = -(k_2/k_0)\dot{y}_0$$

The characteristic equation of the closed-loop system has completely assignable roots (i.e., system eigenvalues)

$$s^3 + k_2 s^2 + k_1 s + k_0 = 0$$

via the appropriate selection of the design constants: $\{k_0, k_1, k_2, k_3\}$. In particular, we set $k_1 k_2 > k_0$ and $k_0, k_2 > 0$.

Summarizing, the system $\ddot{y} = u$, with unknown initial velocity conditions, is exponentially and asymptotically stabilized by means of the (implicit) generalized PI control law (see Figure D.1):

$$u = -k_1 y - \int_0^t [k_2 u(\tau) + k_0 y(\tau)]d\tau$$

The closed-loop behavior of the output satisfies the dynamics described by

$$y^{(3)} + k_2\ddot{y} + k_1\dot{y} + k_0 y = 0$$

The performance of this controller is depicted in Figure D.2. The proposed controller

$$u = -k_1 y - \int_0^t [k_2 u(\tau) + k_0 y(\tau)]d\tau$$

is interpreted in frequency-domain terms (see D.3) as follows:

$$u(s) = -\left(\frac{k_1 s + k_0}{s + k_2}\right)y(s)$$

This is a classical controller of the <u>lead</u> type, since the zero, given by $-k_0/k_1$, and the pole, $-k_2$, satisfy the relation $\alpha = k_0/(k_1 k_2) < 1$. In other words, the compensator zero is closer to the origin than the pole.

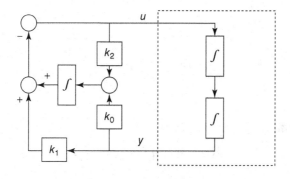

Figure D.1   GPI controller

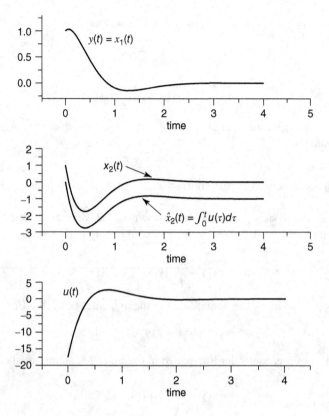

Figure D.2   Performance of GPI controller

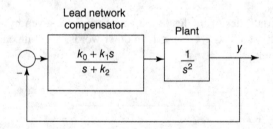

Figure D.3   Frequency-domain interpretation of GPI controller

D.1.3   *Robustness w.r.t. Perturbations*

If the system is known to be affected by *constant* input perturbations, ξ, suddenly appearing at an unknown time τ:

$$\ddot{y} = u + \xi \, 1 \, (t - \tau)$$

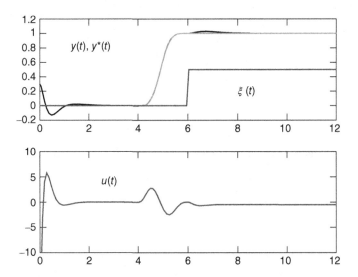

Figure D.4   Behavior with respect to step-input perturbations

then a controller with additional integration action allows for robustness w.r.t. ξ:

$$u = -k_2 y - \int_0^t [k_3 u(\tau) + k_1 y(\tau)]d\tau - k_0 \int_0^t \int_0^\sigma y(\lambda)d\lambda$$

See Figure D.4. The closed-loop system satisfies a dynamics which is equivalent to

$$y^{(4)} + k_3 y^{(3)} + k_2 \ddot{y} + k_1 \dot{y} + k_0 y = 0$$

A globally exponentially and asymptotically stable system can be obtained by suitable choice of the design parameters $\{k_3, k_2, k_1, k_0\}$. In particular, we set $k_3 > 0, k_0 > 0, k_2 k_3 > k_1$, and $k_1(k_2 k_3 - k_1) > k_0 k_3^2$.

Note that the controller is actually the following classic stable filter:

$$u = -\left[k_2 + \frac{(k_1 - k_2 k_3)s + k_0}{s(s + k_3)} \right] y$$

known as the "proportional + integral-lead action" filter.

D.1.4   Relation with Asymptotic Observers

In the example, $s^2 y = u$, consider the *reduced-order* observer for the velocity variable, sy:

$$s\xi = -h\xi + u - h^2 y, \quad \hat{sy} = \xi + hy, \quad h > 0$$

An open-loop observer is obtained by setting $h = 0$. This results in an integral reconstructor of the form

$$s\xi = u, \quad \hat{sy} = sy = \xi \rightarrow \hat{y} = \int_0^t u(\sigma)d\sigma$$

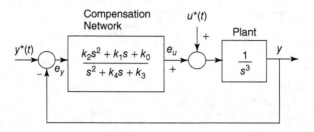

Figure D.5   GPI controller as a classical compensation network

D.1.5   A Third-Order Example

Consider now the third-order system

$$y^{(3)} = u$$

which admits the following integral input–output parametrization:

$$\widehat{y} = \int_0^t u(\tau)d\tau, \quad \widehat{\widehat{y}} = \int_0^t \int_0^\tau u(\lambda)d\lambda$$

The relations between the structural estimates and the real values of the states of the system are given by

$$\ddot{y} = \widehat{y} + \dot{y}_0$$

$$\dot{y} = \widehat{\widehat{y}} + \dot{y}_0 + \ddot{y}_0\, t$$

Suppose that the problem is to have the system output $y(t)$ track a given smooth signal $y^*(t)$. A controller achieving the trajectory-tracking task, when all states are known, is given by

$$u = [y^*(t)]^{(3)} - k_4(\ddot{y} - \ddot{y}^*(t)) - k_3(\dot{y} - \dot{y}^*(t))$$

$$-k_2(y - y^*(t))$$

If we use the structural estimates, plus integral error compensation in the controller synthesis, we have

$$u = [y^*(t)]^{(3)} - k_4(\widehat{y} - \ddot{y}^*(t)) - k_3(\widehat{\widehat{y}} - \dot{y}^*(t))$$

$$-k_2(y - y^*(t)) - k_1 \int_0^t (y - y^*(\tau))d\tau$$

$$-k_0 \int_0^t \int_0^\tau (y - y^*(\lambda))d\lambda d\tau$$

Note that, from the system model, $u^*(t) = [y^*(t)]^{(3)}$. Define $e_u = u - u^*(t)$ and $e_y = y - y^*(t)$. Then we have

$$u = u^* - k_4 \int_0^t e_u(\tau)d\tau - k_3 \int_0^t \int_0^\tau e_u(\lambda)d\lambda d\tau$$

$$-k_2 e_y - k_1 \int_0^t e_y(\tau)d\tau - k_0 \int_0^t \int_0^\tau e_y(\lambda)d\lambda d\tau$$

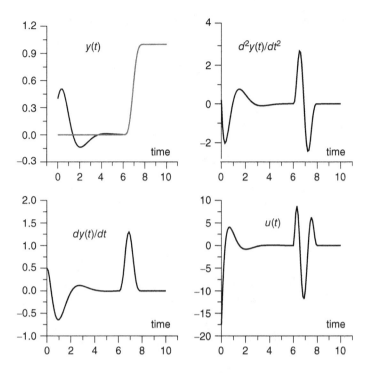

Figure D.6 Performance of GPI controller for third-order system

In the frequency domain, the proposed controller satisfies the following relation:

$$\left(1 + \frac{k_4}{s} + \frac{k_3}{s^2}\right) e_u = -\left(k_2 + \frac{k_1}{s} + \frac{k_0}{s^2}\right) e_y$$

Then, as depicted in Fig. D.5, we have,

$$u = u^*(t) - \left[\frac{k_2 s^2 + k_1 s + k_0}{s^2 + k_4 s + k_3}\right] (y - y^*(t))$$

The performance of this controller is depicted in Figure D.6.

D.1.6   Compartmental Model of a Heating System

GPI control, based on classical state-space representations of the plant, can also be implemented in a more direct fashion by first obtaining an integral reconstructor of the state vector. The following example depicts the procedure for obtaining a state reconstructor in a mono-variable linear system:

The normalized model of a compartmental heating system is given by

$$\dot{\theta}_1 = \theta_2 - \theta_1$$
$$\dot{\theta}_2 = \theta_1 - 2\theta_2 + \theta_3$$
$$\dot{\theta}_3 = \theta_2 - 2\theta_3 + u$$

The system is controllable (i.e., flat), with flat output $y = x_1$, and it is also observable (i.e., constructible) from this output.

From flatness, differential parametrization of all system state variables is readily obtained, as follows:

$$\theta_1 = y$$
$$\theta_2 = \dot{y} + y$$
$$\theta_3 = \ddot{y} + 3\dot{y} + y$$

From observability with respect to the (flat) output, one obtains the following integral parametrization of the state variables:

$$\left(\int\int \theta_1 \right) = \left(\int\int y \right)$$
$$\left(\int\int \theta_2 \right) = \left(\int y \right) + \left(\int\int y \right)$$
$$\left(\int\int \theta_3 \right) = y + 3\left(\int y \right) + \left(\int\int y \right)$$

From the system state equations, we obtain

$$\theta_1 = \left(\int \theta_2 \right) - \left(\int \theta_1 \right)$$
$$\theta_2 = \left(\int \theta_1 \right) - 2\left(\int \theta_2 \right) + \left(\int \theta_3 \right)$$
$$\theta_3 = \left(\int \theta_2 \right) - 2\left(\int \theta_3 \right) + \left(\int u \right)$$

Iterating once:

$$\theta_1 = 2\left(\int\int \theta_1 \right) - 3\left(\int\int \theta_2 \right) + \left(\int\int \theta_3 \right)$$
$$\theta_2 = -3\left(\int\int \theta_1 \right) + 6\left(\int\int \theta_2 \right) - 4\left(\int\int \theta_3 \right)$$
$$+ \left(\int\int u \right)$$

$$\theta_3 = \left(\int\int\theta_1\right) - 4\left(\int\int\theta_2\right) + 5\left(\int\int\theta_3\right)$$
$$-2\left(\int\int u\right) + \left(\int u\right)$$

Eliminating the double integrals of the state variables, we obtain

$$\theta_1 = y$$
$$\theta_2 = -4y - 6\left(\int y\right) - 5\left(\int\int y\right) + \left(\int\int u\right)$$
$$\theta_3 = 5y + 2\left(\int y\right) + 11\left(\int\int y\right) - 2\left(\int\int u\right) + \left(\int u\right)$$

which is an integral input–output parametrization of the state, modulo initial conditions.

D.2   Generalization to MIMO Linear Systems

The results presented here are easy to particularize for SISO systems.

Consider the observable, time-invariant, linear system of m inputs and p outputs:

$$\dot{x} = Ax + Bu, \quad x(0) = x_0, \quad y = cx$$

Integrating the system in the operational calculus sense (or, in the sense of Mikusiński), we have

$$x = A\frac{x}{s} + B\frac{u}{s}$$

Iterating on this functional relation, we have

$$x(t) = A^2\frac{x}{s^2} + AB\frac{u}{s^2} + B\frac{u}{s}$$

Iterating $n - 1$ times, one obtains

$$x(s) = A^{n-1}\left(\frac{x(s)}{s^{n-1}}\right) + \sum_{i=1}^{n-1} A^{i-1}B\frac{u(s)}{s^i}$$

In contrast, consider the output and its successive derivatives:

$$\begin{pmatrix} I \\ sI \\ \vdots \\ s^{(n-1)}I \end{pmatrix} y(s) = \begin{pmatrix} C \\ CA \\ \vdots \\ CA^{n-1} \end{pmatrix} x(s)$$
$$+ \begin{pmatrix} 0 & 0 & \cdots & 0 \\ CB & 0 & \cdots & 0 \\ \vdots & \vdots & \vdots & 0 \\ CA^{n-2}B & \cdots & \cdots & CB \end{pmatrix} \begin{pmatrix} I \\ sI \\ \vdots \\ s^{(n-2)}I \end{pmatrix} u(s)$$

Integrating $n - 1$ times, we have

$$\begin{pmatrix} \frac{I}{s^{n-1}} \\ \frac{I}{s^{n-2}} \\ \vdots \\ I \end{pmatrix} y(s) = \begin{pmatrix} C \\ CA \\ \vdots \\ CA^{n-1} \end{pmatrix} \frac{x(s)}{s^{n-1}} + \mathcal{M} \begin{pmatrix} \frac{I}{s^{n-1}} \\ \frac{I}{s^{n-2}} \\ \vdots \\ \frac{I}{s} \end{pmatrix} u(s)$$

Thanks to the observability of the system,

$$\frac{x(s)}{s^{n-1}} = [\mathcal{O}^T \mathcal{O}]^{-1} \mathcal{O}^T \left[\begin{pmatrix} \frac{I}{s^{n-1}} \\ \frac{I}{s^{n-2}} \\ \vdots \\ I \end{pmatrix} y(s) - \mathcal{M} \begin{pmatrix} \frac{I}{s^{n-1}} \\ \frac{I}{s^{n-2}} \\ \vdots \\ \frac{I}{s} \end{pmatrix} u(s) \right]$$

where $\mathcal{O}$ is the observability matrix of the system.

We can now combine this expression with the preceding one to obtain

$$x = A^{n-1} \left(\frac{x(s)}{s^{n-1}} \right) + \sum_{i=1}^{n-1} A^{i-1} B \frac{u(s)}{s^i}$$

Finally, we have

$$x(t) = \mathcal{P}(s^{-1})y(t) + \mathcal{Q}(s^{-1})u(t)$$

We address the expression, which does not take into account the influence of the initial states,

$$\hat{x}(t) = \mathcal{P}(s^{-1})y(t) + \mathcal{Q}(s^{-1})u(t)$$

the *integral state reconstructor* based on iterated integrals of inputs and outputs.

The integral state reconstructor may be used, in principle, on any linear state feedback control law

$$u = -k^T x(t)$$

as long as it is complemented by additional compensation which absorbs the effect of the neglected initial conditions, that is,

$$u = -k^T [\mathcal{P}(s^{-1})y(t) + \mathcal{Q}(s^{-1})u(t)] + v$$

Such a compensator only requires iterated integrations of the outputs (or output tracking errors), or of the inputs (or input tracking errors).

D.2.1 Example: Planar Vertical Take-off and Landing Aircraft (PVTOL)

Consider the following normalized model of the PVTOL aircraft:

$$\ddot{x} = -u_1 \sin\theta + \epsilon u_2 \cos\theta$$
$$\ddot{z} = u_1 \cos\theta + \epsilon u_2 \sin\theta - g$$
$$\ddot{\theta} = u_2$$

Linearization of the system equations around the equilibrium point

$$x = \bar{x}, \quad z = \bar{z}, \quad \theta = 0, \quad u_1 = g, \quad u_2 = 0$$

yields the following simpler, decoupled, model:

$$\ddot{x}_\delta = -g\theta_\delta + \epsilon u_{2\delta}, \quad \ddot{z}_\delta = u_{1\delta}, \quad \ddot{\theta}_\delta = u_{2\delta}$$

Note that x_δ is a non-minimum phase output, since the associated zero dynamics is given by

$$\ddot{\theta}_\delta = (g/\epsilon)\theta_\delta$$

Consider the following set of flat outputs:

$$F = x_\delta - \epsilon\theta_\delta, \quad L = z_\delta$$

The input-to-flat outputs system is equivalent to two independent chains of integration:

$$u_{1\delta} = \ddot{L}, \quad u_{2\delta} = -\frac{1}{g}F^{(4)}$$

A compensator based on pole assignment is readily found to be given by

$$u_{1\delta} = -\gamma_2\dot{L} - \gamma_1 L$$

$$u_{2\delta} = -\frac{1}{g}[-k_5 F^{(3)} - k_4\dddot{F} - k_3\dot{F} - k_2 F]$$

which requires the availability of the phase variables associated with the flat outputs.
   In order to circumvent the need to build an observer, one then takes the following system outputs for a GPI controller design:

$$y_{1\delta} = x_\delta, \quad y_{2\delta} = z_\delta$$

The integral state reconstructor associated with these outputs is

$$\theta_\delta = \left(\int\int u_{2\delta}\right), \quad x_\delta = y_{1\delta}, \quad z_\delta = y_{2\delta}$$

$$\dot{\theta}_\delta = \left(\int u_{2\delta}\right), \quad \dot{x}_\delta = -g\left(\int\int\int u_{2\delta}\right) + \epsilon\left(\int u_2\right),$$

$$\dot{z}_\delta = \left(\int u_{1\delta}\right)$$

A control law based on the integral reconstructor approach is thus given by

$$u_{1\delta} = -\gamma_2\left(\int u_{1\delta}\right) - \gamma_1 y_{2\delta} - \gamma_0\left(\int y_{2\delta}\right)$$

$$u_{2\delta} = -\frac{1}{g}\left[-k_5\widehat{F}^{(3)} - k_4\widehat{\dddot{F}} - k_3\widehat{\dot{F}} - k_2\widehat{F} - k_1\left(\int y_{1\delta}\right) - k_0\left(\int\int y_{1\delta}\right)\right]$$

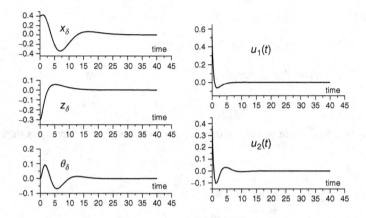

Figure D.7   Performance of multi-variable GPI controller

and the generalized PI controller is, therefore,

$$u_{1\delta} = -\int [\gamma_2 u_{1\delta} - \gamma_0 y_{2\delta}] - \gamma_1 y_{2\delta}$$

$$u_{2\delta} \quad -\frac{1}{g}\left[(k_5 g u_{2\delta} - k_1 y_{1\delta}) - k_2 y_{1\delta} + \int\int [(k_4 g + k_2\epsilon)u_{2\delta} - k_0 y_{1\delta}]\right.$$

$$\left. + \int\int\int (k_3 g u_2 - k_2 y_{1\delta})\right]$$

D.2.2   Remarks

GPI control, based on the state reconstructor approach, has a theoretical justification in the case of linear time-invariant systems, via the theory of modules, its localizations, and the Mikusiński transform. They lead to robust feedback control schemes that connect, in an interesting way, the classical control theory with that of the state space.

The extension to multi-variable discrete-time linear systems is particularly simple and, in the first case, yields dead-beat, open-loop, reduced-order observers with a state reconstruction synthesized on the basis of inputs and outputs in a finite window over the immediate past.

An extension to linear time-varying systems is also feasible. In particular, that development has importance in the control of nonlinear systems around nominal trajectories.

Mikusiński's operational calculus is the key to extending GPI control to linear time-invariant systems with delays. This results in more robust Smith predictor-based feedback control schemes.

References

Fliess, M., Marquez, R., Delaleau, E. and Sira-Ramírez, H. (2002) Correcteurs proportionnels-intègraux généralisés. *ESAIM, Control, Optimization and Calculus of Variations* **7**(2), 23–41.

Index

Algebraic Identification and Estimation Methods in Feedback Control Systems, First Edition.
Hebertt Sira-Ramírez, Carlos García-Rodríguez, Alberto Luviano-Juárez, John Alexander Cortés-Romero.
© 2014 John Wiley & Sons, Ltd. Published 2014 by John Wiley & Sons, Ltd.

Printed in the United States
By Bookmasters